Expanding the Boundaries of Transformative Learning

Praise for
Expanding the Boundaries of Transformative Learning:

"The rates of social change, the movements of the world's people, industrialization, globalization, and militarization continue to escalate at an ever-accelerating pace. Educators face an unprecedented task. They must support people to become highly creative, collaborative, problem solvers, and critical thinkers. They must cultivate people's capacities to see the world from profoundly different perspectives. They must nourish people's capacities for connection and caring in a fragmented and divisive world. The authors of the essays in this volume have been wrestling with all of these questions. They push our thinking forward. Nothing is more important."

—Mary Field Belenky, co-author of
Women's Ways of Knowing and *A Tradition That Has No Name*

"If we are to prevent global disaster, we need a totally different way of educating our children. The essays in this book take us an important part of the way toward the transformation we need."

—Rabbi Michael Lerner Editor,
TIKKUN Magazine and author,
Spirit Matters: Global Healing and the Wisdom of the Soul

"This is truly a unique book. It challenges us to transform ourselves and our planet by being aware of the creative cosmic power that flows through ourselves and the universe we inhabit. Too often, alas, we abuse this power. These authors ask us to look deeply into our souls through 'others" piercing eyes—those who represent our spectrum of humanity from activists to spiritualists. Such a vision will transform all who read these essays."

—Wm. E. Doll, Jr., Vira Franklin and J. R. Eagles
Professor of Curriculum, Louisiana State University

"*Expanding the Boundaries of Transformative Learning* offers a new vision of global community and new strategies for political struggle. It is a book that does not limit social transformation to the narrow confines of the classroom, but locates learning in the larger arena of mind and spirit. The authors put praxis into the service of rebuilding a world devastated by global capitalism; it is a project that confirms how decolonizing pathways to the human heart can be instrumental in the larger anticapitalist struggle ahead."

—Peter McLaren, Graduate School of Education and
Information Studies, University of California, Los Angeles

Expanding the Boundaries of Transformative Learning

Essays on Theory and Praxis

Edited by

Edmund O'Sullivan
Amish Morrell
Mary Ann O'Connor

palgrave

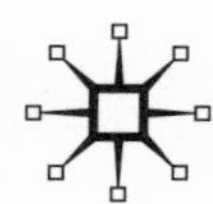

EXPANDING THE BOUNDARIES OF TRANSFORMATIVE LEARNING

First published 2001 by PALGRAVE™
175 Fifth Avenue, New York, N.Y.10010 and
Houndmills, Basingstoke, Hampshire RG21 6XS.
Companies and representatives throughout the world.

PALGRAVE is the new global publishing imprint of St. Martin's Press LLC Scholarly and Reference Division and Palgrave Publishers Ltd (formerly Macmillan Press Ltd).

ISBN 0–312–29507–3 hardback
ISBN 0–312–29508–1 paperback

Library of Congress Cataloging-in-Publication Data

Expanding the boundaries of transformative learning : essays on theory and praxis / edited by Edmund V. O'Sullivan, Amish Morrell, Mary Ann O'Connor

p. cm.
Includes bibliographical references and index.
ISBN 0–312–29507–3 (hard)—ISBN 0–312–29508–1 (pbk.)
1. Critical pedagogy. 2. Postmodernism and education. I. O'Sullivan, Edmund, 1938- II. Morrell, Amish. III. O'Connor, Mary Ann, 1956-

LC196.M85 2002
370.11'5—dc21

2001057531

A catalogue record for this book is available from the British Library.

Design by Letra Libre, Inc.

First edition: May 2002
10 9 8 7 6 5 4 3 2 1

Printed in the United States of America.

Contents

Acknowledgments

When the idea of this volume first materialized as a project for the Transformative Learning Centre, the sense of anticipation that we were about to embark on a journey with many of our colleagues at the Ontario Institute for Studies in Education at the University of Toronto gave an energy and excitement to our undertaking. It was one thing asking your colleagues to contribute an article to the collection of essays that you were planning, it was another thing getting those articles, reading and commenting on them and then asking for the changes that were part and parcel of the responsibilities we had as editors for what has turned out to be a rich and complex set of offerings in a field that is in the making. Indeed in the view of the editors of this volume, Transformative Learning is *a field in the making.* What all of us, as editors of this undertaking, hoped for at the outset was that we would tap into the energy and creativity of our colleagues and to be able to encourage them to write on a topic that would also be on the cutting edge of their own scholarly passions and commitments. We would like to acknowledge and thank our colleagues at the Ontario Institute for Studies in Education for their chapters that they have written in this volume. We would also like to thank the Transformative Learning Centre at OISE/UT, which encouraged our efforts throughout this undertaking. Finally, we would like to thank Luciana Costa for her painstaking copy editing work here in Canada and to Michael Flamini, Amanda Johnson, and Donna Cherry of Palgrave Macmillan in New York for their encouragement of the work and for their editorial and production assistance.

Amish Morrell
Mary Ann O'Connor
Edmund O'Sullivan
Toronto, Canada
February 2002

Contributors

Darlene Clover, Ph.D., coordinated two international research projects for the Transformative Learning Centre from 1993 to 1999. She is currently an assistant professor at the University of Victoria in British Columbia in the Faculty of Education and the School of Environmental Studies. Darlene's research and teaching focuses on environmental adult education, feminist adult education, women and the environment, and women and the arts.

Ardra L. Cole is a professor in the Department of Adult Education, Community Development, and Counseling Psychology and Coordinator of the Centre for Arts-informed Research. She teaches courses in teacher development, teacher inquiry, qualitative research methods, and arts-informed research methods. Her research interests and areas of professional and scholarly work are teacher knowledge and learning, teacher education reform, Alzheimer's disease, and qualitative research methods, especially autobiography, life history, and arts-informed inquiry and representation.

Vanessa Compton is an artist and educator currently pursuing a doctorate in the Curriculum, Teaching, and Learning Department at the Ontario Institute for Studies in Education of the University of Toronto with a focus on holistic and aesthetic education. She recently published an article in the *Journal of Experiential Education* that describes the interplay between the elements of transformative process—fire, brimstone, poetry, and myth—in her bronze-casting workshops. For alchemy of a different kind, she introduced and cofacilitates the Labyrinth at St. Paul's Westdale (Anglican) Church in Hamilton, Ontario, and writes extensively of the signs and wonders arising from community involvement with sacred public art. Her long-term goals are to develop curriculum for a New School of Chartres and to let loose her inner Mahalia Jackson.

George J. Sefa Dei is professor and associate chair, Department of Sociology and Equity Studies, at the Ontario Institute for Studies in Education of the University of Toronto. His teaching and research interests are in the areas of antiracism education, development education, international development, Indigenous knowledges, and anticolonial thought. He is the first president of the Ghanaian-Canadian Union, an umbrella group of Ghanaian-Canadian cultural, ethnic, and religious associations in Ontario. He has served as the president of the Canadian-Ghanaian Organization (now Cross Edge Network) and is on the board of trustees of the Harry Jerome Scholarships Awards of Canada. Currently he is working on a research project on minority education in comparative contexts.

ANNE GOODMAN (Adelson) comes from South Africa, where she was active in the anti-apartheid struggle. She has a doctorate in peace education from the Ontario Institute for Studies in Education of the University of Toronto. Her areas of research, interest, and practice include peace and reconciliation in Africa, especially South Africa, peace education, trauma healing and reconciliation, the culture of peace, nuclear issues, and ethnic identity and conflict transformation. She has taught at the Centre for Peace Studies at McMaster University in Hamilton, Ontario, and at the Transformative Learning Centre of OISE/UT. She has a long history of involvement with the environmental, women's, and peace movements and was a member of the National Working Group for the International Year for the Culture of Peace. She is a founder and active member of an international project called Voice of Somali Women for Peace, Reconciliation, and Political Rights.

BUDD L. HALL is dean of the Faculty of Education at the University of Victoria. He is a cofounder of the Transformative Learning Centre and former chair of the Adult Education and Counseling Psychology Department at the Ontario Institute for Studies in Education of the University of Toronto. Budd's special interests are in participatory research and social movement learning. His most recent book, coedited with George Dei, is *Indigenous Knowledges in Global Perspectives: Multiple Readings of Our World* (University of Toronto Press, 2000). He is also a poet.

J. GARY KNOWLES practices art-making and educational inquiry so that, in his life, the two may become one day soon a seamless fabric of professional work. As codirector of the Centre for Arts-informed Research (located within the Department of Adult Education, Community Development and Counseling Psychology at the Ontario Institute for Studies in Education) of the University of Toronto, he encourages graduate degree candidates to take artistic and intellectual risks in their research and be counted as arts-informed scholars. His educational scholarship is located in educational journals and books as well as found hanging on walls. Recently, with Lorri Neilsen and Ardra Cole, he edited *The Art of Writing Inquiry* (Backalong Books, 2001), and, with Ardra Cole, wrote *Lives in Context: The Art of Life History Research* (AltaMira, 2001).

LISA M. LIPSETT, Ed.D., is a writer, painter, and ecological educator whose work focuses on artful human–Earth reconnection. She is on the faculty of the Muskoka Healing Arts Teaching Center in Huntsville, Ontario, where she teaches ecologically based spontaneous painting and artful self-care. She is also a founding member of the Center for Arts-informed Research at the Ontario Institute for Studies in Education of the University of Toronto. Her chapter in this book is part of her doctoral thesis entitled *On Speaking Terms Again: Transformative Experiences of Artful Earth Connection.* More of her writing and art work can be found at the Artemeter Gallery (*www.artemeter.com*).

MATT MAXWELL is a singer/songwriter/recording artist, educational consultant and teacher based in Bowen Island, British Columbia. He received his doctorate at the Ontario Institute for Studies in Education of the University of Toronto where he worked closely with Jack Miller, David Selby, and Edmund O'Sullivan. He has published articles on second-language acquisition, ecological education, and alternatives to technology-based curricula. He can be reached at mmaxwell@lightspeed.ca.

BRIAN MILANI is a writer, researcher, student, teacher, carpenter, and activist interested in the theory and practice of making change—both individual and social. He teaches courses on green economics at the Metro Labour Education Centre, at York University's Faculty of Environmental Studies, and at the Ontario Institute for Studies in Education of the University of Toronto's Transformative Learning Centre. He is author of *Designing the Green Economy: the Postindustrial Alternative to Corporate Globalization* (Rowan & Littlefield, 2000). His doctoral work in the adult education department at OISE/UT and at the Institute for Environmental Studies (also at the University of Toronto) is on building materials in a green economy. He is also a student of internal martial arts.

ANGELA MILES teaches in the Adult Education and Community Development Program at the Ontario Institute for Studies in Education of the University of Toronto, focusing on feminist theory and practice in a global perspective. She is a member of Toronto Women for a Just and Healthy Planet, author of *Integrative Feminist Perspectives: Building Global Visions, 1960s–1990s* (Routledge, 1996). She is also a member of the executive and editorial boards of Inanna Publications and Education Inc. (operating as *Canadian Woman Studies/les cahiers de la femme*).

JACK MILLER has worked in the area of humanistic/holistic education for thirty years. He is currently professor and coordinator of the Holistic and Aesthetic Education Focus at the Ontario Institute for Studies in Education of the University of Toronto. He is author of eleven books, including *Education and the Soul* (SUNY Press, 2000) and *The Holistic Curriculum* (OISE Press, 1996) which have been translated into Japanese and Korean. Jack has made several trips to Asia to work with holistic educators there.

AMISH MORRELL is a doctoral student in the Department of Adult Education, Community Development and Counseling Psychology at the Ontario Institute for Studies in Education of the University of Toronto. In his thesis research, titled *Reclaiming the Real: Photographic Literacy and Historical Memory in Big Intervale, Cape Breton,* he is exploring ways of using photography in critical social research and education. He is also interested in social movements and subcultures. His master's thesis (1999) is titled *Imagining the Real: Theorizing Cultural Production and Social Difference in the Cape Breton Back-to-the-Land Community.*

MARY ANN O'CONNOR is a doctoral candidate at the Ontario Institute for Studies in Education at the University of Toronto and an associate of the Transformative Learning Centre. She is the director of the Women's Educational Center at the University of Michigan-Flint.

EIMEAR O'NEILL is a psychotherapist, community activist, and educator as well as a member of the Transfomative Learning Centre at the Ontario Institute for Studies in Education of the University of Toronto. Her doctoral work, *Holding Flames: Illuminating Women's Learnings Around Transformation,* integrates deep dynamic understandings from relational therapy and antioppression/antiviolence education with the transformative power of community art installations.

EDMUND O'SULLIVAN has been a professor at the Ontario Institute for Studies in Education of the University of Toronto for over thirty-two years, and has taught courses in child development, educational psychology, critical mass media studies, critical pedagogy, and cultural studies. He currently heads the program for Transformative Learning. He is the author of eight books and has over a hundred articles, chapters in books, and refereed journals. His books include *Critical Psychology and Critical Pedagogy* (Greenwood Press, 1990) and more recently *Transformative Learning: Building Educational Vision for the 21st Century* (St. Martin's Press, 1999). Ed recently was recipient of the OCUFA Teaching Award for Excellence in Teaching at the university level.

CAROLE ROY is a doctoral student in the Transformative Learning Specialization in the Department of Adult Education, Community Development and Counseling Psychology at the Ontario Institute for Studies in Education of the University of Toronto. She is a former educator in nonformal experiential and international youth educational exchanges and has an interest in women's resistance, social justice, and the use of creativity, wit, and humor in protests. She has long been concerned with peace, environment, and social justice, walking for peace and nuclear disarmament from New Orleans to New York, from Bonn to Vienna, and from Victoria to Nanoose Bay (BC) in the 1980s. She also joined the Central America Peace March from Panama to Mexico City to support human rights and social justice.

DANIEL SCHUGURENSKY is an assistant professor in the Department of Adult Education, Community Development and Counseling Psychology at the Ontario Institute for Studies in Education of the University of Toronto. His current research deals with the interactions between citizenship education and participatory democracy. His teaching interests include pre-service and graduate courses in the sociology of adult education, politics of adult education, community development, adult education and social change, research methods, Latin American education, and adult education in comparative and international perspective.

DAVID SELBY is professor of education and director of the International Institute for Global Education, Ontario Institute for Studies in Education of the University of Toronto. His latest book, coedited with Tara Goldstein, is *Weaving Connections. Educating for Peace, Environmental and Social Justice* (Sumach Press, 2000). He is currently writing a book on the implications of quantum and ecological relationships and processes for the nature of learning and for the dynamics of change in learning communities. He can be contacted at dselby@oise.utoronto.ca.

RENEE SHILLING is an Anishinaabe woman from Manitou Rapids, Rainy River First Nations, in northwestern Ontario. Her interests revolve around the role that adult education can play in the decolonization of Indigenous nations. She believes that it is essential to develop critical education programs that encourage dialogue regarding the developmental issues of Indigenous communities by addressing the historical circumstances of colonization. She is currently a candidate in the Master of Education program at the Ontario Institute for the Studies of Education of the University of Toronto.

NJOKI NATHANI WANE is an assistant professor at the Ontario Institute for Studies in Education of the University of Toronto and teaches in the areas of Black Canadian feminism, cultural knowledges, Indigenous knowledges, and antiracist studies. Her most recent publications are *Indigenous Knowledges: Lessons from the Elders; Reflections on Mutuality in Mothering* (University of Toronto Press, 2000) and a coedited book: *Equity in School and Society* (Canadian Scholars Press, 2000). She is currently working on a book on Black Canadian feminisms.

Introduction

AMISH MORRELL AND
MARY ANN O'CONNOR

This group of chapters represents the unique perspective of the Transformative Learning Centre in a growing body of theory and research in contemporary education. Although this Centre was not formally recognized by the University of Toronto until 2000, it has taken different forms over the past three decades, bringing together a group of scholars, both students and faculty, who have learned from, collaborated with, and influenced and inspired each other. The idea for this anthology on transformative learning came from the realization that, while each of us at the Transformative Learning Centre of the Ontario Institute for Studies in Education of the University of Toronto (OISE/UT) is doing her/his own research and writing, as a whole group we are also doing exciting and unique work. Our approach to transformative learning is distinctively *ours*—related to, but also distinguishable from, the approaches of other educators, activists, and theorists.

OISE has provided a unique space for this work, itself emerging from a particular set of historical conditions to provide a rich ground for critical scholarship. It was founded in 1965 during a period of unprecedented economic expansion. The former premier of Ontario, Progressive-Conservative party member William Davis, who was at the time Ontario's provincial minister of education, saw the educational sector as a key force in an expanding economy and created provincial endowments for programs that provided graduate training for educators. With increased funding in the educational sector came a widening of the scope and vision of what educators and educational institutions were expected to accomplish. Also emerging in the late 1960s and the early 1970s was an ethos of education as development. Proponents of this ideology perceived that education, writ large, could be both a catalyst for human personal and cultural development and an instrument by which to enter the arena of global politics by addressing issues of development and underdevelopment. Under these ideas an "undeveloped" third world came under the educational tutelage of the "developed" first world.

In the mid-1970s, economic expansion began to wane. The educational sector was deemed responsible for not supplying the training necessary to uphold the previous period of growth. Funding was revoked from the educational sector at both the federal and provincial levels, and in this new climate of economic austerity, the government sought to replace educational innovation with back-to-basics curriculum.

Since the end of the 1970s, and now into the twenty-first century, Canada's educational institutions have become victims of conservative educational policies that go under the name of neoliberalism. Within the neoliberal mandate is a commitment to remove the government from the key sectors of health, welfare, and education, opening them up to the vicissitudes of the market. In Canada and elsewhere, education is increasingly being delivered by governments into the hands of market forces and its directives. With the view that education must move toward exclusive support from market forces comes the imperative that educational institutions must, in turn, serve the needs of the market. Despite these trends, OISE has provided a place for its very critics.

Earlier efforts at OISE to create an alternative to liberalist paradigms in education resulted in the emergence of critical pedagogy as an educational subdiscipline by a group of scholars working together in the late 1970s. The intention of this group, which included Philip Corrigan, David Livingstone, dian marino, Paul Olson, Roger Simon, Edmund O'Sullivan, and Jack Quarter, was to create a North American version of the field of cultural studies that had been emerging in the United Kingdom with the work of radical adult educators including Stuart Hall, Raymond Williams, and Richard Hoggart. Edmund O'Sullivan and Madan Handa (who founded an earlier incarnation of the Transformative Learning Centre called the Centre for Global Peace) saw cultural studies and critical pedagogy as limited in that they focused only on human issues. They strove to integrate environmental issues into a new paradigm that was both critical and holistic. Three years later, with the work of Budd Hall, Edmund O'Sullivan, and Margrit Eichler, the Transformative Learning Centre was founded, and a departmental learning specialization called global transformation and community development (later renamed transformative learning), was created. A key influence in the definition of transformative learning, in terms of how it is conceptualized from within the Transformative Learning Centre, was a talk entitled "A Moment of Grace" given in November 1993 by Thomas Berry as part of the Madan Handa Memorial Lecture Series. Berry made crucial links between cultural history and ecology, highlighting the necessity of developing creative responses to environmental destruction and social inequality. The focus of this volume is a critical integration of issues of peace, social justice, and ecology in such a way as to present a fundamental challenge to market-driven education.

Many of us, and especially those who are members of the Transformative Learning Centre, already have in common the basic principles of critical theory; namely, that systems of power and privilege structured around class, race, and gender are key to understanding human relationships and society; that capitalism, racism, and sexism function as interlocking, mutually reinforcing systems of oppression to effectively marginalize, manipulate, exploit, and impoverish many (in fact, most) women, children, people of color, people of the so-called third world, and low-income people of the industrialized world; and that contemporary policies of economic restructuring for "global markets" are increasingly oppressing these same majorities while enormously benefiting the already wealthy global elite. We also share critical understandings of physical and economic violence in families and within nations. All of us have deep commitments to social justice, peace, and preservation of the biosphere. Most of us are social and political activists, along with being members of the academy (both faculty and students). It could be said that many of us share the same basic frameworks of radical critique, although our emphases would be different. And we share some visions, at least, of alternative social, economic, and ecological relationships.

This is not to say that every contributor to this volume has the same ideas about transformative learning. On the contrary, there are many different points of entry.

The passion each of us has for her/his own work has made for lively dialogue among us and will undoubtedly make for more. We look forward to dialogue with larger communities of teachers and learners through this volume.

Given the diversity of people and approaches that we knew would be represented here, we provided our contributors with a rough working definition of transformative learning. We offer it to our readers as well—not as a fixed definition, but as a way to stimulate discussion.

> Transformative learning involves experiencing a deep, structural shift in the basic premises of thought, feelings, and actions. It is a shift of consciousness that dramatically and permanently alters our way of being in the world. Such a shift involves our understanding of ourselves and our self-locations; our relationships with other humans and with the natural world; our understanding of relations of power in interlocking structures of class, race, and gender; our body-awarenesses, our visions of alternative approaches to living; and our sense of possibilities for social justice and peace and personal joy.

The fascinating thing to us, as editors, is how differently each of the contributors comes at his/her own work and how we take our learning and teaching in so many new directions. When we are prepared to engage educational questions in terms of new ways of being, of deep structural transformation, all sorts of hitherto unimagined possibilities arise. And this is a key point for the distinction between OISE's approach to transformative learning and the approaches taken elsewhere.

We are most interested in the generation of energy for radical vision, action, and new ways of being. If humans are going to survive on this planet, we need new connections to each other and to the natural world. Changing political and economic relationships is part of the larger project of reconstituting and revitalizing all of our relationships. Our purpose in transformative learning is not to delineate abstract principles about how adults learn, and we are not interested in theoretical "generalizability"—at least not in the sense in which this term is ordinarily used. We are asking ourselves why transformative learning matters. And when we speak of transformation, we need to know: from what to what? These chapters collectively expand on and problematize the purposes of transformative learning. At the same time, these questions also require us to understand the content of learning in ways that other approaches to transformative learning have shied away from. As reflected in this collection, we aim to leave much more open space for the marginal, the liminal, the unconscious, and the embodied. In the following chapters, the authors explore many different sites and contexts in which learning occurs, reflecting broad definitions of what constitutes education.

Thus we have, for example, contributors whose praxis includes artful inquiry and arts-informed research methods, contributors who have meta-analyses of the media and global capital, and contributors who draw from various types of Indigenous knowledge to shape their pedagogical strategies. We anticipate that such diversity will encourage readers to develop and refine their own critical analyses and, at the same time, to find for themselves new points of entry into alternatives. We intend this as a partial inventory of transformative learning within OISE, not as a definitive text, in the hopes that readers critique and expand on what we present in these pages. We would also note that for each of us as editors, there are break points with different authors, places where we personally would not go or approaches that we would not necessarily agree with. Yet we embrace the diversity represented here.

Our particular vision(s) of transformative learning also differs from other versions with respect to the role and potential of dialogue or rational discourse. There is no question that challenging discussions can stimulate change, particularly when there is safety, trust, and respect. In our view, however, there are also other routes to transformative learning. We do not insist on the primacy of reason or of articulation for transformative learning. We understand that crucial learning often takes place nonverbally, in the inarticulate dimensions of our bodies. We even would claim that essential transformative learning takes place unconsciously and that there is no need to attempt to bring everything into our consciousness, no need to try to pin a name on every experience. When we dance, for example, or spend a night outside under a star-filled sky, or examine a photograph, we learn. We learn in ways that change us and give us the vision and compassion and strength to work for both personal and social change.

One of the consistent features of our OISE approach to transformative learning is that it is interdisciplinary. The contributors to this reader are without exception people who have multiple skills and areas of expertise. As researchers, most of the contributors are multimodal, using a variety of cognitive and noncognitive approaches to engagement with their material.

We have read a particular thematic trajectory into these essays and have organized them accordingly. You may have a different way of conceptualizing and organizing these pieces. This volume begins with chapter 1 by Edmund O'Sullivan, "The Project and Vision of Transformative Education: Integral Transformative Learning." O'Sullivan has been teaching a course at OISE titled "Transformative Learning" that has provided a starting point for many new scholars affiliated with this field. In chapter 2 Matt Maxwell critiques modernist frames of educational curricula, providing the reader with a conceptual framework that can help in placing some of the ideas contained within the book. He calls for pedagogical methods that allow us to see outside of the parameters of existing curriculum. Angela Miles, in chapter 3, develops a definition of transformative learning as integrating progressive personal and social change. She calls for a transformative learning practice that allies itself with new social movements concerned with challenging and presenting alternatives to global capitalism. Similarly, in chapter 4 Budd Hall looks at the role of adult education in the establishment of the new market utopia and asserts the importance of building utopian visions within civil society that counter trends toward privatization and neoliberalism. With chapter 5 by Brian Milani we begin to see a transition between a critique of dominant educational models and their relation to the market to an analysis of their impact on our very sense of "being" in the world. Milani has produced an ardent analysis of the impact of industrialization, capitalism, and modernism on human consciousness. Eimear O'Neill and Edmund O'Sullivan, in chapter 14, look at the relationship among domestic violence, religion, and capitalism. They assert that the central role of patriarchy in both Judeo-Christianity and industrialism has resulted in a culture of violence. They advocate a redefinition of power, not as exploitative but as mutually enabling and about self-definition, equal access to resources, and conscious participation in the decisions affecting one's own life. Read together, these pieces give us social, cultural, economic, as well as psychological contexts for understanding contemporary educational and everyday life practices. These pieces also give us a ground for constructing alternative social visions and pedagogical strategies by which to address the social and ecological issues that concern transformative educators.

A problem often noted by OISE graduate students is that having a critical analysis of the operation of power within society can lead to a sense of disempowerment. Learning intended to be transformative can lead to inaction rather than action. In chapter 6, Daniel Schugerensky writes that "many educators (including adult educators) began to identify themselves as agents or instruments of the dominant groups, and decided that continuing to work in education was either an act of complicity or a waste of time, because the little relative autonomy of educational institutions was merely symbolic and ineffectual." He draws on the example of the participatory budget process in Porto Alegre Brazil as an example of a process of public deliberation through which critical analyses can be translated into transformative social action.

Authors draw from their own personal experiences and from their various cultural backgrounds to describe alternative ways of teaching, seeing, and being. Some authors develop theoretical frames that enable us to recognize different forms of knowledge that they believe will be instrumental in addressing the problems of social inequality and environmental degradation. David Selby, in chapter 7, describes how his childhood experience of a laneway that led from his hometown in northern England into the surrounding fens shaped his sense of identity and his understanding of human-ecological relationships. He proposes a model of radical interconnectedness in which everything is understood as interconnected and mutually constitutive, or everything is, as his title states, "the signature of the whole." Similarly, in chapter 8 Jack Miller writes about spirituality in the classroom, demonstrating ways of being attentive to the interconnectivity that Selby writes about. Miller writes about how to situate ourselves in relation to a broader context that we want to change and how to be attentive to those around us. Selby connects spirituality and social justice in ways that are critical in an era in which the language of spirituality is being evoked to further "family-values" legislation and right-wing political agendas. Vanessa Compton illustrates, in chapter 9, a particular method for regaining a sense of wholeness. She describes the process of walking the labyrinth, through which she believes that it is possible to restore a sense of "balance in the relation between self, soul, society, and world, 'reenchanting' contemporary life."

Postcolonial criticism has been an important aspect of work done by OISE scholars affiliated with transformative learning, leading to interesting and unique hybrid pedagogies. In chapter 10, George Dei looks at the role of Ghanian spirituality in transformative learning and the development of an anticolonial politic. Dei writes that transformative learning needs to "assist the learner to deal with pervasive effects of imperial structures of the academy on the processes of knowledge production and validation; the understanding of indigenicity; and the pursuit of agency, resistance, and politics for educational change." If spirituality is to be part of transformative learning, it must be self-consciously politicized. Similarly, in chapter 11 Njoki Wane writes about African women's spirituality and its role in sustaining African women throughout the violence of colonization. Wane further draws on interviews with Embu women to describe African women's spirituality and how it shapes their sense of place in nature and the universe. Renee Shilling writes, in chapter 12, about how the historical trauma of colonization has been transmitted over generations of Indigenous peoples, and she addresses what a decolonizing educational practice would look like.

Chapter 13 by Darlene Clover looks at forms of centuries-old knowledge that have sustained harmonious relationships between Indigenous peoples and their environment and the role of capitalism in destroying these diverse knowledges and cultures. She calls for an integrative environmental adult education based on people's cultural identities and their relationships with the places in which they live. She writes that "ecological cultural identity is based on the feelings and relationships people develop with the place they live, the patterns and changes in the landscape, and interactions with other forms of life. It is spirituality, the self, collectivity, self-esteem, values, customs, self-respect, and self-direction. It is both a cognitive and an embodied way of knowing." There are many connections between this work and the other essays that focus on spirituality.

The authors in this volume write not only about expanding our definitions of knowledge but about expanding our very senses of self. In chapter 15 Anne Goodman draws upon her extensive background as a peace activist and educator to discuss what it means to develop a "culture of peace." She both talks about the "official" vision of a culture of peace developed by UNESCO and her "unofficial" critical-visionary concept of a culture of peace based on an expanded notion of our humanity. According to Goodman, "The modern western conception of the individual, with its emphasis on autonomy and separation, is only one model of being human. . . . In African culture, for example, relationships are central, and the humanity of each person is seen as integrally bound to the humanity of others."

More subjective forms of inquiry, such as research involving the arts, are integral for transformative pedagogies. Gary Knowles and Ardra Cole write in chapter 16 about their experiences in the academy as arts-based researchers. Amish Morrell describes, in chapter 18, his research on how photographs can shape our understanding of ourselves and our surroundings in a way that is critical and transformative. In chapter 20 Carole Roy discusses the Raging Grannies as a public pedagogy, and theorizes how humor is a critical aspect of their work. Both Lisa Lipsett, in chapter 17, and Mary Ann O'Connor, in chapter 19, discuss ways of accessing the unconscious and exploring the unconscious dimensions of both social and ecological relationships. Lipsett describes her use of spontaneous painting while O'Connor's work is based on fiction writing. They both present us with ways of understanding our relations to nature and to social and cultural forces that they evoke through innovative research paradigms.

For us at OISE's Transformative Learning Centre, inner and outer change are reciprocal processes, necessarily reflecting one another in all patterns of relationship. Social change cannot happen without individual psychic change; physical health cannot be separated from planetary well-being; self-conceptions cannot be carved off from conceptions of *anima mundi*. We make our world, and our world makes us, in some obvious ways and in some very subtle ways. This transformative potential is to be celebrated. Our hope is that this volume is part of the celebration.

CHAPTER I

The Project and Vision of Transformative Education

Integral Transformative Learning

Edmund O'Sullivan

Unless we live our lives with at least some cosmological awareness, we risk collapsing into tiny worlds. For we can be fooled into thinking that our lives are passed in political entities, such as the state or a nation; or that the bottom-line concerns in life have to do with economic realities of consumer life-styles. In truth, we live in the midst of immensities.

—Brian Swimme, *The Hidden Heart of the Cosmos*

Prologue

A discussion of a transformative vision of education will involve a diversity of elements and movements in contemporary education. A vision statement must attend to these diversities. I see my approach to transformative learning as an integral endeavor and thus call my approach *integral transformative learning.*[1] *Webster's Collegiate Dictionary* (1986) gives as part of its definition of "integral" the following descriptors: "essential to completion, formed as a unit with another part, integrated: lacking nothing essential."

While I would like to point out that there are many definitions of transformative learning and that it behooves any author using this term to specify her or his definition, as I will demonstrate, my attempt at defining transformative learning is essentially an integrative process. Without further ado, let me begin this process.

We are living in a major transitional period of history in which there are many contesting viewpoints. To some extent these trends are operating somewhat separate and independent of one another. I would like to name some of those elements because I think they will form part of a weave of a new type of integral education that will contest the vision of education for the global marketplace and move toward a more integral transformative vision. I couch these elements within a broad cosmological framework that I believe is my contribution to the effort of offering an alternative to our present conventions in education. I attempt to distill from my own work on transformative learning some core features of an integral planetary education that can be taken up in adult education. The reader should, however, be aware that my use of the term "transformative learning" covers a much wider frame of reference and addresses foundational issues in the field of education at large.

The New Century as Transformative Historical Moment

Our movement into the twenty-first century is momentous not because it is a millennium turning point or a movement into some kind of postmodern history, nor because we are moving from an industrial age into a new information age. The period in which we are living is not simply a turning point in human history; it is a turning point in the very history of the earth itself. We are living in a period of the earth's history that is incredibly turbulent and in an epoch in which there are violent processes of change that challenge us at every level imaginable. The pathos of the human being today is that we are totally caught up in this incredible transformation, and we have a significant responsibility for the direction it will take. What is terrifying is that we have it within our power to make life extinct on this planet. Because of the magnitude of this responsibility for the planet, all our educational ventures must finally be judged within this order of magnitude. This is the challenge for all areas of education. For education, this realization is the bottom line. When setting educational priorities, every educational endeavor must keep in mind the immense implications of our present moment. This demands an attentiveness to our present planetary situation that does not go into slumber or denial. It poses significant challenges to educators in areas heretofore unimagined. Education within the context of "transformative vision" keeps concerns for the totality of life's context always at the forefront.

Integral Modes of Transformative Learning: The Tripartite Distinction of Survive, Critique, and Create

The idea of transformative vision starts with the notion of transformation within a broad cultural context. In the larger cultural context, transformation carries the dynamism of cultural change. When any cultural manifestation is in its florescence, the educational and learning tasks are uncontested and the culture is of one mind about what is ultimately important. During these periods, there is a kind of optimism and verve that ours is the best of all possible worlds. It is usual to have a clear sense of purpose about what education and learning should be. There is also a predominant feeling that we should continue in the same direction that has taken us to this point. Here one can say that a culture is in "full form" and the form of the culture warrants "continuity." We might say that a context that has this clear sense of purpose or direction is "formatively appropriate." A culture is formatively appropriate when it attempts to replicate itself and the educational and learning institutions are in synchrony with the dominant cultural themes.

Even when a culture is formatively appropriate, at times there seems to be a loss of purpose or a loss of the qualities and features that have led to the florescence of that particular culture. Part of the public discourse, during times such as these, is one of "reform criticism." Reform criticism calls a culture to task for its loss of purpose. It is a criticism that calls the culture back to its original heritage. It accepts the underlying heritage of the culture and seeks to put the culture, as it were, "back on track." When reform criticism is directed toward educational institutions, we call this "educational reform."

Another type of criticism that is radically different from reform criticism calls into question the fundamental mythos of the dominant cultural form and indicates that the culture can no longer viably maintain its continuity and vision. This criticism—"transformative criticism"—maintains that the culture is no longer "formatively appropriate" and questions all of the dominant culture's educational visions of continuity. In contrast to reformative criticism, transformative criticism suggests a radical restructuring of the dominant culture and a fundamental rupture with the past. I suggest that transformative criticism has three simultaneous moments. The first moment I have already described as the critique of the dominant culture's formative appropriateness. The second is a vision of what an alternative to the dominant form might look like. The third moment includes some concrete indications of the practical exigencies of how a culture could abandon those aspects of its present forms that are functionally inappropriate while, at the same time, points to a process of change that can create a new cultural form that is functionally appropriate.

Having looked at the notion of transformative learning in its broad cultural context, it now is appropriate to look at it from the point of view of the knower and known. I draw on general systems theory to give us a specific entry point into the world of the individual learner. In the systems theory viewpoint, in the process of learning the mind organizes itself by virtue of feedback—that is, by monitoring its interactions with its environment. The key term in systems theory is "monitoring." The basic assumption is that open systems, like the mind, self-monitor. In the context of learning, monitoring assumes that the mind watches what it is doing and adjusts. The mind initially operates through preconceptions; these preconceptions not only shape our interpretations of the world but also impinge on the world itself.

In order to comprehend how the world and the mind shape each other, it is necessary to examine the two main ways that feedback works. The first way is homeostatic or negative feedback, a process that brings the world around us in line with our assumption and goals. The second is adaptive or positive feedback, which leads to change in internal codes or presuppositions. Thus, negative feedback indicates that one is on track, with no need for adjustment, while positive feedback signals deviation from the objective and the need to correct or alter course. Positive feedback operates when we act on our environment to make it intelligible in terms of our presuppositions. Adaptive positive feedback occurs when there is a persistent mismatch between perception and code—that is, when we can no longer interpret experience in terms of our old assumptions. The cognitive system then searches for new codes by which novel and confusing perceptions can be made intelligible. This search amounts to the cognitive system exploring new ways to reorganize itself, and it continues until codes and constructs evolve that can deal with the novelty of the new data. In this particular context, living systems adapt by transforming themselves, and learning occurs. Thus, from the perspective of transformative learning, real learning is not something added to a system in adaptation. Transformation means, in essence, the reorganization of the whole system. In this process, the viewed world is different and so is the viewpoint of the viewer. Transformative learning processes are counted as the *creative* function of cognitive crisis. Creativity occurs within a cognitive system when old habitual modes of interpretation become dysfunctional, demanding a shifting of ground or viewpoint. The breakdown, or crisis, motivates the system to self-organize in more inclusive ways of knowing, embracing, and integrating data of which it had been previously unconscious.

I draw on the work of distinguished cultural historian and ecologist Thomas Berry (1993), in a talk he gave to the Transformative Learning Centre, to highlight the historical task of cultural transformation today. Berry sees our present moment in history as a "terminal state" and also as a "moment of grace." In moments of grace we take danger and turn it into opportunity. In moments of grace we take decadence and turn it into creativity. Berry maintains that in order to survive our moment, we must be prepared to take a journey into a new creative "story." In the present we are living between "stories." He suggests that our present cultural story, exemplified in the technical-industrial values of western Eurocentric culture, is dysfunctional in its larger social dimensions even though we continue to firmly believe in it and act according to its guidance. According to Berry, we are in pressing need of a radical reassessment of our present situation, especially concerning those basic values that give life some satisfactory meaning. We need an "integral" story that will educate us, a story that will heal, guide, and discipline us. The transition between stories will be referred in this chapter as the movement from the "terminal Cenozoic" to the "Ecozoic." It is a movement of transformative learning at both the individual and the cultural levels of history.[2]

I contend that transformative learning encompasses an "education for survival," an "education for critical understanding," and an "education for integral creativity." Surviving, critiquing, and creating are, in fact, central to my definition of "integral" transformative learning. Let us look at these three modes in turn.

SURVIVAL EDUCATION (SURVIVE)

The word "survival" comes from the French verb *survivre.* Basically, survival means creating conditions for the continuance of living. The terminal aspects of our historical moment, described above, involve educational concerns at the personal, communal, and planetary level. The sense of all of the locations of survival can be crystallized in the awareness that something is very amiss with the practices of an economic system that has led to both human suffering and environmental disaster. Patterns of devastating destruction that are neither random nor accidental have arisen from a consciousness that fragments existence. The matrix of western culture, originating in modern European culture and transplanted all over the world, that considers human existence and, above all, human consciousness and spirit as independent from and above nature dominate the world's imagination. This dynamism, embedded in the onslaught of the global market, is withering away our sustenance at the personal, communal, and planetary levels of existence. We see its effect in environmental devastation, human rights violations, the hierarchies of race, the prevalence of violence, the idea of technological progress, and the problem of failing economies. Understanding these complex issues as an intricate part of the current ecological crisis is the survival task of transformative learning. Understanding these complex issues in isolation threatens the survival of Earth in its total context—one that encompasses the planetary and cosmological context as well as the human and the larger biotic species context.

In terms of facing these profound issues of survival, there are three learning aspects that we do not ordinarily identify with learning: the dynamics of denial, despair, and grief. In my work on transformative learning, I deal with these dimensions at length; here they must be dealt with all too briefly. Denial is a defense mechanism that prevents us from being overwhelmed by the deeply problematic nature of our times. But in order to solve problems, it is necessary to come out of denial. Once the depth of our problems is allowed in, we must contend with despair. Despair will be one of the major difficulties in facing survival issues. Without the development of a critical understanding and creative vision, despair has the capacity to overwhelm. Finally, grieving is a necessary ingredient in the survival mode. The sense of loss at the personal, communal, and planetary level that is part and parcel of the survival mode demands a grieving process at profound levels. Therefore, transformative learning in the survival mode is a learning process requiring the ability to deal with denial, despair, and grief.

Critical Resistance Education (Critique)

In the survival mode, we are dealing with a profound cultural pathology that requires a deep cultural therapy. Part of this cultural therapy involves a transformative mode of cultural criticism. We need to examine the factors and conditions that have brought us to this devastating historical moment. We are in need of a resistance education that moves in the direction of cultural criticism and that includes moments of resistance and critical pedagogy. A critical pedagogical moment will cover several areas in need of deep critical reflection.

The first dimension to be examined is the matrix of thought that provides the frame of reference and worldview for the forces of the modern world. Here we need to be attentive to the deep ontological basis of western European thinking. In relation to the natural world, the dynamics of the modern scientific-industrial worldview have moved an agenda that has had profound effects on how the modern western European human views the natural world. Modern cosmological systems start with the postmedieval Enlightenment period and take us into the mid-twentieth century. This is the apex of the final phases of the late Cenozoic period. What is characteristic of this period is the major cultural revisioning of thought systems that are now identified as the technical-scientific-industrial worldview. The movement into this postmedieval period represents a very profound change in cultural consciousness; a change from the medieval into the modern temper is one of the most far-reaching cultural changes in human history. This movement from the medieval world to the modern world, as with other great historical junctures, is one that involves a new metaphysical and ideological change that impacts on all major cultural institutions. In essence, it forms a new picture of the cosmos and the nature of the human being. The sense of the world that views nature as a mechanism enmeshed in mechanical forces has led humankind to a profound disenchantment with the natural world. Within these forces of modernism, there is a loss of the sense of a wider cosmology in which human actions are embedded.

The philosopher Stephen Toulmin, in his book *The Return to Cosmology* (1985), gives us a convenient entry point for our discussion of the term "cosmology." He observes that there appears to be a natural attitude taken by humans at all times and in all places when reflecting on the natural world that includes a comprehensive ambition to understand and speak about the universe as a whole. Toulmin notes that, in practical terms, this desire for a view of the universe as a whole reflects a need to recognize where we stand in the world in which we have been born, to grasp our place in the scheme of things, and to feel at home within it.

A second dimension of critical resistance education is dealing with the saturation of consciousness. W. C. Fields once quipped, "I don't know who discovered water, but it certainly wasn't the fish." This image succinctly conveys the problem of submerged consciousness that a critical resistance education must encounter. Like the fish in water, we, too—in relation to our culture—are in what Paolo Freire (1970) calls a submerged state of consciousness. With the incredible saturation of information that our modern consciousness demands we attend to, we are, according to John Ralston Saul (1997), an *unconscious civilization.*[3] We are caught in a situation where our knowledge does not make us conscious. Information, in modern societies, comes to us indiscriminately, and the information is disconnected from usefulness. We are swamped with information and have virtually no substantial control over it. Reflecting back to the earlier discussion of a fractured cosmology, we see a loss of coherence because we no longer have a coherent conception of ourselves, our universe, or our relation to one another and our world. Within the context of a broader cosmological background, information diversity is as critical to our long-term survival as biodiversity.

The third dimension of transformative learning is the critical examination of hierarchical power. Our modern western historical inheritance is deeply embedded in a hierarchical conception of power that comes to us in the structures of patriarchy and imperialism. In its simplest interpretation, patriarchy is a system of power where men dominate. Our culture exists within the matrix of patriarchal power. In reflecting on our western heritage, male dominance can be seen in four establishments that have been in control of western history over the centuries: the classical empires, the ecclesiastical establishment, the nation-state, and the modern corporation. Historically, women have had minimal, if any, role in the direction of these establishments. A critical transformative deconstruction of patriarchy is one learning task that will pick up the feminist movement's challenge to the destructive effects of patriarchy throughout contemporary societies. Simultaneously, with the deconstruction of patriarchy, the deep structure of hierarchical power and violence needs to be critically examined in such areas as race, class, gender, and sexual orientation. The structures of imperialism that have characterized the expansiveness of western culture in the modern era are now being brought into the light of profound criticism.

VISIONARY TRANSFORMATIVE EDUCATION (CREATE)

I develop the idea of creative transformative learning by highlighting the themes of education for planetary consciousness, for integral development, for quality of life, and education and the sacred. The first theme is education for a planetary consciousness. In spite of its powerful dynamics in our world today, the idea of the global market is a small idea. I feel that we must locate our lives in a larger cosmological context much more breathtaking than the market vision of our world. Brian Swimme (1996) points out that we can be fooled into thinking that our lives are passed in political entities such as the state or a nation or the bottom-line concerns in life having to do with economic realities. He invites us to understand that we live in the midst of immensities. In presenting a larger cosmological context for our lives, it is not enough to defend a new cosmology in opposition to the one that has underpinned modernity. In a visionary context for transformative learning, we must articulate a planetary context for learning that can effectively challenge the hegemonic culture of the market vision and that can orient people, in their daily lives, to create an environmentally viable world in our present time. The philosopher Alan Gare (1995) maintains that we need a new grand narrative that helps us to find our way in a larger narrative story than the market. He ventures that such a narrative will help us to participate as learners in addressing the momentous environmental crisis we currently face. We need stories of sufficient power and complexity to orient people for effective action to overcome environmental problems, to address the multiple problems brought about as a result of environmental destruction, and to reveal possibilities available for *transforming* this situation and the role people can play in this project. The scope and magnitude of transformative stories brings many cultural pieces together in creative dynamic tension.

> In order for such stories to "work," to inspire people to take them seriously, to define their lives in terms of them, and to live accordingly, such stories must be able to confront and interpret the stories by which people are at present defining themselves and choosing how to live in an environmentally destructive way. It is also important to reveal how power operates, and show why those individuals who are concerned about the global environmental crisis are unable effectively to relate their own lives to such problems. The new grand narrative must enable people to understand the relationship between the stories to which they define themselves as individuals and the stories by which groups constitute themselves and define their goals, ranging from families, local communities, organizations and discursive formations, to nations, international organizations and humanity as a whole. (Gare, 1995: 140)

In my own work on transformative learning, I suggest that we need to expand our horizon of consciousness to the universe itself. The cultural historian and ecologist Thomas Berry (1989: 1) reveals the scope of a vision where the universe is the matrix:

> The universe in its full extension in space and its sequence of transformations in time is best understood as story; a story in the twentieth century for the first time with scientific precision through empirical observation. The difficulty is that scientists have until recently given us the story only in its physical aspects not in the full depth of its reality or in the full richness of its meaning. The greatest single need for the survival of the earth or of the human community in this late twentieth century is for an integral telling of the Great Story of the Universe. This story must provide in our times what the great mythic stories of earlier times provided as the guiding and energizing sources of the human venture.

I am using Berry's ideas as entry point into a broader cosmological context. This point of entry has opened up a system of larger meaning that can help create an organic planetary context for educational endeavors that transcends the myopic vision of the global marketplace.

I believe that educational vision in the twenty-first century must be accomplished within a planetary context. We live on a planet and not on a globe. When we look at the universe story, we encounter an organic totality and not a cartographic map. We are one species living on a planet called "Earth," and all living and vital energies come out of this organic cosmological context. The globe is a construct of human artifice. Before 1492, cartographic procedures for mapping commerce routes were flat. For Europeans, Colombus moved the mapping systems for commerce from a flat surface to a globe. The globe was made for commerce today, and the language of globalization is first and foremost for commercial purposes. For all of the major issues discussed under "critique," the language of globalization is the background context. The major fundamental shift in our time is that the power structures on the globe have moved from national state business (including the military business) to transnational business. All over the world, at this moment, nation-states are delivering their governments to transnational business. We cannot, therefore, dispense with global language, and it is absolutely necessary that it be the subject of deep-order cultural criticism at a world level.

The second visionary theme is education for integral development. I have chosen the words "integral" and "development" carefully. In spite of the very critical analysis of the concept of development that can be offered against market-driven development, it is nevertheless important to retain a conception of development in a treatment of transformative education. It is one thing to severely criticize our western conceptions of development; it is another to try to conceive of education in the absence of an overarching conception of development. Creative visionary education must include a conception of development that will transcend the limitations of our western ideas on development and its attendant conception of underdevelopment. Therefore, integral development links the creative evolutionary processes of the universe, the planet, the earth community, the human community, and the personal world. It is a development that must be understood as a dynamic wholeness that encompasses the entire universe and a vital consciousness residing both within us and, at the same time, all around us in the world. The endpoint moves toward a deep personal planetary consciousness that we can identify as ecological selfhood.

The third theme to be addressed within the context of transformative vision relates to quality of life as part of the learning frame of reference. We in the minority world (first world) must confront and come to terms with the quality of life that we have created for ourselves and also assume the responsibility of how that manner of living has diminished the manner of living of countless people in the majority world and in our own. The bottom line, in the global market economy, is profit. The singular major goal is economic growth indexed in the gross national product (GNP). We have sold this dream of profit to our world by commodity fetishism. Our western labor force has bought the notion of "standard of living," but this is only a comparative phrase to indicate if buying power has increased or decreased in wage potential to buy market commodities. Standard of living, however, does not add up to quality of life. Our economic market vision has left our whole culture with a crisis of meaning. Michael Lerner (1996) maintains that, in the final analysis, we hunger for more meaning and purpose in life. Our cultural values, fixated on the marketplace, have resulted in a profound cynicism that makes us question whether there is any deeper meaning and higher purpose to life beyond material self-interest. The bottom line of all this materialism and glorification of self-interest is that we find ourselves.

We are living in a period of human and Earth history that is in a state of radical transformation. Some of the habitual patterns that we have inherited have become dysfunctional for our present circumstances. We are being driven, by necessity, to devise new patterns for living to survive in a manner that gives us a sustainable quality of life. I feel that we cannot deal with our present historical moment by surface responses to our difficulties. We are now becoming aware that our western scale of progress and development is not tuned to human scale or, for that matter, the scale of Earth. Our task must be to deepen our understanding of development in a manner that takes into account a much wider spectrum of human needs.

Any examination of quality of life must attend to our deep-seated need for community and sense of place. Rootlessness, transitoriness, and dispossession are the fallout of globalization as, in an increasing number of communities around the world, people are moving to find better jobs and corporations are moving to find cheaper labor. Products for consumption, such as food, are transported thousands of miles to reach global markets while neighborhoods where people grew up (like my own) shift within a generation. Our sense of belonging to a stable community and our security are lost in the shuffle of accelerated change and mobility. This loss of connection to where we live, to the people in our communities, and to the natural world that surrounds us is tangible.

Educational institutions at all levels must play a pivotal role in fostering a community's sense of place. This is accomplished by having, as part of the curriculum, studies of the "bioregion." Bioregional study would encompass study of the land, the history of the community that has occupied a region, and the histories of people in a bioregion. Educating for the purpose of cultivating a sense of history of an area enables people to have loyalties and commitment to the place of their dwelling. In a time when the global economy can no longer be relied on to provide the basic necessities of life, the cultivation of a sense of locality and place has built within it a corrective to the vagaries of globalization. Educating for a sense of place not only has a history to give; it also has a history to make. In the latter context, locality education encourages each self-identified community to build in its educational goal the fostering of an independent local economy capable of providing goods and services for inhabitants of a locality.

Another feature associated with the issue of quality of life is the need for diversity within and between communities. In the global world toward which we are moving, there is an educational imperative for all members of the planet to enter communities of greater inclusion. Inclusion does not entail violation of boundaries. Inclusion means openness to variety and difference. Inclusion means attending to the uniqueness of each and every member of the whole. Educating for an inclusive community is thus open to the fullest sense of differentiation as well as to the deeper mystery of each person where the principle of subjectivity is honored. It is important to understand that inclusive communities operate not on the basis of sameness but on the basis of the creativity of difference. Inclusion in today's world is not created in a vacuum. Most groups and communities present themselves at varying degrees of inclusion. Finally, a transformative vision of education should be built on the foundational processes of the universe—differentiation, subjectivity, and communion. It allows a simultaneous articulation of both difference and the communal. The creativity of the community must be grounded in an awe and respect for the larger biotic community—the web of life.

Quality-of-life concerns also are connected to the development of a "civic culture." The responses to globalization, bureaucratic government, and media-driven community disempowerment are intermediary structures that link the local community to the larger global structures. We call these intermediary structures civil societies or cultures. The need for these intermediary structures is a response to the exigencies of our present global situation. The needs of an alert conscious citizenry become clear as we assess our circumstances in the global world we appear to be moving toward. The notion of citizen comes to the fore again. An alert citizenry is the ultimate check on the activities of politicians and commercial and financial institutions. Effective governance will depend on individuals exerting their rights and responsibilities to monitor the activities of governments and apply pressure to ensure that the rule of international law is not violated. Good "world citizens" will refuse to be influenced by the propaganda of governments or the media. They will be sensitive to the need to match consumerism with sustainable development and to use their voting power to ensure that economic and financial policies reflect proper care of the world's resources.

The final theme, which brings to an end this vision statement on transformation, is the sense of the sacred. I believe that any in-depth treatment of transformative education must address the topic of spirituality and that educators must take on the development of the spirit at a most fundamental level. Contemporary education today suffers deeply by its eclipse of the spiritual dimension of our world and universe. Spirituality, in our times, has been seriously compromised by its identification with institutional religions. It also has been compromised by the vision and values of the market. In a world economy governed by the profit motive, there is no place for the cultivation and nourishment of the spiritual life. Leisure, contemplation, and silence have no value in this system because none of these activities is governed by the motivation of profit. People who attend to their spiritual life are seen as nonproductive or underdeveloped. Our world economy is geared to material wants and needs, and ignores what people have called the hunger of the spirit. Our first and foremost task in life is to take hold of our spiritual destiny—a phrase that is not a household word in education. Nevertheless, we must begin to consider education as a spiritual venture. The sense of the sacred, from my own perspective, encompasses all aspects of transformative vision. It is a dimension that is integral to all three modes of transformative learning and cannot be separated from any one of these modes.

In conclusion, we must understand that transformative education fundamentally questions the wisdom of all our current educational ventures. Our present educational institutions, which are in line with, and feed into, industrialism, nationalism, competitive transnationalism, individualism, and patriarchy, must be put into question. All of these elements coalesce into a worldview that exacerbates the crisis we are currently facing. There is no creativity because there is no viewpoint or consciousness that sees the need for new directions. It is a very strong indictment to say that our conventional educational institutions are defunct and bereft of understanding our present planetary crisis. In addition, a strong case can be made that our educational wisdom suffers from what we have been calling the "loss of the cosmological sense." In truth, something was gained and, I am now just coming to understand, something was lost. I am not talking about shallow changes in fashion. I am talking about a major revolution in our view of the world that came with the paradigm of modernism.

A relevant quote from Susan Griffin (1995: 29) will give the reader an anticipation of the overall pathos of our present historical moment:

> The awareness grows that something is terribly wrong with the practices of European culture that have led both to human suffering and environmental disaster. Patterns of destruction which are neither random or accidental have arisen from a consciousness that fragments existence. The problem is philosophical. Not the dry, seemingly irrelevant, obscure or academic subject known by the name of philosophy. But philosophy as a structure of the mind that shapes all our days, all our perceptions. Within this particular culture to which I was born, a European culture transplanted to North America, and which has grown into an oddly ephemeral kind of giant, an electronic behemoth, busily feeding on the world, the prevailing habit of mind for over two thousand years, is to consider human existence and above all human consciousness and spirit as independent from and above nature, still dominates the public imagination, even now withering the very source of our own sustenance. And although the shape of social systems, or the shape of gender, the fear of homosexuality, the argument for abortion, or what Edward Said calls the hierarchies of race, the prevalence of violence, the idea of technological progress, the problem of failing economies have been understood separately from the ecological issues, they are all part of the same philosophical attitude which presently threatens the survival of life on earth.

Following on the spirit of this quote, my editorial colleagues and I formulated, in the process of developing this volume, a core definition of transformative learning that highlights its integral nature:

> Transformative learning involves experiencing a deep, structural shift in the basic premises of thought, feeling, and actions. It is a shift of consciousness that dramatically and permanently alters our way of being in the world. Such a shift involves our understanding of ourselves and our self-locations; our relationships with other humans and with the natural world; our understanding of relations of power in interlocking structures of class, race, and gender; our body-awarenesses; our visions of alternative approaches to living; and our sense of the possibilities for social justice and peace and personal joy.[4]

In the end it is for the reader to decide, but we believe this anthology reveals a diversity of sites where transformative learning takes place. It is a privilege for us at the Transformative Learning Centre to have colleagues (both students and faculty) at the Ontario Institute for Studies in Education at the University of Toronto who can speak so definitively and eloquently on these matters.

NOTES

1. This phrase is a distillation of some core ideas covered in my book *Transformative Learning: Educational Vision for the 21st Century.* See O'Sullivan (1999).
2. It is very important in the critique of western culture or the critique of modernism and industrialism to not slip into the royal "we." What I would like the reader to be

mindful of, in reading this short summary piece, is that western culture as I am critiquing it must be seen as a complex set of people and institutions. If you refer to chapter 5 on diversities and transformative learning in my book, *Transformative Learning: Educational Vision for the 21st Century,* you will see that my critique focuses on patriarchy, male-centrism, and Euro-centrism.

3. See O'Sullivan (1999) for a detailed discussion.
4. This core definition appears on our website: www.tlcentre.org.

REFERENCES

Berry, Thomas. (1993, November). A moment of grace. Paper presented at the Madan Handa Memorial Lecture Series. Ontario Institute for Studies in Education of the University of Toronto, Canada.

Berry, Thomas. (1989). The twelve principles for understanding the universe and the role of the human in the universe. *Teilhard Perspective, 22* (1): 1–3.

Freire, Paolo. (1970). *The pedagogy of the oppressed.* New York: Seabury Press.

Gare, Arran. (1995). *Postmodernism and the environmental crisis.* New York: Routledge.

Griffin, Susan. (1995). *The eros of everyday life.* New York: Doubleday.

Lerner, Michael. (1996). *The politics of meaning.* Reading, Mass.: Addison-Wesley.

O'Sullivan, Edmund. (1999). *Transformative learning: Educational vision for the 21st century.* London: Zed Books.

Saul, John Ralston. (1997). *The unconscious civilization.* Toronto: Anansi.

Swimme, Brian. (1996). *The hidden heart of the cosmos.* Marynoll, N.Y.: Orbis Books.

Toulmin, Stephen. (1985). *The return to cosmology.* Berkeley: University of California Press.

Webster's Collegiate Dictionary. (1986). New York: Merriam Webster Inc.

CHAPTER 2

What is Curriculum Anyway?

MATT MAXWELL

It is standard practice when trying to tease out the sense of a term that has been given multiple meanings over time to look at its etymology, so as to discern its roots and origins, and to see how much or how little it has been semantically altered. In the case of "curriculum," this can indeed be an instructive exercise. As most every educator knows, it is the noun that comes from the Latin verb *currere,* "to run"; a curriculum is thus etymologically the "running of a course" or a "race course." This analogy proves very interesting when examining technocratic education because of its obvious parallels with a competition. If the (largely implicit) purpose of modern schooling is to socialize young people to accept the paradigm, and to function within the framework, of a competitive world socioeconomic system that we characterize as a "rat race," or more kindly as a "horse race," then presumably the most appropriate training would be to subject students to endless practice in the running of a race course, around and around, under the watchful eyes of their trainer (the teacher) and the audience in the stands (parents, administration, community, corporations, etc.). Those who over the years show themselves to be the most adept (i.e., the most rapid) at running these various races eventually get the opportunity to leave the confines of the educational racetrack for larger, more demanding, but just as circular racetracks in the workplace of the "real world," where the stakes of winning or losing the race are much more significant than a certain letter grade.

There is an irony in all of this, because the whole modern project, as has been pointed out by a host of thinkers (e.g., Berry and Swimme, 1992; Knudston and Suzuki, 1992; O'Sullivan, 1999) has been premised on an ongoing upward and onward linear progress. Yet the lived experience of the day-to-day existence of *Homo modernis* can be adequately described only as one fraught with repetitiveness; the modern person lives a life dictated by tightly circumscribed routine, from the moment s/he wakes to the alarm of the digital clock, and for the balance of the day—as s/he progresses through the morning cup of coffee, daily commute, morning news, stock market reports, work, break, work, break, work, evening commute, microwave dinner, sitcoms, Web surfing, national news, and finally the respite of sleep. The cyclical world of the premoderns, based on the phases of the moon, the changing of the seasons, and other natural rhythms, has been replaced with one that privileges synthetically created cycles of weeks, days, hours, and minutes. With each passing year, and its concomitant explosion of information, data, commerce, consumption, and so on, more energy is added to the functioning of these cycles, making them ever more frenetic. If one can liken them to "dissipative structures" (see, e.g., Capra, 1996 or O'Sullivan, 1999 for clear explanations of Prigogine's work in this area), one can say that these high-entropy, high-energy input structures are becoming increasingly unstable. These cycles cannot continue to operate for long in their present condition—they will either evolve into something else or crash and burn.

Those arguing from a technocratic position will not agree with this analysis; rather they will choose to believe that in spite of the various crises in the world, things are better now than they ever have been and that, with a little luck, at least a good portion of humanity will be able to enjoy the fruits of the emerging *technotopia* (technological dystopia). I would like to suggest that this naïveté stems from the fact that those "running the course" of curriculum and later that of work are able to maintain an interest in participating in these endless circular journeys—within the confines of these cultural, social, and economic structures, which hold that there is no greater meaning than the accumulation of wealth—only by wearing blinkers. These psychic blinds are what keep students' eyes on the course immediately before them and stop them from being "dis-track-ted"—by other participants in the race, by the onlookers, and for that matter, by the blue sky overhead. Such distractions might lead one to question the purpose of the race, or, even more important, to wonder what lies beyond the confines of the racetrack—maybe fields, forests, meadows?

Removing the blinkers and inviting a panoramic perspective is precisely the purpose of education for transformation as it is being articulated by the likes of Selby (1997, 1998), Miller (1996, 1999), and O'Sullivan (1999). One thing should be made perfectly clear: Education that is transformational in character is *not* about replacing one set of blinkers with another, more comfortable set. It is *not* about implementing from on high educational blueprints that must be strictly followed in a prescriptive manner. And it is *not* about simply critiquing the blinders, the racecourse, or the onlookers. To remain solely at the level of critique is to remain committed to an atomistic paradigm that privileges exclusivity (either/or) over complementarity (and/also).[1] The focus is on statements like "This is wrong because . . ." or "This proposal is better than that one because . . ." as opposed to ones such as "These proposals share the following common territory . . ." or "This proposal complements that one in the following ways. . . ." Our parliaments are built on the liberal adversarial model in which the opposition parties designate members of their caucuses as "critics" of defense, education, health, environment, and the like. The modern academy is characterized by a modicum of ground-breaking thought and a plethora of critiques and commentary of this original conceptual trail-blazing in which often more time is spent pointing out the weaknesses of a new idea than its strengths. This type of approach comes, I believe, from our firmly ingrained view that no two objects can occupy the same space at the same time. From this perspective, the rightness of one approach implies the wrongness of another. One position attempts to aggressively take the place of another rather than trying to complement it. Thus, we return to the *ourobouros*[2]-like conundrum of the modern predicament in which our various approaches and positions are simultaneously trying to coexist with and supplant each other, in an ongoing, dynamic, and unresolved self-cannibalistic marriage of convenience.

Education for transformation offers a way out of this unproductive circular voyage by proposing not only to disassemble and to remove these psychic blinkers but to explore, describe, and articulate the nature of the territory lying beyond the confines of the structures of the racecourse. This exploration, at one level, begins with questions such as "How can we move beyond destructive behavior, to Earth and to each other?" but quickly moves to "How do we educate for meaning?" or, in more philosophical fashion, "What is the nature of the *logos?*" Education for transformation is about helping all participants within a "community of learners" to locate understanding, within themselves, within each other, within the world. The educations of Selby (1997, 1998), O'Sullivan (1999), Miller (1996, 1999) and the ecological strand of environmental education (e.g., Orr, 1992; Smith and Williams, 1999) all attempt, in my mind, an exploration, articulation, and expression of meaning, albeit in slightly different ways.

Before I graphically represent the ways in which these educational approaches are both similar and dissimilar, I would like to draw the reader's attention to Wilber's (1995, 1996) four-quadrant model, an epistemological map that can be used to locate different strands of thought that have recurred under different guises at different points in time. (See figure 2.1.) The left side of this schema represents our inter- and intrasubjective territory, our interiority, and those aspects of ourselves that are more readily qualified than quantified. The right side denotes all that appears on the "outside" in what scientism refers to as "objective" reality. The upper half of the model refers to individual activity, and the lower half represents collective phenomena. Thus we have four quadrants: the individual-intentional (upper left), the individual-behavioral (upper right), the collective-behavioral (lower right), and the collective-intentional (lower left). Following Wilber (1996: 86), we can identify different thinkers as privileging one or other of the quadrants.

What should become immediately obvious from a cursory look at this model is that very different entities can coexist within a given quadrant (e.g., Max Weber and

Figure 2.1
Wilber's Four-Quadrant Model

Left-Hand Paths (*Interior*) · Interpretive · Hermeneutic · Consciousness		*Right-Hand Paths* (*Exterior*) · Monological · Empirical, positivistic · Form
(Individual)	Sigmund Freud C. G. Jung Sri Aurobindo Hildegard von Bingen Teresa of Avila	B. F. Skinner John Watson Empiricism Behaviourism Physics, biology, neurology, etc.
(Collective)	Charlene Spretnak Wilhelm Dilthey Max Weber Hans-Georg Gadamer Susan Griffin	Systems theory Talcott Parsons Karl Marx Gerrhard Lenski

Susan Griffin), but the most discrepancies occur in the upper left quadrant, the individual-intentional, because ultimately it is here that the great divide between the *personal* and the *transpersonal* realms becomes evident, as in the chasms separating Freud and Teresa of Avila or Piaget and Hildegard von Bingen. (These differences would become more obvious if we were to represent this schema in a three-dimensional manner, but that could be awkward.) However, for the most part, Wilber's (1996) model does point out some strong commonalities among those who privilege activity in a given quadrant. Indeed, it is fairly easy to situate the thought of most philosophers, psychologists, sociologists, anthropologists, political scientists, and, for that matter, educators within this framework. But before doing so, it is important to acknowledge an important caveat, namely that although various people may privilege a particular quadrant, that is not to say that they do not take activity in other quadrants into account. For instance, Marx was strongly influenced by the dialectic of Hegel (lower left), but just as much so by empiricists like Hume (upper right). He believed that by altering the Collective Behavior (lower right), one could, by way of the Individual Behavior (upper right), effect the necessary positive changes in individual psychology (upper left). The thing to remember is that the starting point is, for Marx, the Collective-Behavioral quadrant. With this in mind, placing different thinkers within the schema becomes almost perfunctory. For instance, we can situate the Frankfurt School in the lower left; Robert Owen and Peter Kropotkin in the lower right, the mystics such as Thérèse Neumann, Milarepa, and Ramakrishna, as well as nearly contemporary contemplatives like Emerson firmly wedged in the uppermost corner of the upper left; some deep ecologists (e.g., Arne Naess) and some ecofeminists (e.g., Susan Griffin or Carolyn Merchant) in the lower left; whereas others, such as Earthfirst!er David Foreman, belong more appropriately in the lower right quadrant. There are some who straddle different territories, like Thoreau, who was sometimes pulled out into the world (lower right) into a passionately embodied pantheism and at other times pulled within (upper left) in the direction of an Emersonian neoplatonism. The work of ecofeminist Vandana Shiva embraces the lower left (cultural deconstruction) and the lower right (political activism). But the main point remains that most thinkers concerned primarily with the right-hand paths have views that are embedded in materialism (both reductionistic and holistic), positivism, and scientism (worldviews that are largely based in a western patriarchal perspective), whereas those on the left-hand paths (which also could be described as more intrinsically feminine, intuitive, and qualitative) are concerned with looking behind appearances, which immediately begins to happen when the observer no longer takes the observation and interpretation process for granted but realizes that it is an infinitely complex, dynamically fluid, ongoing intra- and intersubjective event that resists any accurate description.

It is clear that activity in the right-hand quadrants has been the predominant theme of modernity. But the materialist edifice is, I believe, being eaten away—both from within, with the challenge of postmodern deconstructionists such as Jacques Derrida (1991) and Jean François Lyotard (1989), who mock the notions of progress and the supremacy of scientism as essentially meaningless master narratives; and from without, by New Paradigmers, latter-day followers of the perennial philosophy[3] and ecological constructivists like David Orr (1992) and ecofeminists such as Charlene Spretnak (1991)[4] and Susan Griffin (1995).

I will now apply this model within the context of educational thought, past and present. (See figure 2.2.) For our purposes here, the critical concern is the placement of the various thinkers who are working to articulate an education for transformation, whatever they conceive that to be.

Figure 2.2

	Left-Hand Paths (Interior) · Interpretive · Hermeneutic · Consciousness	*Right-Hand Paths (Exterior)* · Monological · Empirical, positivistic · Form
(Individual)	Rudolph Steiner Ralph Waldo Emerson Jack Miller Ron Miller Maria Tagore Maria Montessori	John Locke Edward Thorndike B. F. Skinner Computer-based learning Physics, biology, neurology, etc.
(Collective)	C. A. Bowers Graham Pike & David Selby Edmund O'Sullivan Paolo Freire Nel Noddings Ecological education	Horace Mann Robert Owen Back-to-basics Educating for the global marketplace Positivist-biased curriculum Traditional environmental education

I will now briefly examine the work of several educators who work in the fields of holistic curriculum, global education, transformative education, and ecological education. I will be doing this within the framework of Wilber's Four-Quadrant model and will highlight some commonalities and differences in their thought.

For obvious reasons, I have situated Jack Miller's (1996, 1999) educational philosophy in the upper-left quadrant, in company with Emerson, Steiner, and Montessori. Miller's interest has always been primarily articulating an education founded on the wisdom of the perennial philosophy and embedded in an experience of that which extends into the transpersonal realm. Although Miller addresses in considerable depth intersubjective territory, such as community connections and social reconstructionism, and even investigates right-hand path specifics (e.g., school environment design), his first concern is always with "education for the soul," "spiritual literacy," "holistic curriculum," the "contemplative practitioner," and "spirituality and education." Miller can thus be described as a *transcendental idealist* who sees personal transformation as a requisite for intersubjective as well as external transformation. This being the case, he tends to privilege positive, visionary work over a critique of present-day injustices and crises. In this sense, Miller opens himself up to criticism from critical theorists, neo-Marxists, radical feminists, and cultural studies scholars of naive spiritual narcissism in the face of ubiquitous human misery. Miller probably would reply that injustices need to be addressed, but this does not mean that one can ever forget one's responsibility to become a more conscious, compassionate, present-to-life, humble, responsible person, and this is something that ultimately must be resolved within the sphere of the individual intentional realm of activity.

I have situated the global education of Pike and Selby (1988) within the lower-left quadrant, because their concept of curriculum is thoroughly embedded in a holistic paradigm that emphasizes the complex, dynamic, fluid web of relationships that exist between the interior and exterior, the whole and the parts, the local and the global, the culture and the individual, the past, present, future, and so on. Moreover, many of the learning activities they suggest are highly transactional in nature, employing cooperative learning strategies, simulations, and the like, which usually involve a near-equal interplay between right- and left-hand paths (collective intentional and behavioral). Nevertheless, the primary focus is on encouraging a paradigm shift; the focus of the activities is not primarily that of data collection (although research often can be a large component), but rather in eliciting deeper understanding through deeper questioning. Pike and Selby (1988) make reference to spirituality but not in the same explicit terms or in the same depth as Miller (1996, 1999). They suggest activities that involve guided visualizations and include developing trust, empathy, and inner exploration (e.g., "Four Hands on Clay," Pike and Selby, 1988: 46), but these tend to be secondary to the many interactive group learning activities that form the core of their curriculum. One could say, therefore, that Pike and Selby's educational philosophy emphasizes mutual inclusivity but privileges the collective-intentional and behavioral spheres of activity to about the same extent that Miller accentuates the individual-intentional realm.

Not unlike Pike and Selby, O'Sullivan's (1999) main focus is on the collective-intentional quadrant, although he also acknowledges the importance of activity taking place in all four quadrants. His *survive, critique, create* model (inspired by the work of Thomas Berry) involves a rigorous taking stock of the present situation. O'Sullivan explores all the political, social, economic, racial, gender, and environmental dimensions of the global problematic but ultimately moves beyond the level of critique to take on the more demanding work of envisioning and articulating an ecological, communitarian, and spiritual cosmology in which transformative learning can take place. O'Sullivan emphasizes the importance of fluidity, tensions, and dynamics as essential components of an evolution toward a new paradigm. He calls for the articulation of a new education, one that is *integral/developmental,* mirroring the creativity of autopoietic dissipative structures. O'Sullivan says: "I use the term 'integral development' rather than holistic or integrated development because of the creative dynamic and evolving nature of the processes. The term 'holistic' places undue emphasis on harmony and integration. My sense of the term 'integral' is that it connotes a dynamic evolving tension of elements held together in a dialectical movement of both harmony and disharmony" (1999: 208–209).

The function of an integral/developmental education is to move the human cultural "dissipative structure" in an upward course, in which we can become ecologically, personally, cosmologically, and transpersonally reintegrated into the universe from which we arise and in which we are embedded. O'Sullivan's point is that the transformative process that moves one to a higher level is one characterized as much by conflict as it is by harmony. Indeed, it is the shaking-up process itself, which comes from a high-entropy, high-energy throughput state, that gives rise to a greater variety of what Selby (1999) has called possible, probable, and desired futures.

Like Miller (1996, 1999), and Pike and Selby (1988), O'Sullivan (1999) is intent upon building bridges and finding shared territory among progressive, transformative, and transformational educations. However, not all of the sources from which O'Sullivan draws are as tolerant of difference as he is; they are not all ready to accept his invitation to sit at the same table. Most neo-Marxists have little time for spirituality; many radical feminists are concerned almost exclusively with deconstructing patriarchal hegemonies and view any transcendental spirituality with suspicion; deep ecologists like George Sessions (1995) have little good to say about Gaia theorists like James Lovelock (1990); and so on. In a sense, this makes O'Sullivan's attempts at reconciliation even more admirable; his endeavors—based like those of Pike and Selby on the principle of mutual inclusivity—run counter to the ingrained culture of critique so prevalent in the academy. Ultimately, the direction in which O'Sullivan points is primarily one of a cultural and paradigmatic nature. In adopting Thomas Berry and Brian Swimme's (1992) vision for a culture that will make possible our passage from the terminal Cenozoic to the Ecozoic epoch, O'Sullivan is placing the most emphasis on the cosmological (intersubjective) aspects of activity. His philosophy is also more politically, socially, and ecologically engaged than that of Miller, but his examination of spirituality, while highly eclectic, is less specific than Miller's. O'Sullivan also is suspicious of the overwrought navel-gazing[5] and the naive idealism of more radical New Agers who eschew all activism in favor of spiritual self-fulfillment.

I have somewhat reluctantly placed "ecological education" in the lower-left quadrant because it is in many respects nearly as equally concerned with specifically right-hand concerns, such as energy and geochemical cycles, food webs, biodiversity issues, and so on. Ecological education, however, unlike more traditional, scientistic environmental education, calls for deeper questioning of taken-for-granted cultural values and ultimately for embedding curriculum in an ethos that accepts the democratic principles and the rigorous scientific methodology of modernity but rejects its reductionism, atomism, anthropocentrism, andropocentrism, and cultural and ecological imperialism. This antimodern sentiment moves in various directions, which have been labeled blue-green (neo-premodern [Bowers, 1995; W. Berry, 1993]), green-green (ecological postmodern [Spretnak, 1991; Capra, 1996; Orr, 1992]), red-green (ecosocialism [Pepper, 1995]). Moreover, some theorists like Noel Gough (1999) include critical theory and postmodern deconstruction in their work, stretching the psychological and sociological boundaries of environmental education in new directions.

In one profound way, the philosophy of ecological education differs from the ideas of Miller (1996, 1999), Pike and Selby (1988), and O'Sullivan (1999) in that, almost by definition, its agenda is more specific and restrained. Orr (1992) is correct in saying that "all education is environmental education," but he neglects to add that all education—by omission or inclusion—is peace education, multicultural education, antiracist education, experiential education, global education, and so on. The point is that the ecological aspects of living and learning have clearly been, and continue to be, pitifully ignored in our learning institutions, and this serious lacuna begs to be addressed. But as Bob Jickling (1994) points out, if we simply do so by, for instance, "educating *for* deep ecology," we are falling into the trap of confusing a subtle form of indoctrination for education. In this sense, ecological education has to walk a bit of a razor's edge: On one hand, it needs to reveal hidden pathologies in the modern worldview and to suggest alternative approaches to doing things, but it also must attempt to avoid prescriptive proposals. Programs like Joseph's Keifer's Common Roots (see Smith and Williams, 1999) are exciting in that they pull together various communities—school, local, bioregional, biotic—into a cohesive curricular whole. But curricula like these do not need to be set in stone; for instance, there is no reason that any curriculum should not contain a measure of ancient history and the study of far-off lands, even if there is no obvious correlation with the local community.

As I have mentioned earlier, ecological education tends to ignore spirituality in curriculum, except in a most peripheral way. There is a general aversion to activity taking place within the upper-left quadrant because it is so easily associated with the separate modern ego or self, which is seen to be at the root of most of our ecological woes. C. A. Bowers (1995) especially views just about all aspects of the individual intentional with extreme suspicion. He virulently attacks all "modern" notions of creativity, individuality, and so on, as being hopelessly flawed, embedded as they are in a worldview that places the interests of the individual above those of the (human and biotic) collective. As vigorously as he attacks everyone—from Rousseau to Howard Gardner, from Freire to Derrida—who has advocated for anything that smacks of individualism or human-centeredness, Bowers waxes poetic about premodern cultures that embedded an ethos of respect for the greater community and the natural world in their cultural schema. Thus, for Bowers, *everything* has to do with *culture,* because culture is such a powerful and ubiquitous (in a relative, intersubjective sense) coding mechanism. Following this logic, it is better to be under the sway of a neo-premodern, ecofriendly, communitarian cultural coding than one that promotes selfishness and ecological destruction. What this line of thinking does not address is the issue of transpersonal evolution, which, by all accounts, allows individuals to access a perennial wisdom that steps beyond many of the particulars of time and place and, therefore, of many aspects of cultural coding itself.

Clearly all these educators are exploring similar terrain, although the main focus for each of them is somewhat different. All are attempting in their own ways to articulate an educational philosophy that balances the inner and the outer, the universal and the particular, the ideal and the material. I believe that it is possible to create a balanced curriculum by keeping in mind the dynamics of these four areas—interior and exterior, collective and individual—and how they are in constant interaction with each other.

The coming decade, I believe, will see a further rapprochement of these radical, transformative educations. It will be exciting to witness this ongoing and dynamic dialogue between those thinkers advocating alternatives to mainstream, technocratic curricula that are embedded in a tired modernist paradigm.

NOTES

1. For an interesting analysis of these two different epistemological approaches, see Zohar and Marshall (1994).
2. A mythological snake that swallows its own tail.

3. This term was first coined by G. W. Leibniz and popularized by Aldous Huxley in his book, *The Perennial Philosophy*—it refers to the transcultural esoteric knowledge common to all world religions.
4. Who incidentally delivers a blistering attack on deconstructionists such as Derrida!
5. This is an appropriate term, because the navel is the approximate site of the third chakra, which, like all chakras below the fourth (heart) chakra, is associated with self-centered activities.

REFERENCES

Berry, Thomas, and Swimme, Brian. (1992). *The universe story: From the primordial flaring forth to the Ecozoic era—A celebration of the unfolding of the cosmos.* San Francisco: Harper.

Berry, Wendell. (1993). *Home economics.* New York: North Point Press.

Bowers, C. A. (1995). *Educating for an ecologically sustainable culture: Rethinking moral education, creativity, intelligence and other modern orthodoxies.* Albany, N.Y.: SUNY Press

Capra, Fritjoff. (1996). *The web of life.* New York: Anchor Press.

Derrida, Jacques. (1991). *A Derrida reader: Between the blinds.* New York: Columbia University Press.

Gough, Noel. (1999). Rethinking the subject: Deconstructing human agency in environmental education research. *Environmental Education Research, 5* (1): 35–48.

Griffin, David Ray. (1998). *Unsnarling the world knot: Consciousness, freedom and the mind/body problem.* Berkley: University of California Press.

Griffin, Susan. (1995). *The eros of everyday life: Ecology, gender and society.* New York: Doubleday.

Jickling, Bob. (1994). Why I don't want my children to be educated for sustainable development. *Trumpeter, 11:* 114–16.

Knudston, Peter, and Suzuki, David. (1992). *Wisdom of the elders.* Toronto: Stoddart.

Lovelock, James. (1990). *The ages of Gaia: The biography of a living earth.* New York: Norton.

Lyotard, Jean François. (1989). In Andrew Benjamin (Ed.). *The Lyotard reader.* Oxford: B. Blackwell.

Merchant, Carolyn. (1980). *The death of nature: Women, ecology and the scientific revolution.* New York: Harper & Row.

Mies, Maria, and Shiva, Vandana. (1993). *Ecofeminism.* London: Zed Press.

Miller, John P. (1996). *The holistic curriculum.* Toronto: OISE Press.

Miller, John P. (1999). *Education and the soul: Toward a spiritual curriculum.* New York: SUNY Press.

Noddings, Nel. (1984). *Caring: A feminine approach to ethics and moral education.* Berkeley: University of California Press.

Orr, David. (1992). *Ecological literacy: Education and the transition to a postmodern world.* Albany, NY: SUNY Press.

O'Sullivan, Edmund. (1999). *Transformative learning: Building educational vision for the 21st century.* London: Zed Press.

Pepper, David. (1995). *Eco-socialism: From deep ecology to social justice.* London: Routledge

Pike, Graham, and Selby, David. 1988. *Global teacher, global learner.* Toronto: Hodder and Stoughton.

Selby, David. (1997). Schooling in sustainability: Towards education that sustains and educational change that can be sustained. Paper delivered at the Sixth International Touch Conference, Centre for Environmental Education and Ethics, Krkonose, Czech Republic, April 26- May 2, 1997.

Selby, David. (1998). Global education: Towards a quantum model of environmental education. Environment Canada Online Colloquium, October 19–30, 1998.

Selby, David. (1999). Global education: Towards a quantum model of environmental education. *Canada Journal of Environmental Education, 4:* 125–141.

Sessions, George (Ed). (1995). *Deep ecology for the 21st century: Readings on the philosophy and practice of the new environmentalism.* Boston: Shambala.

Smith, Gregory A., and Williams, Dilafruz (Eds). (1999). *Ecological education in action: On weaving education, culture and the environment.* New York: SUNY Press.

Spretnak, Charlene. (1991). *States of grace: The recovery of meaning in the postmodern age.* New York: Harper Collins.

Wilber, Ken. (1995). *Sex, ecology, spirituality.* Boston: Shambala.

Wilber, Ken. (1996). *A brief history of everything.* Boston: Shambala.

Zohar, Dana, and Marshall, Ian. (1994). *The quantum society: Mind, physics and a new social vision.* New York: Quill/William Morrow.

CHAPTER 3

Feminist Perspectives on Globalization and Integrative Transformative Learning

ANGELA MILES

INTRODUCTION

Traditionally, humanist adult education has focused on individual personal development. Critical adult education has emphasized collective engagement, critical awareness of, and resistance to, unjust social structures and relationships. Each of these broad approaches has its reductionist proponents who dismiss or ignore the social or personal as the case may be. At their best, neither humanist nor critical approaches deny the element they do not focus on. They do, however, center *either* the individual *or* the social. Transformative Learning[1] in the field of adult education has generally fallen on the humanist side of this divide, referring to individual learning that tranforms the personal worldview of the learner without integral theoretical attention to the core values or social meaning of the changes involved (Mezirow and Associates, 2000). Integrative transformative learning, on the other hand, as understood here, incorporates progressive personal change and progressive social change as mutually constitutive of each other and focuses integrally on both. This defining integration is both necessary to and made possible by another essential characteristic of integrative transformative learning, the full and conscious articulation of alternative life-centered values. Uses in the educational literature closer to the integrative definition proposed here may be found in the work of Kamla Bhasin (1994) and Edmund O'Sullivan (1999).

In the current period of triumphal capitalism, the need for alternative life-centered values is becoming more pressing and more obvious. These values are increasingly being expressed in local communities and, at the global level, by many social movements with integrative transformative learning at their heart. To be worthy of the name, Transformative Learning in the academy must consciously learn from and contribute to this practice and to the political project of imagining and creating new life-sustaining futures. Here I present an analysis of the nature and impact of patriarchal corporate globalization and the dominant rationality that legitimizes it. I hope this will make it clear (1) why Transformative Learning today has to be about transforming the world and not only individuals, and (2) why this means it must explicitly affirm not just equality and justice but also *life* as its defining value.

CORPORATE GLOBALIZATION

Capitalism has grown over centuries through the often violent colonization and exploitation of nature, women, workers, Indigenous peoples, and traditional cultures and communities. At its heart is a process by which land, labor, and nature (indeed all things) are reduced to commodities and exploited. Things come to be valued not for themselves but for their worth on the market and the profit they make for a few. Women's unpaid work and nature remain outside the market, invisible and unvalued but nevertheless essential to market production. Capitalist production is dependent on this much larger uncommodified world, which is nevertheless perceived as marginal and trivialized.

Whatever cannot be used for profit is considered expendable, including life itself. The decimation, displacement, and enslavement of whole populations, from crofters in the highlands of Scotland to Indigenous populations the world over, was the sine qua non of early capitalist development and remains characteristic. Today rural areas in Colombia are sprayed from the air with Roundup, the "Agent Orange" of the "war against drugs."[2] Indigenous and peasant communities in Canada, India, and China, with aggregate populations in the millions are respectively sacrificed to the James Bay, Narmada and Three Gorges dams, to produce hydro-electric power for "development." Farming and fishing communities in the North, whose viability has been destroyed by corporate agriculture and fishery, are deemed expendable.

However, this is not all there is to capitalism. Enormous increases in wealth also have come from the labor of workers in industrial production. In the richer nations, popular, women's, and workers' struggles over many years have won some male workers a share in this wealth and an increased standard of living for their wives and children. People have also gained political voice and statutory and constitutional rights. The "trickle down" of wealth in the affluent industrial nations has provided not only increasing private consumption but expanding public institutions and social services that have greatly enhanced the quality and security of life.

Between World War II and the end of the 1970s, policies supporting capitalist growth were justified in public discourse as a means to improve private incomes and these public services. People's voting options included political parties committed to maintaining, even improving, levels of employment, income, and services. Today all this has changed. In public discourses successfully reshaped by Margaret Thatcher, Ronald Reagan, Brian Mulroney, and their successors in all political parties, economic growth has become an end in itself rather than a means to a better life. People are now told, and have largely come to accept, that all but a tiny minority of "winners" in this high-stakes economy will have to be sacrificed in the battle for global competitiveness and profit. And this has become our reality.

All notion of communal life and values and government responsibility outside the support of economic "growth" is fast disappearing. Controls on the pursuit of profit in the name of social priorities such as equity, health, security, cultural integrity, labor rights, and environmental protection are losing legitimacy and are being progressively removed. As Vandana Shiva (1997: 22) has noted, we are seeing "the replacement of government and state planning by corporate strategic planning and the establishment of global corporate rule."

Free trade agreements such as the North American Free Trade Act (NAFTA) and the Free Trade Act of the Americas (FTAA) and international agreements such as the General Agreement on Trade in Services (GATS) and Trade Related Aspects of Intellectual Property Rights (TRIPS) are being negotiated without democratic scrutiny, debate, or agreement and forced on less powerful nations by the more powerful. The world is being opened not to free trade with a level playing field but to unfettered trade by transnational corporations. Not only goods but also services and capital are now traded "freely," regardless of cost to communities or environments (Barlow, 1996). Governments are not free to withdraw from these agreements once they have signed. Continuous and irreversible "liberalization" is assured by ratchet mechanisms that compel them to "liberalize'" ever more, never less. The provisions of these agreements are interpreted and applied not by national governments but by supranational bodies. These bodies have the power to nullify traditional practices and overrule elected governments' attempts to pursue social, environmental, and other ends that might conflict with trade and profit maximization (Palast, 2001).

Under NAFTA, for instance, traditional collective land-holding patterns and rights become illegal barriers to private ownership and profit. The Zapatistas in Chiapas are partly struggling to defend communally based livelihoods, and therefore Indigenous and peasant survival, from this NAFTA-sanctioned robbery. When the Canadian government banned the use of a gasoline additive hazardous to human health and the environment, its producer, Ethyl Corporation, brought a $251 million suit under the NAFTA for damages to its reputation and future profits. Canada settled out of court a year later, lifting the ban, apologizing publicly, and paying $13 million dollars.

Under GATS and TRIPS, corporations may challenge governments over any of their actions, or the actions of others within their jurisdiction, deemed to have infringed their property rights or "unreasonably" limited their opportunities for profit. Cases are heard by unelected judges at the World Trade Organization (WTO) in Geneva, Switzerland, in closed hearings where neither the public nor the media are allowed into the room, no transcripts are available, and no appeals are possible.

Recently, for instance, the WTO upheld the import rights of Monsanto and U.S. cattle and dairy associations against the European Union's attempt to ban beef with synthetic hormones because of the known health hazards. Canada is using the WTO to prevent countries from banning the importation of asbestos, despite its known dangers.

Until the creation of the WTO in 1995, few countries in the economic South had intellectual property laws. Now, however, U.S. intellectual property rights rules, which extend patent rights for twenty years, have become standard. All one-hundred-and-forty WTO members must conform. When the South African government passed a law allowing cheaper generic drugs to be produced and sold, thirty-nine pharmaceutical giants used international trade agreements in the courts to protect their twenty-year patents and astronomical profits, despite the desperation of people and countries doomed to do without life-saving drugs at these high prices. World outrage at the drug companies' actions in this case have resulted in some face-saving moves on their part to make the drugs they produce available more cheaply in the poorer countries. But the patents, providing vast profits from twenty-year monopolies, remain.

The Group of Seven/Group of Eight (G7/G8) governments are imposing neoliberal agendas nationally as well as internationally, further abdicating their public and social responsibilities in the process. Through the World Bank (WB) and the International Monetary Fund (IMF), they have been forcing brutal "structural adjustments" on poorer nations for decades, insisting that these nations maximize foreign currency earnings above all other production needs or policy goals in order to repay their foreign debt. Because of deteriorating terms of trade and massive interest rate jumps, the foreign debt of these nations continues to rise, despite the fact that they have repaid the principal many times over. Interest rates jumped from 2.2 percent in the 1970s, when the debts were originally incurred, to 16.6 percent in 1982. The World Watch Institute (2000) reports that in 1971 the debt of developing countries was $277 billion; by 1997 it had reached $2.171 trillion.

The debt of developing countries provides huge and continuing windfalls to the nations' private and public creditors in the economic North—$50 billion annually in transfers from the South to the North since 1987, according to the IMF (1996). It also enables the WB and the IMF to require neoliberal reform of the domestic economies of these countries as a condition for receiving the loans they need to service their debt (Isla, 1993). Each indebted nation of the South must accept a tailor-made Structural Adjustment Program (SAP) proposed by the IMF if it is to receive the WB loans it depends on.

More recently, the specter of domestic debt and deficit has been used to legitimize the imposition of the same neoliberal agenda on the populations of the G7 nations as well (Isla, Miles, and Molloy, 1996). Government subsidies to capital remain intact and even are augmented as a panoply of restructuring "austerity" measures are introduced in the economic South and North.[3] These include selling off emergency food stores; ending price controls on staples; liberalizing trade; privatizing state enterprises; "downsizing" government offices; cutting back and privatizing social service; reducing corporate taxes; and weakening labor and wage protections including unemployment insurance, minimum wage, and old age pensions.

Broadcasting, healthcare, child care, home care, public housing, welfare, unemployment insurance, education and research, transportation, environmental protection, garbage collection, public parks, and amenities are cut, to name only some of the affected areas. The resulting deterioration is weakening people's confidence in state provision. Accompanying talk of "crises" is fostering the idea that public services are always and necessarily "inefficient" and "unaffordable," and ultimately not viable. Privatization is offered as a panacea. Railroads, mines, airlines, local transportation systems, postal services, and water and power systems are being sold off to private owners, despite disastrous consequences.[4] Public services and government responsibilities such as education, healthcare, air traffic control, environmental monitoring, correctional services, even military "services"(!)[5] are being farmed out to private companies with less skilled, nonunionized, lower-paid workers and lower (sometimes dangerous) standards of performance.[6]

At the same time, new freedom of movement for capital, goods, and services, although not labor, has allowed companies to shift their operations to little-regulated, union-free, low-pay locations, undercutting wage levels and worker security in labor forces already threatened by aggressive business and government "downsizing" and reductions in workers' rights and benefits. Even though profits were at a forty-five-year high, between 1980 and 1993, the Fortune 500 companies cut their payrolls by more than 25 percent, eliminating nearly 4 million secure well-paying jobs (At Home, 2001). Most new jobs are low-paid, insecure, often part-time jobs with no benefits. Even in rich nations real earnings for average workers are falling and families increasingly require two adult earners, sometimes with more than one job each.

The impact of these business practices on the growing number of economic losers is heightened by the simultaneous shredding of social safety nets in an ideological climate that denies all communal life and redefines all public wealth as personal impoverishment ("tax theft").[7] Margaret Thatcher's famous and extreme dictum that there is no such thing as "society," only "individuals," has become the defining orientation of governments that construct every group except business as a "special interest group" and entrench corporate rights over human rights. The antihuman presumptions that follow from this logic have become so pervasive as to be no longer remarked: "Lay off workers in Britain and move your factory to the other side of the world—where labor is cheaper, unions are weaker and regimes are more brutal—and you are hailed as an entrepreneur. Arrive in Dover on the back of a lorry with the intention of working long hours for low pay and you will be branded 'bogus' and labelled a scrounger" (Young, 2001: 11).

It should not be surprising that in a system so skewed toward their interests, the power and wealth of large corporations has come to far outweigh all but the largest national economies. If the gross sales of corporations are considered as equivalent to the GDP of a country, we find that fifty-one of the world's hundred largest economies are internal to corporations (Korten, 1999).

Vulnerability to social disintegration and ecological destruction varies by region, race, class, and gender, and the poor and powerless suffer disproportionately. But everywhere today, people are impoverished and environments threatened by processes of economic growth (Waring, 1988; Douthwaite, 1999).[8] A few are managing to ride the wave that is swamping so many and they are getting richer as the rest get poorer.[9] Between 1979 and 2000 . . . the wealthiest 1 percent of Americans saw their share of the country's assets double, from a fifth to approaching one half (Beckett, 2001). Little wonder, then, that in 1994 there were 51 percent more poor children in Canada than in 1989 (Carey, 2000; Yalnitzyan, 1998).[10]

On a world scale the polarization is obscene. In 1997, 450 billionaires had assets equal to the combined annual income of the poorest 50 percent of the world's population (Korten, 1999). Worldwide, the number of people living on $1 a day or less increased from 1.2 billion in 1987 to 1.5 billion in 2000—22 percent of the world's population (Millennium Forum, 2000). As a result of this poverty, 17 million people die annually of malnutrition and preventable diseases (Russell, 2000). Millions of others must sell themselves or their body parts or their children to survive miserably even for a short while. "Trade" is growing in blood, body parts, babies, brides, domestic workers, child soldiers, and sex workers. Two million girls between five and fifteen years old are sold/entrapped/recruited into the "commercial sex market" every year (United Nations Population Fund, 1977).

Sociological and ecological damage are cause and consequence of each other in processes of scarcity creation that enrich corporate capital as they impoverish people and threaten the very survival of the planet. Increasing infertility provides large markets for pharmaceuticals and new reproductive technologies, at least in the richer nations. AIDS, heart failure, angina, and other diseases are also huge sources of profit in these nations. Soil depleted by agribusiness monocropping requires more and more fertilizers, while crops require more pesticides. Water, polluted or used up by large-scale agriculture, industrial production, and tourism, becomes a valuable commodity for those who can pay,[11] which does not, of course, include most of the one billion people who lack access to clean water. The depletion of fish stocks heralds fish farming with all its attendant and costly inputs. Oil spills offer profitable opportunities for cleanup, at least in those privileged countries and contexts where cleanup is required. The displacement of vast populations whose waged or subsistence means of livelihood have been undermined provides cheap and convenient (legal and illegal) migrant labor,[12] and profits in transporting this labor.

Social disintegration spawns massive prison industrial complexes;[13] large domestic and foreign markets for arms, police, and military equipment; and lucrative contracts for military service provision and training.[14] Each year $280 billion is spent guarding borders against desperate migrants; waging wars, civil wars, rebellions, invasions, and occupations; "protecting" dictators, oil companies, and G7, WTO, FTAA, and European Union meetings against mass protests; fighting over scarce resources; orchestrating a "war on drugs" against poverty-stricken farmers with no other way to survive; conducting United Nations (UN) peace-keeping activities and North Atlantic Treaty Organization (NATO) campaigns. Most of these conflicts are caused by the activities of the very corporations that profit from the sale of arms. Eighty percent of the profit from this trade is reaped by companies based in the five permanent member countries of the UN Security Council whose governments are complicit. The UN Summit on Social Development in 1995 estimated that the cost of the absolute eradication of poverty would be $80 billion per year over a period of twenty years, a small fraction of worldwide military spending.

Human and nonhuman life has always been expendable in this mad race for profit. Today capital's parasitical relationship to life is intensifying and expanding in significant ways as processes of corporate globalization dismantle democratic controls built over many decades of struggle. Not only are threats to life increasing everywhere, life itself is being controlled and commodified in entirely new ways through new reproductive technologies, genetic engineering, and biotechnologies, the new frontier of patriarchal capitalist development.

New intellectual property rights (IPRs), allowing the patenting of seeds, plants, animals, and human genes, transform the very basis of life into private property and legalize corporate theft of knowledge, seeds, and plants from local populations who have known, used, and developed them over centuries. Vast areas of land are being expropriated as nature reserves, with local populations expelled to "preserve" biodiversity for bioprospecting/biopiracy by or on behalf of corporations. In India the plants and seeds of basmati have been claimed as inventions by Ricetec, a U.S. corporation.

Monsanto Corporation has developed and patented genetically modified seed designed to withstand spraying of its proprietary weed-killer, Roundup. It is illegal for farmers to reuse patented seed or to grow these seeds without signing a licensing agreement to pay royalties. This policy is strictly policed by toll-free snitch lines and private police who check farmers' fields and crops. On March 29, 2001, in a case followed the world over, a judge ruled that a Canadian farmer, Percy Schmeiser, whose fields had been contaminated by Monsanto's genetically modified canola seed must pay the company thousands of dollars for violating its patent on genetically modified canola seed (Rural Advancement Foundation International, 2001).

It is a small step from making it illegal for farmers and peasants to use and reuse seeds without paying a corporation, to developing a terminator seed, as Monsanto has done, whose sterility makes this impossible. Although this "terminator technology" has been disavowed in the wake of international horror, it clearly reveals that corporations are not interested in owning life in order to protect it or to overcome scarcity or feed the world.[15] Quite the reverse, in fact. Transnational corporations are actively removing the means of livelihood from individuals and communities all over the world. Processes of globalization are dangerously concentrating power in corporate hands, power not just over the means of life but over life itself. However, these processes are also revealing, beyond a shadow of a doubt, the destructive nature of patriarchal capitalism and the urgency of tranformative change.

TRANSFORMATIVE LEARNING AND TRANSFORMATIVE POLITICS

> *Part of the fanaticism of the economic system that we now call globalization, part of its bigotry, is that it pretends that no alternative is possible. And it's simply not true.*
>
> —John Berger, "Tragedy on a Global Scale"

International communication and cooperation grounded in alternative values is fast increasing. This is partly a by-product of shrinking distances and new communication technologies brought by globalization. But it is driven mainly and most significantly by the growing need everywhere to resist the neoliberal agenda. Indigenous, women/feminist, Black, youth, farmer, fisher, worker, lesbian and gay, anticolonial, environmental, peace, antiglobalization, community, health, literacy activists, and many others are naming and refusing the costs of this agenda and (implicitly or explicitly) articulating alternative life-sustaining values and possibilities. A multifaceted, multicentered global movement is emerging built on the struggles of the many varied groups around the world (some facing direct threat of extinction) who are affirming life in the face of corporate devastation.

When the Zapatistas undertook a historic 3,400-kilometer march from their home area to Mexico City, their slogan was "Never again a Mexico without us." This was not a plea for the protection of one specific group in a small corner of a world ruled by corporate capital. It was a powerful assertion of their commitment to make a new world in which diversity of culture and biology is honored, all life is protected, and power is decentralized. In a speech to a crowd of 150,000 who met the marchers when they entered the capital, Subcomandante Marcos specifically honored a motley list of potential allies who are not usually associated with each other or with revolution, inviting Indigenous brothers and sisters, workers, peasants, teachers, students, farm workers, housewives, drivers, fishermen, taxi drivers, office workers, street vendors, gangs, the unemployed, journalists, professionals, nuns and monks, homosexuals, lesbians, transsexuals, artists, intellectuals, sailors, soldiers, athletes and legislators, men, women, children, young people and old, brothers and sisters to join with the Zapatistas in this movement (Campbell and Tuckman, 2001: 1)

A speech by Shima Das from Bangladesh to the Canadian Research Institute for the Advancement of Women (CRIAW) expresses a similar spirit:

> We believe in the concept of spinning local threads, weaving global feminism. We may have different threads both in colours and texture from many local areas. But we are rich in experiences from various parts of the world, from Africa, Latin America, Asia and Pacific, Australia, Europe and America. We have experiences where women are

> fighting against poverty, against external domination, against the business corporate exploitation, against racism, against patriarchal social and cultural systems. The experiences are diverse and sometimes far from each other even in terms of understanding. We do not think that there can be only one design to be followed as an "answer." The diversity in design and colour but woven in a single cloth will make us united and strong. (1991: 17)

Celebration of solidarity in diversity is characteristic of the many centers of visionary opposition to corporate capitalism. These groups are affirming their own value and identity as well as their connection to others; seeking their own good as well as the good of all; selectively building on and honoring traditional culture and contributions while welcoming change; articulating new senses of self and others and of possible futures grounded in the value of all life. And they are doing this *against* a hegemonic antilife worldview backed by the fast-increasing power of patriarchal capital in collusion with the state.

In this context Transformative Learning in the academy that does not actively and consciously join the search for alternative futures will, knowingly or unknowingly, serve corporate interests. Current transnational corporate ascendancy is reflected in government policies prioritizing transnational trade and economic growth. These policies are resulting in a rapid concentration of wealth and power and the continuing impoverishment of communities and degradation of nature. Adult education is being called on to play an expanding dual role in supporting this neoliberal agenda—enhancing individual and corporate competitive competence *and* containing the social costs of competition pursued at the expense of people and the planet. In the emerging "information economy," adult education is increasingly required to serve the lifelong learning needs of thriving individuals in demanding jobs. It also must respond to the growing need for English-as-a-second-language classes for migrants and refugees displaced by the consequences of SAPs and the militarization required to force these policies on unwilling populations and for "retraining" fishers and farmers whose traditional livelihood has been destroyed, for workers who have lost their jobs, and for young people who have never had jobs.

Integrative Transformative Learning, which consciously and explicitly affirms human and nonhuman life rather than profit and allies with transformative social movements, can do this ameliorative work in ways that challenge rather than serve the destructive and still largely unquestioned neoliberal growth agenda. This agenda currently is restructuring education along with the rest of the world. It includes harsh funding cutbacks; increasing commercial influence on curriculum and school culture; growing central control of education budgets and decisions; threats to privatize public education; tighter alignment of educational goals with the market; rising fees and decreasing access to higher education; research agendas set by business and market goals; ever tighter academic and business "partnership."

Protecting space for Integrative Transformative Learning in the academy requires active resistance to this educational agenda. A critical understanding of the dynamics of corporate globalization is an essential contribution to the growing global capacity to build more positive alternatives.

Notes

1. In order to distinguish between Transformative Learning in the academy and more general references to transformative learning, I have capitalized the former.
2. "Roundup (a broad-spectrum herbicide patented by Monsanto) destroys almost everything it touches, wiping out legal crops alongside illegal ones, poisoning rivers, shattering one of the most fragile and biodiverse forest ecosystems on Earth, precipitating both acute and chronic human diseases. Now the U.S. administration wants to take this ecocide a step further by spraying the jungle with a genetically engineered fungus that produces deadly toxins" (Monbiot, 2001: 11).
3. The Ontario government, for instance, recently relieved Ontario Hydro's two main successor companies of $21 billion of debt (Martin, 2001).
4. In Cochabamba, Mexico, recently local people rioted to force the government to take back newly privatized water services which had left them without water. The privatized rail service in Britain has cost lives and ruined the rail system. Privatization of power provision in California has led to energy blackouts and interrupted service.
5. When a Peruvian air force jet shot down a small plane on April 20, 2001, in the mistaken belief that it was carrying drug smugglers, it was revealed that the plane had first been spotted and wrongly identified by a U.S. surveillance aircraft carrying employees of a private firm with a CIA contract. The incident, in which a young mother and her child died, cast light on the privatization of the drug war.
6. In Ontario, the provincial government's hasty offloading of the water monitoring function to private labs has been implicated in a serious outbreak of e-coli, which caused a number of deaths and much serious illness (Harris, 2001).
7. In Ontario, Premier Mike Harris reduced social assistance by 22 percent as soon as he was elected in 1995. In Canada, the value of the minimum wage declined 48 percent between 1972 and 1992. In 1989, 87 percent of the unemployed were eligible for unemployment benefits; by 1996, tightening criteria had reduced eligibility to 40 percent.
8. In the past decade, Canada has seen increasing poverty despite a 25 percent surge in the economy.
9. A new fifteen-country study by the international organization Social Watch documents widening income gaps in every country since the advent of free trade (www.socialwatch.org).
10. For extensive documentation of increasing economic polarization in Canada, see Yalnitzyan (1998).
11. Environment Canada has just put out a call for tenders on a contract to put a dollar value on Canadian water, all the while denying that this move has anything to do with the eventual sale of the water (Williams, 2001).
12. Seven million Filipino women have left their country in search of work in Asia, Europe, North America, and the Middle East.
13. In the United States the prison population increased from 200,000 in 1970 to 2 million in 2000 (World Watch, 2000).
14. A great deal of the $1.3 billion allocated for Plan Colombia, the mixed program of military and development aid intended to fund the "drug war" in Colombia, is going to commercial ventures. DynCrop has a five-year, $200 million contract to fly lethal crop-dusters over Colombia. Other private businesses conduct aerial surveillance and have trained Colombian officers (Monbiot, 2001).
15. In order to counter the costly negative publicity of this terminator seed, corporations moved very cynically and very quickly (with public funding) to produced rice engineered with vitamin A precursors. Extensive publicity billed this "Golden Rice"

as a solution for widespread vitamin A deficiency in the third world. Free licenses "for humanitarian use" were granted for all intellectual property rights, and the rice is free to farmers earning under $10,000 per year (Cockcroft, 2001). Unfortunately, in order to gain the necessary vitamin A from this rice, people would have to eat 3 kilograms (uncooked weight) of rice every day, whereas the normal ration is only 100 grams (Bové, 2001). The reintroduction of traditional diverse intercropping, originally displaced by expanded monocropping of rice, would improve people's nutrition far more effectively than this genetically modified rice but would be much less profitable.

References

At Home. (2001). Website, run by retired senior citizens in Portland. *http://web.pdx.edu/~psu0135/wealth.html.*

Barlow, Maude. (1996). Global pillage: NAFTA has become a charter of rights and freedoms for corporations. It's time to create a moral vision. *This Magazine* (March/April): 8–11.

Beckett, Andy. (2001). When corporate love just isn't cool. *Guardian Weekly,* March 15–21: 16. (Review of *One market under God: extreme capitalism, market populism and the end of economic democracy* by Thomas Frank.)

Berger, John. (2001). Tragedy on a global scale. *Guardian Weekly,* June 7–13: 19.

Bhasin, Kamla. (1994). Let us look again at development, education, and women. *Convergence 27* (4): 5–14.

Bové, Jose. (2001). On the front line of a new world war. *Guardian Weekly,* June 8-July 4: 30.

Campbell, Duncan, and Tuckman, Jo. (2001). Zapatistas march into the heart of Mexico. *Guardian Weekly,* March 15–21: 1.

Carey, Elaine. (2000). Toronto getting poorer while 905 thrives: Report. *Toronto Star,* June 9: B1, 4.

Coates, Barry. (2001). Big Business at your service. *Guardian Weekly,* March 15–21: 28.

Cockcroft, Claire. (2001). *Guardian Weekly,* (June 28-July 4): 23.

Das, Shima. (1991). Women with diverse designs: Feminism from the perspective of real experiences of women in Bangladesh. *CRIAW Newsletter 12* (1): 15–17.

Douthwaite, Richard. (1999). *The growth illusion: How economic growth has enriched the few, impoverished the many and endangered the planet.* Gabriola Island, B.C., New Society Publishers.

Harris, Kate. (2001). Accountable but blameless. *The Saturday Star,* June 30: A1/6.

IMF (International Monetary Fund). (1996). *World economic perspectives.* Washington: IMF.

Isla, Ana. (1993). The debt crisis in Latin America: An example of unsustainable development. *Canadian Woman Studies/les cahiers de la femme, 13* (3): 65–68.

Isla, Ana, Angela Miles, and Sheila Molloy. (1996). "Stabilzation/structural adjustment/restructuring: Canadian feminist issues in a global framework. *Canadian Woman Studies/les cahiers de la femme 16* (3): 116–121.

Korten, David C. (1999). *The post corporate world: Life after capitalism.* West Hartford: Kumaian Press.

Mezirow, Jack, and Associates. (2000). *Learning as Transformation: Critical perspectives on a theory in progress.* San Francisco: Jossey-Bass.

Martin, David. (2001). Harris has it wrong on energy plan for Ontario. *Toronto Star,* May 17: A3

Miles, Angela. (1996). *Integrative feminisms: Building global visions, 1960s–1990s.* New York: Routledge.

Millennium Forum. "Facing the Challenges of Globalisation: Equity, Justice and Diversity." Report from Millennium Forum posted on www.millenniumfourm.org by Women's International League for Peace and Freedom, June 6, 2000.

Monbiot, George. (2001). Bush's dirty little war. *Guardian Weekly,* May 31-June 6: 11.

O'Sullivan, Edmund. (1999). *Transformative learning: Educational vision for the 21st century.* London : Zed Books.

Palast, Gregory. (2001). Necessity test is mother of GATS intervention. *The Observer,* April 15.

Rural Advancement Foundation International. (2001). Monsanto vs. Percy Schmeiser. *Geno-Types,* April 2 (www.rfi.org).

Russell, Graham. (2000). Want to help the poor? Attack poverty at its roots. *Miami Herald,* September 28. Available on www.rightsaction.org.

Shiva, Vandana. (1997). Economic globalization, ecological feminism and sustainable development. *Canadian Woman Studies/les cahiers de la femme* 17 (2): 22–27.

United Nations Population Fund. (1997). *Annual Report.*

Waring, Marilyn. (1988). *If women counted: A new feminist economics.* San Francisco: Harper & Row.

Williams, Lori. (2001). "National alert: New threats to your water." Sierra Legal Defence Fund fund-raising letter.

World Watch. (2000). Matters of scale: Earth day, thirty years later (Regular fact sheet). *World Watch* (March/April): 25.

Yalnitzyan, Armine. (1998). *The growing gap: A report on growing inequality between the rich and poor in Canada.* Toronto: Centre for Social Justice.

Young, Gary. (2001). Penalising the poor. *Guardian Weekly,* March 22–28:11.

CHAPTER 4

The Right to a New Utopia

Adult Learning and the Changing World of Work in an Era of Global Capitalism

Budd L. Hall

We make, by art, in . . . orchards and gardens, trees and flowers to come earlier or later than their seasons; and to come up and bear more speedily than by the natural course they do. We make them also by art greater much than their nature: and their fruit greater and sweeter and of differing taste, smell, colour and figure, from their nature

—Francis Bacon, *New Atlantis*

I think economic life is for teaching our species it has responsibilities to the planet and the rest of nature.

—Jane Jacobs, *The Nature of Economies*

Introduction

In this chapter, I will do four things: examine economic globalization as a form of global market utopia; discuss the impact of this project on our lives and on our work; explore responses to the form that this market vision takes; and finally make a case for New Utopian visions.

John Carey, in his introduction to *The Faber Book of Utopias* (1999), tells us that "Anyone who is capable of love must at some time have wanted the world to be a better place, for we all want our loved ones to live free of suffering, injustice and heartbreak." He goes on to say of Utopian ideas that "their imaginative excitement comes from the recognition that everything inside our heads, and much outside, are human constructs and can be changed." He ominously notes as well that "They [utopian projects] aim at a New World, but must destroy the old" (1999: xi). I have found Carey's views on Utopias and his encyclopedic anthology of visions from all ages and from all parts of the world fascinating and helpful. The idea of utopian projects helps me to understand both the power of the current market dreams and the importance of recovering our own right to create new utopias. Globalization is a Utopian vision. The creation of an integrated twenty-four-hour-a-day economic system that allows for total freedom for investors to find cheap money to borrow and high returns on investment anywhere in the world is a dream. That all limits on corporate and individual profits would be removed is a dream. That all workers in all countries would be integrated into global networks of production and consumption, which produce untold profits for investors, is a dream. Evidence of the power and excitement of this dream is that multinational corporations have joined with political leaders to promote this as the only dream that has a possibility of making this a better world. Not only is globalization the way for individual investors to accumulate riches beyond their wildest fantasies, but the market utopia is also, its architects would argue, the best hope we have ever had to reduce poverty and create a better world at home and abroad. The global market utopia is a dream, but it may not be our dream. It may not be a dream for all. And keep in mind, as I have earlier noted, that like all Utopias, this one "must destroy the old" (Carey, 1999: 14).

The Global Market Utopia

Globalization is being experienced in a variety of forms and practices. Important dimensions of globalization also include the economy, the state, communications, movements of people, sales of arms, and violence and crime. The most dramatic financial figure, which illustrates the contemporary global market, is that each day, about 1.5 trillion disconnected dollars change hands for financial transactions totally apart from funds needed for global trade purposes. These transactions have to do with currency speculation by private and public banks, with investments of all kinds through the computerized stock markets of the world, with bond undertakings at both private and state levels. The political leadership in most parts of the world has joined the call for each of us to play our part in the competitive global market. Products are assembled everywhere, sold everywhere, and cross borders sometimes scores of times before finding an ultimate place of rest or sale. The movement of durable goods does not stop with sale. Within days, weeks, or years most of the goods produced in the contemporary world will be discarded, and our goods then will rejoin the global search for another resting place. If we live in cities, we send our waste to rural areas. If our waste is poisonous or toxic, we will send it to the farthest reaches of our countries or, failing that, to the poorest parts of the world where countries fight over the right to become a dumping ground for the waste of the rich. Jobs, health and safety conditions, environmental regulations, human rights, and immigration policies are thrown out as deregulation on a global basis strips national legislation of its force.

The state itself has taken on global forms. The richest states of Europe now work together in a powerful economic union where the restrictions and limitations of individual governments are giving way to regional forms of state control. In Asia serious economic decisions are not taken by a single state government without direct or indirect talks with governments of trading partners such as Japan, China, Indonesia, Taiwan, South Korea, Australia, New Zealand, and increasingly Thailand and Malaysia. The United Nations system and related regional banking and development agencies are a further layer of an internationalized state function. These multilateral bodies have more power and influence in the medium and smaller states with institutions such as the International Monetary Fund and the World Bank taking on nearly full control of the economies of the least powerful states.

Crime and violence are also disturbing features of our globalizing world. The complex combination of rich-country drug use and poor-country weak economies creates patterns of international activity that take advantage of all the modern means of communications and money transfers. All of the world's people are caught in vicious patterns of cruelty and violence, that spills over into each and every one of our homes (Commission on Global Governance, 1995). The arms trade is another dimension of globalization. While the overall world expenditures on the military have declined since the 1989 accords between the former Soviet Union and the United States, the arms trade itself has taken on a new life. The United States in particular has accelerated its sales from roughly $9 billion in 1987 to over $22 billion in 1992 to over $50 billion today. According to war historian John Keegan (1994), those "who have died in war since 1945 by cheap, mass-produced weapons have, for the most part, been killed by small-calibre ammunition costing little more than the transistor radios and batteries which flooded the world during the same period" (1994: 331).

Among the low-cost weapons that cross our borders each day are land mines, which can be produced for several dollars each and which can kill or maim a person with ease. Today, there are an estimated 100 million land mines distributed in roughly sixty countries around the globe (Grimmett, 1994).

While money flows with the speed of light and goods and services at the speed of air and sea transport, people are also more mobile than ever before. The combination of economic destruction, civil conflict, and positive inducements to move has created global movements of people. On global terms, over 100 million people are refugees, living against their choice in countries in which they were not born (UNHCR, 1995). Of course, people do not move as easily as either goods or finance capital. Yet much of the contemporary movement of people is involuntary movement as economic and political refugees are forced to shift from their homes in search of security of a means to survive. In Canada, money can move in and out between Mexico, the United States, and our financial institutions with ease, but people have much more difficulty. The open capital market has not produced an open labor market. In spite of the legal restrictions against movement among our three nations, which are partners in the North American Free Trade Agreement (NAFTA), there are very large movements of people from Mexico into the United States illegally, some illegal Canadian movement into the United States, as well as vast displacements of persons from the rural areas of Mexico, Canada, and the United States to places where their chances of finding jobs are better. As Miguel D'Arcy de Olievera and Rajesh Tandon observe, "the weaker, the more vulnerable, the powerless, those who do not produce or consume anything of value for the world market, those who can hardly be privatised or internationalised are becoming expendable." (1995: 7)

FURTHER IMPACT OF THE GLOBAL MARKET UTOPIA

On Children

Canadian scholars John McMurtry and Teresa Turner (2000) in a paper on the impact of globalization on the world's children argue that "In truth what unifies global corporate investment patterns today is a ruling principle which is blind to the well-being or ill-being of children" (7). Dependency on GNP and similar economic indicators is covering over a rising tide of social disintegration and exclusion. "The cry is expressed on the ground with the unprecedented horror of families selling their own children in exponentially rising numbers as sex slaves into the new free market economies, with tens of millions of children across the world now driven not only into forced prostitution, but street beggary, slave labour and military enslavement as a consequence of the impoverishment of their families in the globalization dynamics" (McMurtry and Turner, 2000: 8).

On the Rise of Social Exclusion

The impact of globalization is not, of course, only on children or only in the majority world countries. In 1999, the Centre for Educational Research and Innovation (CERI) of the Organization for Economic Cooperation and Development (OECD) came out with a study that examines the role of adult learning in overcoming exclusion. In this study they draw attention to the report provided for the meeting of Social Policy Ministers in 1998 which notes that "The labour market has turned against low-skilled workers, who in all countries are more likely to find themselves unemployed, non-employed or earning lower wages than their better educated colleagues. . . . Unemployment remains high—35 million or 7 per cent of the work force . . . over half of all unemployed persons had been so for more than 12 months . . . households where children are present are much more likely to have low incomes than they were 10–20 years ago" (CERI, 1999: 18).

Accumulated evidence from the "rich" countries, which includes Australia and Canada, points to widening income inequalities, worrying levels of unemployment and inactivity, and growing poverty often amid a general increase in affluence. In the entire group of OECD countries, a growing number of families have no wage earners as well as a vast majority where *all* adult members of the family are obliged to work. Indeed, one in five of all households in OECD countries are now considered "work-poor"; people are working at income levels below the poverty lines in their respective countries.

On the Environment

On November 24, 2000, we learned that the Global Conference on Climate Change meeting in The Hague had failed to reach an agreement on measures to be taken to reduce greenhouse gas emissions. This is just over one year after we learned that two islands in the Kiribati Archipelago in the South Pacific were the first to be submerged by rising sea levels due to global warming and that others in the area are also in danger (*Globe and Mail*). Canada, along with the United States, Japan, and Australia, was one of the countries reluctant to make an agreement that may in any way slow down the increasing production of automobiles or cut back other industrial processes that threaten the rest of the world. Canada has the highest per capita consumption of energy of any country on earth. But rather than take a creative look at what we might do to provide leadership for greener options, our government prefers to hold out for recognition of our vast forest areas to be considered as credits in the global warming game for their role as global air filters.

In fact, the kinds of lifestyles and consumer patterns that fuel the global market utopia are a cancer for the planet. In the insightful work entitled *Our Ecological Foot-*

print, William Rees outlines a method for determining the percentage of the world's resources that we use as individuals, as communities, or as whole nations (Wackernagal and Rees, 1994). His complex formula points out that if the entire world were to achieve the same levels of development and growth that characterize most lives in the rich countries, we would need four entire planets' worth of energy resources to satisfy these demands. Clearly we are on an ecological collision path between a Utopia of the rich and the carrying capacity of a still-fragile planet.

Effects on the Structure of Work

As capital has loosened itself from the restrictions of the nation-state, so too has production. In Canada, we produce "American" automobiles from parts that are in turn made in Europe and Asia and subsequently flown in for just-in-time assembly in the auto plants of southern Ontario and Quebec. Women workers in Mexico grow the tomatoes we buy in Canada (except at the end of our short growing season); in Mexico regulations on the use of toxic chemicals for production are lower. Few clothes available for purchase in Canada are made, to use that old-fashioned concept, "at home." On the North American continent, as in other parts of the world, production has been growing dramatically in the border area between the United States and Mexico. There, in what are called the *maquiladoras,* lower-paid Mexican workers, a majority of whom are women, live in company dormitories, work in U.S.-owned factories, and send money home to their families in poverty-stricken northern Mexico. In a newer trend we are also seeing that New York City immigrant sweatshops can, for example, be as competitive on the global market as similar workplaces in Thailand (Sassen, 1991).

I am indebted to Brian Milani for his recent book *Designing the Green Economy* (2000) and its insights on the restructuring of work. Throughout the age of industrialization there has been a deskilling of direct production. Jeremy Rifkin (1994), in his study of work change in the United States, estimates that 90 million jobs in that country are vulnerable to replacement by machines. More and more we are finding that creative, fulfilling jobs are becoming rarer and rarer "islands in a sea of work degradation" (Milani, 2000: 39). The explosion in producer services has fed into a polarization not only between nations but also within nations as the fastest-growing categories of workers are building attendants, sales clerks, fast-food workers, and security guards. Contrary to popular opinion or policy documents from governments, most new jobs are and will continue to be low-paid jobs. At the same time the state continues to push for more "flexibility" in labor conditions. In the province of Ontario, after nearly fifty years of a forty-hour work week tradition, the Conservative Government has recently introduced legislation that would make it legal for workers to work up to sixty hours per week if asked by their employers.

Those fortunate to work in the larger globalizing companies of rich countries are seeing a dizzying variety of new practices in the workplace. As Griff Foley (2001) notes in his book *Strategic Learning,* "As corporations and national economies compete for market share, it is increasingly recognised that what makes the difference is the quality of organisations' human resources—its workers and managers" (9). It follows that there is a dramatic need to engage workers and managers in the goals, vision, and ambitions of the new enterprises if a given sector or company is to tap the larger reservoir of worker-manager knowledge to advance in the global market. Owners of twenty-first-century businesses are attracted by various approaches from high-performing teams, to learning organizations, to spirituality in the workplace are increased performance and productivity—the bottom line. If the workforce can create more collective knowledge for company competitiveness, feel more involved in the destiny of the company, and generate massive profits for the shareholders, so much the better.

Responses to Globalization

> *All present understood that the free market was another name for God, but then again, when one got to thinking about it, the market, like God, didn't always answer everybody's prayers.*
>
> —Lewis Lapham, *The Agony of Mammon*

Lewis Lapham is the editor of *Harper's Magazine,* a liberal U.S. publication. He had an opportunity to take part, as a journalist, in the Davos World Economic Forum, the club of 1,100 corporate and world leaders who collectively drive the Global Market Utopia. His account of his experience has been written up in a book subtitled *The Imperial World Economy Explains Itself to the Membership in Davos, Switzerland.* What is particularly insightful about this account of corridor conversations and witness to presentations by the most powerful of our world are the huge layers of uncertainty. The overwhelming topic of conversation, aside from remarks about the Monica Lewinsky affair, which had just been revealed, was the search for security on one hand and the search for a moral grounding on the other.

On November 6, 2000, *Business Week,* an influential U.S. business magazine, published an issue entitled "Global Capitalism: Can It Be Made to Work Better?" The writers therein note that "There is no point denying that multinationals have contributed to labour, environmental and human rights abuses." They quote John Ruggie, the assistant secretary general of the United Nations, as saying that "The current system is unsustainable." And, in one of the most chilling admissions I have seen, they report that "The downside of global capitalism is the disruption of whole societies" (Engardio and Belton, 2000: 8).

Eric Hobsbawn, one of the most respected western historians, in his seminal history of the twentieth century notes that, "for the first time in two centuries, the world of the 1990s entirely lacked any international system or structure" (1994: 559). He goes on to say that "In short the century ended in a global disorder whose nature was unclear, and without an obvious mechanism for either ending it or keeping in under control" (562). He closes out his work with a plea to look towards something new: "The forces generated by the techno-scientific economy are now great enough to destroy the environment, that is to say, the material foundations of life. The structures of human societies themselves including even some of the social foundations of the capitalist economy are on the point of being destroyed by the erosion of what we have inherited from the past. Our world risks both explosion and implosion. It must change" (584–85).

Jorge Nef, a global security political scientist, in a study published by the Canadian-based International Development Research Centre cautions us that "the seemingly secure societies of the North are increasingly vulnerable to events in the lesser secure and hence underdeveloped regions of our globe, in a manner that conventional international relations theory and development have failed to account for" (1997: 13). Nef continues, adding that "In the midst of the current crisis, the established flow of information, ideas, science and worldviews is being shattered" (1997: 94).

Benjamin Barber, a U.S.-based political scientist, has outlined his views in a fascinating book called *Jihad vs. McWorld:* "Jihad forges communities of blood rooted in exclusion and hatred. . . . McWorld forges global markets rooted in consumption and profit, leaving to an untrustworthy, if not altogether factitious invisible hand issues of public interest and common good that once might have been nurtured by democratic citizenries and their watchful governments" (1995: 7). "Unless we can offer," he continues, "an alternative to the struggle between Jihad and McWorld, the epoch on whose threshold we stand is likely to be terminally post-democratic" (7).

Herman Daly and John B. Cobb, Jr., in their powerful plea for an economics that reflects the environmental and human realities of our planet, note that "At a deep level of our being we find it hard to suppress the cry of anguish, the scream of horror . . . the wild words required to express wild realities. We human beings are being led to a dead end all too literally. . . . The global system will change during the next forty years, because it will be physically forced to change. But if humanity waits until it is physically compelled to change, its options will be few indeed" (1989: 13).

Responses from the Worlds of Adult Education and Training

Adult education and training are intimately linked to our daily lives, to our work lives, to our legislative and professional possibilities, and obviously to the larger political economic frameworks of our times. Adult educators and trainers have a complex approach to the changing world that we are facing. Faced with the dominance of the global market utopia, we might look at responses from the communities of adult education and training as facilitating the global market utopia, as making the best of global market implications, or transforming relations of power. In fact, our adult education and training policy frameworks quite often contain elements of all three educational responses.

Facilitating a Market Utopia

The British secretary of state notes in his forward to the 1999 *Green Paper on Learning to Succeed,* that "[t]he skill needs of the future will be different from those of today and it is clear that we will not keep pace with the modern economies of our competitors, if we are unable to match today's skills with the challenge of the developing information and communication age of tomorrow. As labour markets change, we must develop a new approach to skills and to enabling people, and businesses, to succeed" (2).

Indeed there has been a dramatic growth in adult education participation in the past ten to fifteen years. According to Paul Belanger and Paolo Federighi, the percentage of the adult population participating in some form of adult learning activities in 1994–1995 in the industrialized nations range from 14 percent in Poland to 53 percent in Sweden with Australia and Canada at 39 percent and 38 percent respectively (2000: 6). The vast majority of this provision is for labor market learning in areas of new information technologies, restructuring of larger enterprises, and training related to lower-end employment opportunities. National adult education policies in Australia, Canada, the United States, South Africa, Slovenia, Bulgaria, India, Ireland, and Venezuela have emphasized this important new requirement for global survival.

The provision of vocational education is, of course, also the area where the private sector itself has moved into with the most vigor. Adult learning is not only needed for the global market, but in many ways has become part of the global market itself. Adult learning has been variously estimated to represent a huge potential global market, if it could be fully privatized. Prior to 2001, in the province of Ontario in Canada, private universities were not legal. As a result of a new openness in the current government, there are now some sixty proposals in the Ministry of Education seeking endorsement from private for-profit universities waiting to move into the field.

Making the Best of the Market

Belanger and Federighi's recent transnational study of adult learning policies, *Unlocking People's Creative Forces* (2000), makes perhaps the most eloquent case for an adult education framework that responds to both the economic as well as the individual and collective demand for adult learning. Written in the spirit of the 1997 Hamburg Declaration, of which Belanger himself was the architect, this book makes the case for a broad vision of adult learning that meets a variety of needs: "In order to survive and improve their lives, adult men and women on every continent are striving to develop the means to enhance their capacity to act and the understand the ways of the world. Adult learning also has strategic importance for economic actors today. Risk management strategies, economic policy, environmental and health policies all invariably rely on continually raising people's competencies and skills. Similarly, traditional and new types of popular movements and national liberation projects call for strengthening and spreading the capacity for initiative so that people can deal with the challenges facing them, bring about change and take an active part in economic and social development" (2).

One of the most creative approaches to taking up the global market challenge from an adult learning and organizational change perspective that I have come across is the recent work done by the Australian adult educator Griff Foley. Better known for his work on the teaching and learning processes and for his widely respected writings on radical or social movement learning, Foley has in the past several years entered fully into the world of workplace learning and change. In book on what he calls "strategic learning," Foley (2001) outlines a vision for an approach to organizational learning that he feels is both strategic and emancipatory: "A 'strategic learning' approach rejects the attempt to recast adult education and learning simply as an instrument for improving performance and productivity. It sees learning as complex (formal and informal, constructive and destructive, contested and contextual). It assumes that critical and emancipatory learning is possible and necessary. It asserts that a first step to their realisation is an honest investigation of what people are actually learning and teaching each other in different sites—workplaces, families, communities, the mass media, social movements" (20).

Foley also calls for a new foundation of workplace learning. He considers an appropriate analytic framework for workplace learning to include: the nature of and reasons for the long-running global economic crisis; the character and logic of the capitalist political economy; the character and logic of work in capitalism; the experiential learning of people in changing workplaces; the unconscious dimension of workplace learning; and gender relations in workplaces (63).

Resisting and Transforming

It may be argued that more adults learned about the nature of global market structures and the problems generated by them in the several days of the Seattle demonstrations before the World Trade Organization meetings than from any adult education conference yet organized. There is a dramatic increase in social movement learning linked directly to the antiglobalization struggles. Young people, trade unionists, and activists from the majority world came together in Seattle, as they had in lesser numbers in other cities before and as they have since then in Prague, Washington, D.C., and Windsor, Ontario, Canada. A new global civil society is a reality. The same computer networks that support global flow of capital are beginning to support a more active flow of ideas for resisting and transforming the global market frameworks. I would urge those interested to log onto the global websites of the International Confederation of Free Trade Unionists (www.icftu.org) for a rich array of educational materials and ideas for strategic thinking about all aspects of globalization and its transformation. You will find language for policy documents and collective bargaining that go beyond the adaptation to global markets. Similarly, LabourStart, a project of the Labour NewsWire Global Network (www.labourstart.org), is a sophisticated educational and information service for trade union educators and others interested in a labor perspective on the changing workplace.

The world of adult education networks itself is undergoing substantial changes. The International Council for Adult Education (ICAE), a global network of some 700 local and national nongovernmental organizations (NGOs) interested in adult learning, organized a World Assembly of Adult Education in Jamaica, August 9–13, 2001, that called for global advocacy toward adult learning for individual and collective transformation in the age of globalization. One of the most effective regional groups affiliated with the ICAE is the Asia and South Pacific Bureau of Adult Education (ASPBAE). Its website (www.aspbae.org) is another vital resource for a transformative vision of adult education.

THE RIGHT TO A NEW UTOPIA

The most powerful instruments to transform the world that we have are our own minds. The 1985 UNESCO Right to Learn Declaration spoke, among other things, of the right to imagine. We have the right to a New Utopian vision, a vision that responds to the collective needs of the majority of people in the world, not simply the few. We need to grasp the power of the utopic vision for ourselves. The global market utopia is held in place by coercion and force, but it is most firmly supported because we share at least a part of the dream that this global utopia speaks to but does not deliver. In part we want to believe that by following the strictures of the global market we will find our way to a more secure world.

Linda McQuaig, a Canadian journalist, notes that "We have become convinced that we are collectively powerless in the face of international financial markets. And with the widespread acceptance of this view the rich have proceeded to create a world in which the rights of capital have been given precedence over and protection against interference from the electorate"(1998: 283). She argues that in the case of Canada, we have been "sold" a myth of powerlessness because it serves the interests of the current ruling alliances, not because we in fact do not have any power as citizens. I believe that McQuaig's arguments hold true for many parts of the world. We have both the ideas and the means of implementing these ideas if only our gaze could begin to be focused in new directions.

It is a time to claim back the power of the Utopian vision. In the claiming back of the power of a vision of the world we want, as opposed to the world we do not want, adult learning has an important role to play. In surveys in Canada, it has been found that adults already engage in, on average, eight hours of autonomous informal learning on their own each week. Individual and social demand for learning is a transformative force of the greatest power. The adult learning and training world is rich with networks, websites, newsletters, conferences, and other structures of communications that already exist to support the social demands for adult learning. It is time for the resources and capabilities of the adult learning communities to support the search for New Utopian visions. New Utopian visions are found in local community gardens, in community shared agricultural schemes, in individual and family choices to live more simple lives, in the large and still-growing movement for "green economic development," in social economies of varying kinds, and in the literally millions of creative ideas that women and men are engaged in as ways to survive in a world that they do not like yet know not how to change. It is time for us to claim the right to a New Utopia in our workplaces. Adult education and training can support the release of our creativity and imagination. We have the right, as the late Paulo Freire says, to become agents in our own history.

For those of you who doubt the power of citizens working together, consider the fascinating account in *The Economist*'s publication "The World in 2001": "The anti-capitalists have been winning the battle of ideas . . . despite having no ideas worthy of the name. In 2001, their influence on governments and in boardrooms is going to increase" (15). Equally worrying is the influence that they are having with big western companies that nowadays feel obligated to bow down before bogus nostrums of "corporate responsibility" (15). And in another article in the same issue: "Activists have already seized the initiative on global trade. They succeeded in scuttling both the OECD's Multilateral Agreement on Investment in 1998 and the launch of the WTO's new global trade talks a year later. . . . This means that at the very best *trade liberalisation is stalled* [emphasis added]" (Wooldridge, 2000: 96).

In *Which World: Scenarios for the 21st Century* (1998), Alan Hammond of the World Watch Institute reports on a research project that extrapolates current global trends in three different scenarios. This research suggests that, roughly speaking, the contemporary context provides evidence for what Hammond calls Market World, Fortress World, and Transformed World.

In Market World, we see the extension of the global free trade agenda, what I call the global market Utopia. It is, however, a vision that still contains nearly all prevailing economic and political powers and influence. Fortress World arises in the scenario that sees market-led forces failing to redress social disparities and eventually spreading stagnation and fragmentation. Resources are shifted rapidly to security issues to contain growing violence and conflict. In Transformed

World, also drawn from existing trends and beginnings of new global civil society movements, we see a society with transformed values and cultural norms in which power is more widely shared and in which new social coalitions work from the grassroots up to shape what governments do. The market still exists, but it is balanced by other needs. In reality, of course, the future, as the present, will contain aspects of all three scenarios, but in which proportion, in which combination, and for whom?

Today, as has also been the case in the past, the conviction for changing the world exists, as do many of the ideas necessary for making these changes. Gandhi, according to ecofeminist Vandana Shiva (1998), elaborated the concept of *Swadeshi,* an economic self-reliance based on the conviction that "people possess both materially and morally what they need to evolve and design their society and economy and free themselves of oppressive structures" ("Drive to Nurture Swadeshi Spirit," 2001: 13). Indigenous knowledges, long surpressed, offer ideas for the future as well. In a recently completed book that I coedited with George Sefa Dei and Dorothy Goldin-Rosenberg, we suggest that "We need to understand and move beyond the often genocidal effects of decades of colonialism and maldevelopment practices. . . . We need to call for locally defined models of sustainability which will prevail the lived realities of local peoples with all their social, cultural, political, spiritual, moral and ecological goals and aspirations" (2000: 11–12).

Samir Amin, the veteran majority world political economist, notes that "a humanistic response to the challenge of globalization inaugurated by capitalist expansion may be idealistic but it is not utopian: on the contrary, it is the only realistic project possible" (1997: 10).

In closing I am reminded of the late Julius Nyerere, the first president of Tanzania, who had so much faith in adult learning. He set a goal for us in 1976 at the first World Assembly of Adult Education when he noted that the first goal of adult education was to convince people of the possibility of change. All other goals can come if we believe that change is possible.

REFERENCES

Amin, Samir. (1997). *Capitalism in the age of globalization: the management of contemporary society.* Atlantic Highlands, N.J.: Zed Books.

Bacon, Francis. (1627). *New Atlantis.* London.

Barber, Benjamin. (1995). *Jihad vs. McWorld.* New York: Ballantine.

Belanger, Paul, and Federighi, Paulo. (2000). *Unlocking people's creative forces: A transnational study of adult learning policies.* Hamburg: UNESCO Institute of Education.

British Secretary of State. (1999). *Green Paper on Learning to Succeed.* London: Secretary of State Policy Paper.

Engardio, Pete, and Bekton, Catherine. (2000). Global capitalism: Can it be made to work better? *Business Week,* November 6: 8.

Carey, John. (1999). *The Faber book of Utopias.* London: Faber.

Centre for Educational Research and Innovation (CERI). (1999). *Adult education and the allieviation of social exclusion.* Paris: OECD.

Commission on Global Governance. (1995). *Our global neighbourhood.* Oxford: Oxford University Press.

Dei, George Sefa, Hall, Budd L., and Goldin-Rosenberg, Dorothy (Eds). (2000). *Indigenous knowledges in global contexts: Multiple readings of our world.* Toronto: University of Toronto Press.

Daly, Herman E., and Cobb, John Jr. (1989). *For the common good: Re-directing the economy towards community, the environment and a sustainable future.* Boston: Bacon.

D'Arcy de Olievira, Miguel, and Tandon, Rajesh. (1995). *Citizens: Strengthening global civil society.* Washington, D.C.: Civicus

Drive to nurture Swadeshi spirit. (2001). *Times of India,* November 1: 13.

Economist, The. The world in 2001. London.

Foley, Griff. (2001). *Strategic learning: Understanding and facilitating change,* Sydney: Centre for Popular Education, University of Technology.

Grimmett, John. (1994). *Conventional arms transfers to the third world.* Washington, D.C.: Library of Congress.

Globe and Mail. November 24, 2000.

Hammond, Allen. (1998). *Which world? Scenarios for the 21st century.* Washington, D.C.: Island Press.

Hobsbawn, Eric. (1994). *Age of extremes: The short 20th century.* London: Abacus.

Jacobs, Jane. (2000). *The nature of economies.* New York: Modern Library.

Keegan, John. (1994). *A history of warfare.* Toronto: Vintage Books.

Lapham, Lewis H. (1998). *The agony of Mammon: The imperial world economy explains itself to the membership in Davos, Switerzland.* New York: Verso.

McMurtry, John, and Turner, Teresa. (2000). The effects of globalization on the world's children. Unpublished paper. University of Guelph.

McQuaig, Linda. (1998). *The cult of impotence: Selling the myth of powerlessness in the global economy.* Toronto: Viking.

Milani, Brian. (2000). *Designing the green economy: The postindustrial alternative to corporate globalization.* New York: Rowman and Littlefield.

Nef, Jorge. (1997). *Human security and mutual vulnerability.* Ottawa: International Development Research Centre (IDRC).

Sassen, Saskia. (1991). *The global city: New York, London, Toyko.* Princeton, N.J.: Princeton University Press.

Shiva, Vandana. (1998). Swadeshi. *Weekend Observer,* April 3.

United Nations High Commission on Refugees (UNHCR). (1995). *Annual report on refugees.* Geneva: United Nations.

Wackernagal, M., and Rees, W. *Our ecological footprint: Reducing the human impact on the earth.* Gabriola Island, B.C.: New Society Publishers, 1994.

CHAPTER 5

From Opposition to Alternatives

Postindustrial Potentials and Transformative Learning

Brian Milani

The possibilities for social movement learning, and most aspects of transformative learning, are historically specific. The cultural, economic, and technological potentials of society clearly affect the goals of social change. Transformative learning even includes possibilities for personal-spiritual change, but this is a topic I cannot deal with here. In this chapter, I want to look—from a political-economic point of view—at how postindustrial development potentials have affected political strategies, and the educational agenda, for radical social change today.

In the wake of the anti–free trade protests in Quebec City (April 2001), debate among activists and supporters has raged about possible strategies for the antiglobalization movement. Most of the debate has centered on the relative value of nonviolent protest versus more militant direct action, the latter tact unfortunately attracting more media coverage. But some activists are beginning to point out that, whatever option is chosen, so long as the movement remains preoccupied with strictly *oppositional* activity, eventually it will burn out or deflate. The movement, they insist, must begin to clarify a positive vision and increasingly define the alternatives to corporate globalization. A strategy more geared to generating social and economic alternatives might provide a context to allow the violence/nonviolence debate to resolve itself or perhaps even fade into superfluousness.

This is an important debate because the antiglobalization movement expresses a high-water mark of social activism dating back to the decline of the last mass tide of activism in the early 1970s. The decline of those movements was due to many factors, but a major one was undoubtedly that they were unable to define satisfactory alternatives to what they opposed. This is understandable since they were relatively new movements, without a lot of time to find common ground and make their (largely implicit) alternative visions concrete. The environmental movement in particular was very young, and this is very significant in view of how central the ecological dimension is to economic alternatives. Besides this, the economic boom of postwar Fordist capitalism was just cresting, and the hollowness of the promises of the Welfare State and economic growth were just starting to be acknowledged.

Another reason why it would be unrealistic to expect the new social movements of the 1960s to provide comprehensive alternatives is that such an alternative project was historically unprecedented. The new social movements of the Fordist era—for peace, women's liberation, the environment, African American, and Aboriginal self-determination, and so on—were themselves expressions of emerging human potentials and new productive forces (NPFs). These potentials were first visible during the Roaring Twenties with its vibrant cultural experimentation. But after a harsh dose of material deprivation in the Great Depression, and then wartime mobilization, it would be the 1960s before new sensibilities would reemerge in new social movements.

These new sensibilities were largely the product of a major movement of industrialization into the realm of culture. Old roles and identities, based on labor and on gender, began to break down as work, social organization, and consumption changed. The process intensified throughout the twentieth century. The so-called information revolution is often seen to be the driving economic force behind this upheaval, but actually it has been only one small aspect of it. The most important thing is not the new role of information but a new role for human creativity in general—expressed in a fundamentally new relationship of culture to the economy. It is a transformation that also has profoundly changed the nature of politics.

I will come back to elaborate on these political-economic-cultural dynamics, but my basic point is that in this new context, alternatives play a much more important role in social transformation than in the classical era of industrial capitalism. During the early period of industrialism, progressive social movements were concerned primarily with the *distribution* of society's wealth. In the current period—marked by postindustrial potentials—the new social movements are more concerned, at least implicitly, with the *redefinition* of wealth: from quantity to quality, from accumulation to regeneration.

Not only have the concerns of popular movements changed over time, but the NPFs also have fundamentally changed the relationship between opposition and alternatives in progressive movement strategy. The old labor and socialist movements of the past needed to have control not only of the means of production but also of state power in order to implement any substantial alternatives. Today it is possible to begin to create these alternatives directly without having prior control of the state (Roberts and Brandum, 1995). This is not just because ecotechnologies are more developed but because even much mainstream technological development tends toward decentralization. Think, for example, of fuel cells and "distributed generation" in the energy sector. Even though market globalization is trying politically and economically to offset these technological potentials, market globalization (based in giant loops of production and consumption) is very wasteful, and many decentralizing tendencies are still powerful. This can work in our favor, and we are in a very different situation from the one that, for instance, Gandhi faced when he advocated self-reliant community economies.

The tasks of redefining wealth and of defining and creating specific alternatives in every sector of the economy are core responsibilities of transformative learning today. Before getting any more specific about them, however, we should look a little more closely at the role of culture in postindustrial development.

CULTURE AND QUALITATIVE DEVELOPMENT

The new movements that have occupied the progressive political stage since World War II have been much more culturally defined and more concerned with quality of life than the older labor and socialist movements. They sprang from a new form of capitalism in which industrialization had moved into the realm of culture and quality, with both inputs and outputs of production becoming more cultural. The rise of intellectual labor, new kinds of services, cultural industry, mass education, and, eventually, the growing concern with lifelong learning are all manifestations of this new importance of culture. Industrial capitalism has more or less integrated these elements into its forms of production and exchange, but the strains this has put on the system have not generally been appreciated.

The system has been strained because, compared to material products, culture is not so easily commodified and accumulated. Culture is largely a qualitative phenomenon. Industrialism, by contrast, is essentially a system of quantitative development, based in money and matter. More is always better. The system prioritizes accumulation above everything else, satisfying people's needs only indirectly as a by-product, spin-off, side effect, or trickle-down. For example, it produces as many cars as is profitable, assuming people's transportation needs will be taken care of. It produces any food commodities that will sell, assuming nutritional needs will be satisfied. The state is charged with filling in the gaps when the spin-off is intolerably insufficient, but real needs, be they social or environmental, still take a backseat to accumulation. This in fact is the very definition of capitalism: "exchange value" must always come before "use value" or social need.

During an earlier phase of industrialism—when the primary end markets were overwhelmingly for products to satisfy primary needs (food, shelter, clothing, etc.)—this one-sided focus on accumulation made more sense. Primary needs are pretty standard, and the goal of overcoming scarcity was uncontroversial. Socialists and labor activists generally had no quibble with the industrial definition of wealth—money and matter. They were concerned primarily with the distribution of this wealth and the conditions in which workers produced it (Paehlke, 1989). Except for some communitarian and utopian socialists, few argued with the benefits of economic growth or what I will call quantitative development.

Things change, however, when both the inputs and the outputs of production become less material. In the conventional industrial system, the key factors are cog labor and vast amounts of physical resources. With the industrialization of culture, human creativity becomes the key factor; it can begin to displace both drudge labor and resources from production.

This is a major threat to capitalism for a couple important reasons. First, cultural production is not really compatible with capitalist markets geared to accumulation and the "allocation of scarce resources." Industrial capitalism is a mode of quantitative development, based in matter and money; its supposedly self-regulating markets do not work properly when faced with nonstandardized needs and products. This is one reason behind the Great Depression—a market failure that dramatized the historical limits of quantitative development. At this stage, some kind of conscious intervention is needed.

Qualitative wealth defies commodification for a variety of reasons, but one of the most important ones is that it requires the direct and specific targeting of human need. It does not just happen as a spin-off or trickle-down of accumulation. In fact, as we see so clearly in terms of environmental health, an excess of quantity (e.g., economic growth) can destroy quality (e.g., ecological balance or community). Increasingly real quality requires dematerialization. This is not an argument against matter and money—only that they be dethroned as humanity's economic gods and become strictly means to the end of real qualitative development.

SCARCITY, POWER, AND WASTE

The second and related reason why the industrialization of culture has been a latent threat to capitalism is because the NPFs embody the possibility of moving beyond scarcity. Capitalism is a class society, and class society is based in relative scarcity, both material and cultural. Class power involves the control of scarce resources—the economic surplus—by a minority. Absolute abundance undermines class, because when people have all their basic needs taken care of, they are not so compelled to take orders.

In this sense, industrial capitalism always has been a living contradiction, because its open-ended productiveness and its constant economic growth were destined from the first to eventually undermine the scarcity basis of class rule. This inevitability came to fruition with the Great Depression, which has been referred as a structural crisis of overproduction. The post–World War I technological explosion of the 1920s generated a productive output far beyond the capacity of worker-consumers to purchase. After the 1929 crash, a chronic crisis of "effective demand" and "business confidence" ensued and put a final end to classical capitalist free markets (Block, 1987; Guttmann, 1994). In this sense, the Great Depression was a spontaneous system shutdown in response to the threat of abundance, and capitalism would henceforth need various forms of state intervention not simply to perpetuate economic growth but to maintain scarcity.

The threat to scarcity and class was, however, just as much cultural as economic. Class society always has been based on a monopoly of "high culture" by a privileged minority. The maintenance of a gap between high and folk culture has been just as essential to ruling groups as control of the state or the economic surplus. By industrializing culture, however, industrialism gives workers new cultural and intellectual capacities that might undercut their cultural dependence. The twentieth century saw the virtual elimination of the distinction between "high" and "folk" culture—as we see so clearly in the sophistication attained by folk arts like jazz. Capitalism has had to reproduce class relationships in other more subtle ways, but always there has lurked the latent specter of a classless society.

For capitalism, the crucial means by which it has perpetuated scarcity—both material and cultural—has been waste. Quantitative growth has been kept going, and yet this has not begun to meet all the basic needs of the people. After World War II, the key elements of the Fordist waste economy were the arms economy and the privatized consumer economy based in suburbanization. The latter was a fragmented landscape, populated by bungalows and cars, and powered by oil, which maximized the consumption of virtually every material. The blatant wastefulness obviously maintained material scarcity. But just as important is the fantastic waste of human potential implicit in the extension of cog labor to all kinds of unnecessary or alienated production—from cars, to finance, to TV, to junk foods, to pornography. Alienated forms of consumption have been just as suppressive of cultural development. Such a capital-intensive form of development also inevitably worked to limit the power of organized labor—since it channeled the information revolution in the direction of displacing labor (rather than resources) from production. Even with the legitimization of collective bargaining in mass production, eventually labor's strength would be eroded. This has been very obvious since the mid-1970s.

This role of scarcity in maintaining class and quantitative development is one reason why certain environmentalists are wrong when they see our environmental crisis as resulting from affluence in the developed countries. The problem is not affluence but *effluence*—waste that artificially reinforces scarcity relationships. The compulsion to get money is a constant distraction from more regenerative goals. If we were able to free people and firms from the compulsion to get money or accumulate capital, we could more easily target real human and environmental needs and substantially dematerialize the economy.

Appreciating how important waste is in maintaining class today shows how environmental questions are absolutely central to questions of economic development, political power, and social justice. Waste has been the crucial structural means by which industrial capitalism has maintained anachronistic relationships. The alternatives that must be implemented are not simply new forms of distribution or governance but substantial new designs for every economic sector to fit within natural processes. Ecological design helps establish the economic basis for justice, equality, and democracy.

The New Ecology of Politics

Real postindustrialism is all about actualizing potentials for qualitative wealth creation, for putting human development first, for dematerializing the economy, and for integrating economic processes within natural cycles. Besides spawning new potentials for economic development, the industrialization of culture also has created the possibility of new forms of regulation and political action. A green economy does not simply require a new politics of ecology (as the mainstream green parties tell us) but a new *ecology of politics,* featuring the decompartmentalization of politics as a separate realm.

It is no surprise that the new social movements have not focused primarily on elections, party politics, or the state. Intuitively they know that the old political forms are designed to narrow and fragment important issues and relationships. It is not even enough to democratize the state, because the state itself is a problem, serving to keep politics out of daily life and everyday activity. The role of the state as a rule-maker is a very important one. But there are all kinds of ways to create pressure to change the rules while building parallel grassroots power.

The original separation of politics, economics, and culture in classical industrial society was to some extent inevitable. Perpetual economic growth was a totally new phenomenon, and industry was the engine of progress. As political economist Karl Polanyi (1957) showed, the new market economy and the associated property relationships were created deliberately by the state. But it is also true that the new industrial economies raged on like runaway trains without anyone at the controls. The new importance of production gave the working classes a strategic position to exercise power they never had in agricultural societies. Workers struggled for the right to vote, but this unprecedented political power was something that ruling groups could concede. On one hand, electoral democracy could act as a kind of social feedback mechanism for the elite, providing some stability for the runaway economy. On the other hand, the political realm could be isolated from the economic realm, where the real power lay. Class interests used the state to institutionalize property relationships that insulated the economic realm from the political (Montgomery, 1993). But the propertied interests also were protected by the real difficulty of controlling the market economy at this level of economic development. Even if the state was intent on providing a conscious alternative to market distribution, in the absence of sophisticated information technologies and management systems, it was certainly at a disadvantage.

This was a temporary situation, however. As the industrialization of culture gathered stream early in the twentieth century, markets became increasingly incapable of performing their assigned tasks of efficiently distributing resources. The cultural-economic revolution also began to provide the state with the organizational and informational tools to manage the economy: especially white-collar labor, the substance of bureaucracy. The erosion of the industrial separation among politics, economics, and culture had begun—an erosion that would threaten class power if left to continue. Bureaucracy might be seen as something of a threat to markets, but it also could be a new means of reproducing class relationships. It depended on how the new political-economic forms of integration would take place.

State socialist regimes began to appear. Corporations—which are actually large political-economic organizations (or "industrial governments," as David Bazelon [1963] put it)—came to replace individual entrepreneurs and family businesses. New forms of Welfare State capitalism emerged that permitted political-economic integration at the highest levels but kept politics and economics well fragmented in the daily life of ordinary citizens. Again, especially in North America, waste production would provide the economic content for the Fordist economy that could pay more attention to mass consumption than any previous form of capitalism while keeping people chained to alienating work.

In this context, it is no surprise that the new social movements would move beyond strictly distributional struggles, especially those focused on the workplace and the state. And while the new movements would use conventional political avenues, they would not subordinate themselves to these processes. They would insist on a broadening of politics.

Politics and Identity Beyond Cog Labor

The fragmentation of industrial society, along with working-class cultural dependence, dictated a dual strategy of change for the nineteenth-century working class. It had to get control of the means of production, or at least establish some power at the (paid) workplace and on the labor market, which it tried to do through trade unionism. But it also had to exert some power in society at large, by influencing or getting control of the state. Whether peaceful or otherwise, substantial change required action on these two fronts, economic and political.

Underlying it all was cog labor, or what Marxists would call "simple labor power," which defined the role of the human being in the industrial system, as it shaped working-class consciousness and identity. Industrial machines were not, like craft tools, extensions of the worker; but rather, the worker was a cog in the machine system as well as a commodity on the labor market. The essential unpaid domestic work that reproduced this labor, while uncommodified, was equally routine and drudgelike.

Workers immersed in cog labor, be it paid or unpaid, clearly had neither the time nor the skills to carry out the dual strategy alone. Politically, the working class needed representatives—advanced workers and sympathetic intellectuals—to carry on the political struggle via socialist, social democratic, labor, or democratic parties. This was the role of the organized left.

During an era when the working class was shaped by cog labor and cultural dependence, the organized left served as the proverbial head on the working class body. It was the workers' shadow state and carried on the struggle in the separate political realm. It was after state power, and no substantial economic alternatives could be implemented until after its attainment. And, as discussed earlier, the focus of the alternatives was on redistributing material wealth, not redefining it.

Today a very different situation exists. The working class is large and almost synonymous with "the people." It is not a homogenous mass, however, and its progressive elements are constantly moving to express the diversity and complexity of human experience. So-called identity politics is the expression of people looking not only beyond class but also beyond cog labor for a deeper sense of selfhood, integrating their cultural heritages with new developmental possibilities. The new social movements have attacked all restrictions on human potential development as expressed in racism, sexism, ageism, anthropocentrism, and the like. They have worked to undermine not simply class society but all the interrelated forms of domination that date back to the beginnings of civilization. Identity politics is by no means peripheral to economic issues but, at its best, is our means of taking control of the "people production" that is at the core of the new productive forces.

Today the crucial productive forces are not machinery but human creativity and self-development that exist everywhere. We have moved beyond the era of *thing production* to one of *people production,* but the industrial system attempts to disguise this by producing people-as-things and creating so much material waste.

Industrialism always has kept key forms of human and ecological productivity invisible—for example, domestic labor and ecosystem services. Today's casino capitalism has maintained and extended this invisibility via the undervaluing of natural resources and many human activities. But the competitive struggle in capitalism means that the system must deploy at least certain aspects of the NPFs. So it selectively cultivates and channels this creative activity into destructive and antisocial forms of production: financial industry, defense industry, genetic engineering, advertising. Not only is this deployment of human creativity, as expressed in high-skill jobs, destructive, but it is also restricted to a narrow band of the workforce. As sociologist David Livingstone (2001) has shown, the idea of the "knowledge-based economy" is a myth. Most people in developed countries are far more educated and skilled than they need to be for the available jobs. It is *civil society* that is knowledge based, suggesting latent but unrealized potentials for truly regenerative economic development.

REVOLUTION: THEN AND NOW

For the working class of today, seizing the means of production essentially means *seizing ourselves.* While institutional change is important, there is a crucial internal dimension to change. People must not only reclaim control over their creative capacities, but they must begin to cooperatively establish productive economic outlets for this creativity in the form of grassroots community-based ecological alternatives. Many of the existing mainstream forms of production are not worth saving. They are invariably wasteful and inefficient. Whether we look at agriculture, the energy, manufacturing, or even finance, socially and ecologically responsible production would be completely different in form and content.

This strategic importance of alternatives, along with the cultural capacity of the working class, has great implications for both the form and content of revolutionary strategy today. First the historic organizational role of the left is obsolete. The working class does not need to be led or represented—it can do that itself, in many diverse forms. Second, the division among politics, economics, and culture has no material, social, or technological justification as it did in nineteenth-century capitalism. This division is artificially maintained for political reasons. The realm of real politics is everyday life—everyday culture and economics, and the crucial forms of political action are in these areas.

Not even technology can be invoked when justifying world-scale division of labor. As mentioned earlier, the direction of most technological development is toward decentralization. Even in manufacturing, industrial ecologists point out that reuse-based industry with extended producer responsibility demands more local-regional production.

All this means big changes in revolutionary social change strategy. Karl Marx lived and wrote at a time before the emergence of postindustrial productive forces. He believed, correctly, that it would take conscious popular political direction for these new productive forces to be applied comprehensively. He saw working-class revolution as a prerequisite for the creation of the New Human and the establishment of direct democracy. He saw it happening through its representatives, the organized left, whom he apparently believed could be made accountable enough to eventually allow the "withering away of the state" as the gradual emergence of postindustrial productive forces facilitated the blossoming of direct democracy. Marx, clearly, was an optimist.

The actual situation is, of course, that the NPFs have emerged, making possible both working-class autonomy and direct democracy, but prior to the revolution. Put another way, the revolution was to be made in order to create the New Human and implement direct democracy. But today we need to create the New Human and establish direct democracy *in order to make the revolution.* The withering away of the organized left is the internal reflection of the need to move beyond state-focused politics. We do not need a shadow state basically because we do not need the (industrial) state.

THE ECOLOGICAL SERVICE ECONOMY AS POSTINDUSTRIAL SOCIALISM

The kind of state we do need is very different from the industrial state. It is far less autonomous from civil society, community networks, bioregional processes, and the alternative forms of production at the heart of a green economy. By and large, the state must become more of a coordinator and less of a policeman. It would be able to do this not simply because of its advocacy of a community economic vision but because of new rules that could revamp the driving forces of economic life.

Even among those who acknowledge the need for substantially different forms of production and consumption in a green economy, it is not generally acknowledged that we also need radically different forms of regulation. I do not simply mean a more democratized state and greater levels of popular participation. That is certainly important, and I deal with new forms of local democracy, particularly Green Municipalism, in my book (Milani, 2000). But here I want to emphasize that we need not just simple controls on private greed and power but rule changes so fundamental that private enterprise becomes intrinsically social and ecological.

Perhaps the most crucial concept for understanding this possible new role of the state is that of the *ecological service economy.* This idea is antithetical to the conventional notion of the "postindustrial service economy" in which manufacturing is shipped out of the developed nations to cheap-labor countries. In the ecological service economy, manufacturing stays local/regional but becomes more geared to service. In this economy, social need—rather than production for production's sake—would be prioritized, and material substances become simply means to the end of satisfying the "service need." Examples of service needs are nutrition (rather than food commodities), transportation or access (rather than cars), entertainment (rather VCRs), and so on. This kind of service-based economy is the logical outcome of rules that enforce *extended producer responsibility* (or EPR). When producers must take responsibility for substances over their entire life cycles, they get very creative about conserving materials and making them safe, so that lots of things get reused, designed for reuse, and designed for compostability. In a reuse-based economy, even manufacturing work tends to become more like service work (e.g., shoe repair). Compared to the unhealthy drudge labor associated with conventional recycling, reuse-based work is also more creative and highly skilled.

The notion of the ecological service economy has been popularized and elaborated by industrial ecologists like Walter Stahel (1994). But it really goes back to Amory Lovins's (1977) soft energy path analysis of the 1970s, when he argued for an "end-use" approach to energy, where we would aim for "hot showers and cold beer" rather than power plants and fossil fuels. Energy supply would be matched in both quality and scale to the task at hand, *after* the real need had been carefully considered.

Because most of the current discussions about a green service economy come from industrial ecologists, people have associated this mainly with green business and philosophies of "natural capitalism." But the logical implication of making end use the starting point for all economic design is the creation of a kind of postindustrial socialism because capitalism, by definition, does not and cannot prioritize social need. Even if capitalism might be radically democratized, it is geared to accumulation, with social need still a spin-off or by-product. It is possible for businesses to sell services, as Xerox, Interface, and many others do. But it is another matter to base an entire economy on extended producer responsibility and related end-use incentives. It puts the economic spotlight on human (and environmental) need, which ultimately will even force us to distinguish between "wants" and "needs."

This kind of state action is geared to redefining wealth, changing the relationship between means and ends in economic life. Industrial forms of state socialism, except perhaps for very brief historical moments, did not really work to redefine wealth qualitatively. They did not attempt to overcome the split between the social and individual, between exchange value and use value, or between politics and economics. They simply identified with the other side of the division. In doing so, they ended up, as political economist Immanuel Wallerstein (1979) argued, running whole societies like giant corporations.

The point of postindustrial socialism would be not to limit the self-aggrandizing individual with regulatory boundaries but to transform individual enterprise altogether, to infuse it with ethical/ecological action. It would do this by changing the rules of the economic game with such systems as extended producer liability. But such systems would be insufficient in themselves to create a comprehensive environment of regenerative incentives and disincentives. Here I will mention just a few more crucial elements in the economic design mosaic that the state can have some role in facilitating if not directing. Again, it should be clear that we can begin to move on these things even before we have popular control of the state:

- *Appropriate scale for the economy:* This would tend to be much smaller than at present, geared to make the most efficient use of local skills and (largely renewable) resources. Appropriate scale is absolutely essential to build accountability into economic decisions.
- *Community account-money systems:* Such systems employ forms of money that not only support local activity but eliminate accumulation altogether and focus exchange on use value. The most well-known example, the Local Employment Trading System (LETS), already exists in over 700 communities worldwide (Greco, 1994; Douthwaite, 1996; Dobson, 1993; Walker and Goldsmith, 1998; Milani, 2000).
- *People's financial system:* The financial system itself could become an important form of community self-regulation if the principal lenders—credit unions, development banks, and the like—make loans primarily for projects that correspond to the community's green development vision.
- *A green city, or bioregional, development plan:* This would be the vision, developed in a participatory fashion, that would support all the above, outlining potentials and objectives in every sector of the economy.
- *Community indicators:* Indicators are essential tools in generating qualitative wealth. They measure and monitor this wealth, and help to displace money as the primary conveyer of value. Many kinds of indicators can be helpful—for example, alternatives to the national gross domestic product like the well-known Genuine Progress Indicator. But because qualitative wealth is very specific to people and places, the most important indicators are "sustainable community indicators." They combine objective factors (like ecofootprints and social statistics) with more subjective preferences of the community. They provide ways of measuring progress toward achieving the green plan, but they are also the ways that communities discuss and decide what is valuable.

The Transformative Educational Agenda

A visionary agenda is one that can be, and is being, implemented today. The alternatives movement is growing rapidly. It has a very large educational component—

not just a critique of the existing system but positive visions and techniques intrinsic to green community development.

Today there is raging debate about whether education should be geared to a liberal education or training for jobs. The debate totally ignores the nature of the jobs that currently are being created, whose primary purpose is to create profit by destroying community and the environment. There is no question that education should be directed to serve the community and economy. But the key to serving the community and economy is developing whole human beings. Conversely, whole human beings need a social context whereby their energies can help regenerate their communities.

Green and community-development alternatives are, par excellence, knowledge-based activities, much more so than mainstream development. But the educational resources currently at their disposal are woefully inadequate. Permaculture, ecodesign, industrial ecology, and the like are generally not taught in our educational system. At best, they exist as fringe courses at especially progressive faculties. More commonly they are privately organized as weekend workshops and weekly seminars.

The movement to make green community education mainstream would be helped tremendously by the development of large community visions or green development plans. They could serve as both planning and educational tools. Such plans could show how alternatives in the food system, in energy, in manufacturing and resource use, in communications, in healthcare, tie together into a cohesive paradigm of economic development that constitutes a realistic alternative to corporate globalization. These plans could tie into the community indicator projects, which exist in many places already, and focus both local and regional government on real problems and opportunities.

Green plans and indicators are also fertile ground for important research, especially for students and academics who would like to see some positive social outlet for their research activities. These plans also would provide guidelines for new educational initiatives that could provide regenerative skills for community and ecological development. They would provide legitimacy and encouragement for the creative disciplines that desperately need to be expanded and made available to enthusiastic young people.

References

Bazelon, David. (1963). *The paper economy.* New York: Vintage.

Block, Fred. (1987). *Revising state theory: Essays in politics and postindustrialism.* Philadelphia: Temple University Press.

Dobson, Ross V. G. (1993). *Bringing the economy home from the market.* Montreal: Black Rose Books.

Douthwaite, Richard. (1996). *Short-circuit: Strengthening local economies for security in an unstable world.* Dublin: Lilliput Press.

Greco, Thomas A. (1994). *New money for healthy communities.* Tucson, Ariz.: Thomas A. Greco Publishers.

Guttman, Robert. (1994). *How credit-money shapes the economy: The United States in a global system.* Armonk, N.Y.: M.E. Sharpe.

Livingstone, David. (2001). Work and learning in the information age. Paper presented at the Ontario Institute for Studies in Education at the University of Toronto, April 26.

Lovins, Amory B. (1977). *Soft energy paths.* New York: Harper Colophon.

Milani, Brian. (2000). *Designing the green economy: The postindustrial alternative to corporate globalization.* Lanham, M.D.: Rowman and Littlefield.

Montgomery, David. (1993). *Citizen worker: The experience of workers in the U.S. with democracy and the free market.* Cambridge, Mass.: Cambridge University Press.

Paehlke, Robert C. (1989). *Environmentalism and the future of progressive politics.* New Haven, Conn.: Yale University Press.

Polanyi, Karl. (1957). *The great transformation.* New York: Beacon Press.

Roberts, Wayne, and Brandum, Susan. (1995). *Get a life! How to make a good buck, dance around the dinosaurs, and save the world while you're at it.* Toronto: Get a Life Publishers.

Stahel, Walter R. (1994). The utilization-focused service economy: Resource efficiency and product-life extension. In Deanna J. Richards, Braden R. Allenby, and Robert A. Frosch (Eds.), *The greening of industrial ecosystems* (pp. 178–190). Washington, D.C.: National Academy Press.

Walker, Perry, and Goldsmith, Edward. (1998). A currency for every community. *The Ecologist, 28* (4) (July/August).

Wallerstein, Immanuel. (1979). *The capitalist world-economy.* Cambridge, Mass.: Cambridge University Press.

CHAPTER 6

Transformative Learning and Transformative Politics

The Pedagogical Dimension of Participatory Democracy and Social Action

Daniel Schugurensky

Introduction

Since its beginnings in the late 1970s with Mezirow's (1978) publication of a pioneer study of women returning to community college, the field of transformative learning has grown considerably, inspiring a large body of research and scholarly debate. In the last two decades, transformative learning theory has attracted the attention of a great variety of adult educators who have come to the field from different traditions. With the fading of the concepts of "andragogy" and "conscientization" (popularized by Malcolm Knowles [1970] and Paulo Freire [1970], respectively, during the early 1970s), the notion of transformative learning captured the interest of many adult education scholars and practitioners alike. At least in the United States, it has been claimed that by the mid-1990s, transformative learning had replaced andragogy as the primary theory of adult learning (Hanson, 1996). During the late 1990s, the field continued consolidating and growing, including the completion of a critical mass of doctoral dissertations (Taylor, 2000) and the successful organization of an annual conference on the topic, starting with the first one held in 1998 at Columbia's Teachers College. At the moment of this writing the organization of the fourth conference (held in November 2001 in Toronto) is in full motion, and the amount and nature of the papers that are coming from various parts of the world show an increasing interest in both the explanatory and the emancipatory potentials of the concept.

Transformative learning, however, is not a homogeneous field. Whereas there are some common agreements that hold the field together, there are some areas in which ideas are highly contested. These disagreements and exchanges (whether over epistemological, psychological, pedagogical, or political dimensions of the theory) are a good sign, because they reflect a community of scholars and practitioners that is vibrant and alive. If debate and controversy is healthy in any field, as it contributes to the development and refinement of theories and practices, it is particularly important in the field of transformative learning, which is predicated upon the notions of challenging taken-for-granted assumptions, creating spaces for open dialogue and developing critical perspectives.

Transformative learning theory undoubtedly has made a substantial contribution to adult education, and this contribution must be recognized and celebrated. Nevertheless, further elaboration is needed in several problematic areas (see, e.g., Merriam and Caffarella, 1999; Belenky and Stanton, 2000; Brookfield, 2000; Kegan, 2000; Taylor, 2000). In order to address these challenges, future studies on transformative learning theory should, among other things:

1. Incorporate more systematically the influence of context.
2. Reexamine its emphasis on rational discourse and rational forms of knowing, at the expense of other ways (e.g., affective, emotional) of knowing and communicating.
3. Examine more closely the long process and the many steps that people take before reaching the transformational stage, exploring the connections among transformative, expansive, and assimilative learning.
4. Explore further the role of adult educators in promoting transformative learning, and the training of "transformative educators."
5. Make more explicit whether transformative learning is a specific theory of learning applicable only to adults or to all learners, regardless of age.[1]
6. Broaden the outcomes of perspective transformation, and explore the occurrence of "undesirable" outcomes.
7. Distinguish more clearly between critical reflection and transformative learning, which are often conflated.[2]
8. Explicate more clearly the nature of the connection (and eventually the lack thereof) between individual and social transformation.
9. Examine more carefully the claimed relationship between transformative learning and participatory democracy, exploring the possibility of reciprocity.
10. Pay more attention to the diversity of catalysts of the transformational process, including those situations in which transformative learning occurs without the intervention of adult educators or even without an intentional educational project.

In this chapter I would like to touch on some of these points by looking at the relationships between transformative learning and two related areas, participatory democracy and social action. The background for this discussion is the participatory budget of Porto Alegre (a city in southern Brazil that has been governed by a popular coalition led by the Workers' Party since 1989) and the involvement of Rosa Parks in the U.S. civil rights movements. The participatory budget, which has received an award from the United Nations for best practice in urban governance, is characterized by a high presence of women and low-income groups in the processes of deliberation and decision making. This is something that should not be taken lightly, because frequently these groups are underrepresented in the real practices of participatory democracy. Moreover, the high participation of people with low levels of formal schooling is particularly relevant, because most writings on transformative learning are about middle-class learners and in many cases about students enrolled in postsecondary institutions. The real case of Rosa Parks (unlike the myth of Rosa Parks) shows that learning in and for social action is the result of a combination of assimilative, expansive, and transformative processes, in which both the emotional and the rational dimension are at play. It also shows the impact that a particular adult education institution (in this case Highlander) can make to social transformation.

Critical Reflection, Transformative Learning, and Social Transformation: An Ongoing Debate

Among the many issues that are subject of debate in transformative learning theory, two are particularly relevant for this chapter: the nature of critical reflection and the relationship between personal and social transformation. In both debates there is a tension between the more psychologically grounded, process-oriented traditional branch of transformative learning theory and the more sociologically inspired, outcome oriented radical adult education.

To begin with, a conceptual confusion, derived from the overuse of the word "critical," is the tendency to conflate critical reflection and transformative learning. Whereas transformative learning often is understood as critical reflection on our own assumptions and the assumptions of others, it is pertinent to distinguish between critical reflection and transformative learning. As Brookfield (2000) remarks, critical reflection is a necessary but not a sufficient condition for transformative learning to occur. In other words, transformative learning cannot take place without critical reflection, but critical reflection can happen without an epistemological transformation. Moreover, in the radical adult education tradition, critical reflection is understood as ideology critique. In the sense attributed to the concept by the Frankfurt School and by the Freirian/Gramscian tradition, this is the process of unveiling the social, economic, and political dynamics of oppressions that are embedded in everyday situations and practices and of becoming aware of the socially constructed nature of our beliefs, values, ideas, and tastes. In this particular tradition, it refers specifically to the examination of the dynamics of capitalism and to the ideological processes of hegemony building that are put in motion to ensure the reproduction of the system. Hence, in the radical tradition of adult education, "critical reflection" is fundamentally emancipatory since it involves social critique, addresses oppressive social structures, and results in a transformation of a comprehensive worldview and eventually in social change (Brookfield, 2000; Tennant, 1993).

When we talk about transformation, we can make an analytical distinction among transformation of individual consciousness, transformation of individual behavior, and social transformation. The three often are related, but it cannot be assumed that the first necessarily leads to the second and this to the third. An example of this distinction can be observed in smoking. Many smokers may become conscious of the detrimental impact that smoking has on their health and on the health of people who surround them, but such awareness does not necessarily lead to cessation. If awareness was directly correlated to changes in behavior, it would be impossible to explain why, with all the evidence available, many doctors and scientists still smoke. Likewise, a person who becomes aware of the negative effects of smoking may quit subsequently, but this does not mean that this individual is going to become involved in advocacy groups confronting the tobacco industry or proposing legislation. In other words, we cannot assume that there always exists a causal relationship among changes in individual consciousness, changes in individual change, and social change.

Recognizing this situation, transformative learning theory distinguishes transforming habits of mind and transforming structures. Although this theory expects that significant personal and social transformation may emerge from critical reflection, it admits that after critical reflection, a person can decide to change or not to change a particular behavior, depending on factors such as the context, the need for additional information, and the required skills and emotional commitment to proceed (Mezirow, 1992: 251; 1998: 186). Context is especially relevant in explaining the connections between individual and social transformation. For instance, a supportive social environment, a social reality that is susceptible of transformation (i.e., a viable collective project), and a sense of community are important elements in creating the conditions for social transformation.

In the absence of these conditions, critical reflection alone is not only unlikely to lead to transformative social action, but in some cases it may even lead to the opposite situation, which is cynicism, paralysis, and a general feeling of helplessness. Paradoxically, critical reflection, without an accompanying effort of a social organization and without concurrent enabling structures to channel participation in democratic institutions, can nurture the development of individuals who become more enlightened than before but who (because of their realization of the immense power of oppressive structures) may become more passive and skeptical than before. In the words of Brookfield: "It is important to acknowledge that critical reflection's focus on illuminating power relations and hegemonic assumptions can be the death of the transformative impulse, inducing an energy-sapping, radical pessimism concerning the possibility of structural change. It is undeniable that raising the consciousness of adult educators can easily produce negative effects on their practice. Knowing about the forces that use adult education to preserve and transmit dominant cultural values can leave its practitioners feeling puny and alone. Knowing that challenging the dominant ideology risks bringing isolation and punishment down on our heads is depressing and frightening" (2000: 145).

This issue resonates well with the experience of many progressive Latin American educators in the 1970s and 1980s. After reading the works inspired by strong structural Marxism, they reached the conclusion that the educational system was nothing else than an ideological state apparatus set in place and controlled by the dominant class to provide the conditions for capital accumulation and political legitimation. Without discussing here the merits of this analysis, the implication of that conclusion was that many educators (including adult educators) began to identify themselves as agents or instruments of the dominant groups and decided that continuing to work in education was either an act of complicity or a waste of time, because they perceived the little relative autonomy of educational institutions as merely symbolic and ineffectual. They found themselves not only helpless in the face of a perceived all-encompassing power but also guilty for promoting the capitalist agenda. These teachers became more aware of structures of domination and the role of educational institutions in reinforcing them, but such awareness, in the absence of a coherent social movement to promote an alternative educational project, led to paralysis, pessimism, and cynicism. To avoid their feelings of guilt and helplessness, several of those educators left their teaching careers without recognizing that both the state and educational institutions are not monolithic structures but arenas of negotiation and contestation among different social groups and projects.

Hence, involvement in social transformation is not something that arises automatically or naturally from a "new critical consciousness." Conceptualizing individual and social transformation as a causal process in which the former leads to the latter is not always helpful, either to understand change dynamics in the real world or to guide the work of adult educators. Alternatively, from a Freirian perspective, it can be argued that transformative learning is really transformative when critical reflection and social action are part of the same process. In Freire's approach (1970), education is understood as praxis: reflection and action upon the world in order to transform it.[3] The process of transformation as praxis implicit in this model suggests a dialectic movement in which reflection and action constantly interact to produce a different consciousness and a different action upon the world.

Transformative learning theory has been criticized for its emphasis on individual transformation at the expense of social transformation. In this regard, several authors have contended that if the social action dimension is missing, critical reflection easily can become an irrelevant and egocentric exercise (Hart, 1990; Cunningham, 1992; Tennant, 1993; Newman, 1994; Taylor, 1997; Collins, 1998; Spencer, 1998). As Brookfield (2000) notes, critical reflection alone can be perceived as "liberal dilettantism, a self-indulgent form of speculation that makes no real difference to anything" (143). For Newman (1994), traditional transformative learning has no connection to the debates on community development and social action (and in general with radical adult education) because of its psychotherapy foundations, its individualized approach, and its removal from political action. The main charge to transformative learning from this perspective is that individual transformation by itself does not ensure social action, which for some is the raison d'être of adult education, and not just the development of autonomous individuals.

This issue was raised several decades ago by Lindeman (1926) when he pointed out that true adult education is social, not just in its process but also in its purpose. If one accepts this proposition, it can be argued that transformative learning, by its focus on individual changes, is not so much a theory of adult education but a theory of adult learning (Spencer, 1998).[4] While I subscribe to Lindeman's position and am sympathetic to the points raised by radical adult education, I also believe that both individual transformations and social transformations are equally important in building the foundations of a better world. Adult education is about promoting a healthy, democratic, caring, and just society (and this implies dealing with structures of power), but it is also about developing autonomous, critical, creative, democratic, and caring individuals (and this implies dealing with cognitive and attitudinal changes). In this sense, an emancipatory adult education is a political-pedagogical project; to emphasize one dimension at the expense of the other denies the complex nature of transformation and promotes a narrow practice with limited scope.

TRANSFORMATIVE LEARNING, REFLECTIVE DISCOURSE, AND DEMOCRATIC SPACES

If transformative learning implies a critical awareness of the tacit assumptions and expectations that we and other people hold and an assessment of their relevance for making choices, this process cannot occur in isolation. Such a process involves participation in constructive discourse in which participants deliberate about the reasons for their actions and get insights from the meaning, experiences, and opinions expressed by others. Thus one of the main goals of transformative learning is the development of more autonomous thinkers who can justify their choices or reasons, but this development can take place only in relation to others, which makes it a collective, relational process. In this respect, transformative learning theory has been influenced by the writings of Paulo Freire (1970) on the development of critical consciousness and also by the work on Jurgen Habermas (1984) on dialogue and discourse.

Indeed, in addition to reflection, discourse constitutes an important element in the process of learning identified by transformative learning theory. Following Habermas (1984), discourse, or human communication, must involve freedom, tolerance, equality, education, and democratic participation in order to achieve ideal conditions of learning. In the Habermasian model, an ideal speech situation is characterized by a deliberation in which certain basic conditions are met. It is a discourse in which all participants have complete information, are free from self-deception, are able to examine arguments objectively, have equal opportunity to participate, and so on. Transformative learning, thus, requires the presence of different viewpoints (especially those that challenge prevailing norms) and must allow (even encourage) the expression of dissent. These types of situations are expected to promote the development of socially responsible citizens who can participate effectively in decision making processes.

It is pertinent to note that discourse does not need to be a confrontation; ideally it is just the opposite: a conscious collective effort to find agreement, to search for common ground, to resolve differences, and to build a new understanding. To do so implies an effort to be open, to set aside bias and prejudice, and to listen to our own purposes, feelings, values, and meanings in the context of those expressed by others (Mezirow, 1995: 53, 1996: 170, and 2000: 8; Tannen, 1998). This is a particularly important point for the practice of participatory democracy as we live in a culture of polarized argumentation in which dialogue often is reduced to a debate between two opposing sides in which reality is dichotomized and in which winning an argument seems to be more important than understanding different ways of thinking and reaching consensus.

From the practice of adult education, the question is how to move toward those ideal speech situations and toward that active and conscious community of decision makers. A first challenge is to replace oppositional with collaborative dialogue, transforming self-serving debates to careful listening and informed, constructive discourse. A second challenge is to find the most appropriate strategies and locations to promote the development of active, socially responsible, democratic, and caring citizens who have the competencies to engage in collective decision making.

I submit that some of the answers for the development of this new civic culture lie in the twin fields of participatory democracy and civic education. Both intimately related to the project of transformative learning, these two fields have immense potential for the creation and promotion of public spaces of deliberation and decision making that are linked to self-government. Participatory democracy, as the Porto Alegre experiment shows, can sow the seeds that allow societies to generate public spaces of social interaction in which discourse is based on finding agreement, welcoming different points of view, identifying the common good in the myriad of competing self-interests, searching for synthesis and consensus, promoting solidarity, and ultimately improving community life. Civic education, which until recently has been perceived with disdain by radical adult educators and critical theorists, is now being seen as a cornerstone for expanding and enabling political agency. As Giroux (2001) points out, this is happening because of civic education's potential for critical thinking, bridging the gap between learning and everyday life, understanding the connection between power and knowledge, using history to extend democratic rights and identities, and operating across a wide variety of public spheres. In his words, "the struggle over politics and democracy is inextricably linked to creating public spheres where individuals can be educated as political agents equipped with the skills, capacities and knowledge they need not only to actually perform as autonomous political agents, but also to believe that such struggles are worth taking up" (23).

Likewise, in his recent book on transformative learning and visions for the twenty-first century, Edmund O'Sullivan (1999: 252) calls for the development of a new civic culture in which a sense of community and place are the basic empowering infrastructures for more extended involvement in wider communities of participation. He also calls for an alert, conscious citizenry that constitutes the ultimate check on the activities of politicians and of commercial and financial institutions. O'Sullivan admits that this is a particularly difficult challenge given the high level of disenfranchisement and the disconnection from the political processes that can be observed even in advanced democratic societies. Based on my own observations and interviews with Porto Alegre neighbors who have been involved in the participatory budget, I feel confident to say that participatory democracy is a privileged space in which ordinary citizens can acquire the knowledge, skills, attitudes, and values to become more critical, open, effective, and caring political agents

In summary, the basic questions of civic education and civic participation in deliberation and decision making are central to the struggle over political agency and democracy itself. Civic education takes place in a multiplicity of public spaces in which the state and civil society interact in different ways. Participatory democracy constitutes one such a space, and its relevance for transformative learning lies not so much on its high effectiveness for accountable and transparent governance but on its educative potential both for ordinary citizens and for government officials.

THE CONDITIONS FOR REFLECTIVE DISCOURSE: FROM THE UNIVERSITY SEMINAR TO PARTICIPATORY DEMOCRACY

Reflective discourse, then, can be understood as a process in which we actively dialogue with others to better understand the meaning of an experience. It involves assessing the reasons and the evidence advanced to support an argument. This in turn promotes a better understanding of issues by tapping into collective experience and knowledge and allows all participants to find their own voice in light of alternative perspectives (Mezirow, 2000).

A genuine reflective discourse is more likely to take place if certain conditions are present. These conditions relate closely to Habermas's (1984) ideal speech situations mentioned earlier as well as basic feelings of solidarity, empathy, trust, and safety among participants. Mezirow (2000: 16) adds certain preconditions for full participation in discourse, noting that transformative learning groups require certain elements of maturity, education, safety, health, economic security, and emotional intelligence. Given these preconditions, he argues that hungry, homeless, desperate, threatened, sick, or frightened adults are less likely to be able to participate effectively in discourse to help them better understand the meaning of their own experiences. A place where most of these preconditions and conditions for dialogue are met is the graduate seminar, which Mezirow identifies as a model of reflective discourse.

A graduate seminar, however, is an experience enjoyed by less than 1 percent of the world population. Indeed, Belenky and Stanton (2000) remind us that the university seminar is the result of a long selection process that leaves out many people. For this reason, issues of unequal power are seldom addressed. A process that excludes many people begs some fundamental questions for an emancipatory adult education project: Can a genuine transformative learning process take place only among the most privileged members of our societies? What are the conditions that can promote transformative experiences among the most disadvantaged sectors of the population? What happens with those who are not university students and are unlikely to become university students in the near future—are they necessarily precluded from engaging in critical deliberation and reflective discourse? Belenky and Stanton contend that people who have been silenced all their lives have few of the tools needed for participating in the kind of discourse community described in transformative learning theory. They suggest that to bring them into an ongoing dialogue requires the creation of an extremely safe and caring community where people draw each other out and listen to each other with attention.

Again, the Porto Alegre experiment, in which local participatory democracy and a progressive model of adult citizenship education permanently feed each other, constitutes an interesting example for the development of these caring and listening communities, especially among the most marginalized groups in society. In his piece on the transformative learning experiences of Nelson Mandela, Laurent Parks Daloz (2000) outlines several key conditions of successful discourse that can be present in local democracy. Among them are an emphasis on empowerment and self-determination of the participants, a participant-based agenda, validation of emotions as part of the process, and a recognition that confusion is inevitable and legitimate. To a large extent, many of these conditions can be observed in the neighborhood meetings of the participatory budget, and this may explain part of its success. As Bush and Folger (1994) note, when participants who hold different perspectives meet in accordance with certain common principles (e.g., respect, tolerance), they emerge with a transformed capacity to understand other perspectives (a prerequisite for genuine dialogue and for seeking agreement), while still retaining their own uniqueness.

In transformative learning theory, the connection between transformative learning and participatory democracy frequently is established as a unidirectional relationship, in which a successful participatory democracy is contingent upon a successful process of transformative learning. The argument usually goes as follows: Transformative learning, by developing individuals' competencies for engaging in critical yet respectful dialogue with others, nurtures the necessary subjective conditions for a genuine participatory democracy. In the words of Mezirow: "Transformative learning inherently creates understanding for participatory democracy by developing capacities of critical reflection on taken-for-granted assumptions that support contested points of view and participation in discourse that reduces fractional threats to rights and pluralism, conflict, and the use of power, and foster autonomy, self-development, and self-governance—the values that rights and freedoms presumably are designed to protect" (2000: 28).

I argue that it is possible to go one step further and say that a relation of reciprocity exists between transformative learning and participatory democracy. This means that transformative learning can promote participatory democracy but also that participatory democracy has the potential to nurture transformative learning. In other words, transformative learning can improve the quality of citizens' participation in democratic institutions, and at the same time democratic participation itself creates powerful opportunities for self-transformation.

On one hand, transformative learning improves citizens' participation in democratic institutions by developing capacities for critical reflection of taken-for-granted assumptions, and it does so through open and pluralistic dialogues in which all participants can freely express their viewpoints and in which all viewpoints are subject to examination. Learning accrued through these "ideal speech situations" can equip people with the communicative competencies to seek the truth and to arrive to consensus through a process of argumentation and mutual persuasion that is not distorted by cultural and power differences. It can be argued that when these critical and communicative capacities are applied to public spheres of political decision making, the democratic process can be improved dramatically. This is essentially a reformulation of the key arguments advanced by the progressive tradition of citizenship education. (See Habermas, 1984; Dewey, 1966; Freire, 1970.)

On the other hand, as posited recurrently by supporters of participatory democracy, democratic participation itself creates spaces for significant learning and for self-transformation. Although many of them acknowledge that participatory democracy may not be the most efficient model of democratic governance, they stress that it certainly is the most educative one (Pateman, 1970; Berry, Portney, and Thomas, 1993). When people have the opportunity to actively participate in deliberation and decision making in the institutions that have most impact on their everyday lives (i.e., schools, workplaces, and local governments), they engage in substantive learning and can experience both incremental and sudden transformations. The transformative effects are usually more significant when this institutional participation provides empowering experiences. For instance, Warren (1992: 8) states that through those experiences people can become more public spirited, more tolerant, more knowledgeable, more attentive to the interest of others, or more probing of their own interests. Through these type of collective experiences people are more likely to consider another's point of view and eventually change their own perspective in order to promote the common good (Lohman, 1992).

The participatory budget provides an interesting case of the potential of participatory democracy to foster this type of situation. A recent example can help to illustrate this. In one of the meetings held in a poor neighborhood, there was a consensus that the main priority for that year's budget was street paving. The problem was that each group of neighbors in the community wanted its street paved first, but the budget allocation for that year allowed for only a few blocks to be paved. Each group was trying to persuade the rest that its street had to be paved before the others, but it was painfully obvious that no group had a more convincing argument than the rest, because everyone was in the same situation. At one point the debate reached a stalemate and participants felt disappointed because they were not clear on how to proceed. The discussion continued, and suddenly someone suggested a compromise. If resources were indeed limited, and all streets in the community needed to be paved, it made sense to pave first the streets that led to the school and to the hospital, which were the ones most people had to use on a daily basis. In subsequent years they could make new agreements on priorities, until all streets in the community were paved. Through this exercise of mutual persuasion and constructive dialogue, people were able to move from their own narrow self-interest to an understanding of the common good. This process, replicated at larger levels eventually to encompass the entire city, can be frustrating at times, but in the long run creates a new sense of community, solidarity, and mutual collaboration. It also has the potential, as the evidence that is emerging from the participatory budget clearly suggests, to nurture the development of those more "socially responsible, clear-thinking decision makers" among ordinary men and women (Alfonsin, 1999; Abers, 2000; Moll and Fischer, 2000; Pontual, 2000; Schugurensky, 2001).

The Progressive Education Legacy: Learning Democracy by Doing It

In *The School and Society,* first published in 1899, John Dewey advanced an argument that can be summarized in a phrase that would be adopted by many progressive educators and parents during the twentieth century: "Children learn by doing." This has certainly become a cliché today, but the fact that it has become a catchphrase does not mean that is less true. Perhaps this has become too obvious, but as Freire (1970) used to say, sometimes it is helpful to explicit the obvious once again.

While many educators, philosophers, and scholars have been concerned with the relationship between education and democracy throughout the centuries, Dewey's (1966) ideas were particularly insightful. For him, education could prepare people to live in a democracy only if the educational experience were also democratic. Because Dewey focused his work on schools, it is understandable that he advanced this principle using children as its subject. However, it is possible to argue that this principle is valid across the life span, and hence it is as relevant to children as it is to adults. Indeed, Dewey never argued that the connection between learning and doing became weaker with age. Moreover, as he regarded the dialectic between knowledge and action to be a philosophical premise, there is no reason to assume that the general principle applies only to children (Menard, 1997). To argue that people learn by doing does not mean that we cannot learn through other means (e.g., attending a lecture or watching a movie). The problem is that learning by doing, while recognized as one of the most powerful means of learning, is the one that has been most marginalized by the education system.

The guiding principle that we can extract from Dewey, even a century later, is that the best and most effective way to learn democracy is simply by doing democracy. Several of Dewey's (1966) ideas on this topic were inspired by Jane Addams, the founder of Hull House and the first American woman to be awarded the Nobel Peace Prize. Addams (1930) believed in experiential forms of learning, what she called "education by the current event," and her work with adult learners constituted a precursor of what today is called "learning in social action" and of methodologies such as participatory action research. Addams believed in problem-solving education rather than in discipline based education and contended that social institutions are educational institutions. In summary, the work of Dewey, Addams, and others in the progressive education movement provides at least three basic interrelated principles that can guide the theory and practice of progressive adult educators interested in the connections between transformative learning and democracy:

1. Learn by doing (in this case, learn democracy by doing it).
2. Social institutions are educational institutions.
3. Democratizing institutions is a political-pedagogical project.

We know today that people can learn and do learn (and, some may argue, learn best) through real and meaningful projects. As Mary Belenky pointed out in the 1998 Transformative Learning Conference, significant learning happens through engagement in action projects, and social action is both a goal and a means of learning in preparation for further action (cited in Mezirow, 2000: 335). Unfortunately, however, in spite of its relevance for the understanding of adult learning, little research exist on informal learning that occurs in social action. Social action, especially when it leads to transformations, promotes not only a new learning experience but considerable personal change. In the words of Maxine Green, "it is actually through the process of effecting transformations that the human self is created and re-created" (1988: 21).

Assimilative, Expansive, and Emotional Learning

Transformative learning theory suggests that once a personal transformation has occurred, most people seldom return to their old perspective, but at the same they are unlikely to experience a consistent forward movement. A learner can get stalled at any phase of the process, but particularly at the beginning (because there is a threat to a long-established sense of order) and later when a new awareness calls for commitment and action that imply certain degree of uncertainty and risk. The question that has not been sufficiently explored is what happens before that personal transformation takes place. In this regard, Belenky and Stanton (2000) claim that "although Mezirow's important theory provides an elegant, detailed description of one important endpoint of a long developmental process, it does not trace the many steps people take before they can know what they know in a highly elaborated form" (72). This is particularly important for adult educators, because transformations usually occur through long incremental processes. Belenky and Stanton suggest that focusing narrowly on the culmination point of development may be problematic for a theory that appeals to adult educators, because practitioners might overlook the reality of their students' lives.

Hence the need to recognize the role of assimilative learning the transformations of individuals. Indeed, research on this topic suggests that transformative learning can be propelled by a disorienting dilemma (which often is a dramatic, abrupt, and even traumatic experience), but it also can be propelled by an "integrating circumstance." An "integrating circumstance" is usually the culmination of a relatively long process (conscious and/or unconscious) of searching and exploring for something that is missing in one's life. The encounter with the missing element can catalyze a transformative learning process. In this case, perspective transformation can be the result of a process of assimilative learning, not necessarily the product of a critical reflection on assumptions (Clark, 1997; Courtenay, Merriam, and Reeves, 1998; Mezirow, 1998).

Consider, for instance, the case of Rosa Parks, the heroine of the U.S. civil rights movement. The public myth around her figure suggests that she spontaneously refused to give up her bus seat in Montgomery to a white man and by so doing she single-handedly gave birth to the civil rights efforts. The reality is that before sitting in that bus, Rosa Parks had spent more than a decade as a community organizer at the local NAACP chapter. In the summer of 1965, a few months before the bus incident, she attended a workshop at Highlander where she expanded her realm of experience by meeting an older generation of civil rights activists of different races. Although her action on the bus was not prearranged and she acted on an individual basis, Rosa Parks knew that she was contributing to the civil rights cause when she refused to move, and her action was the result of many years of assimilative learning. At the same time, although she was not new to the civil rights movement when she attended the Highlander workshop, the experience made an impact on her and provided her with new insights. This is how Rosa Parks described her first Highlander workshop: "At Highlander, I found for the first time in my adult life that this could be a unified society, that there was such a thing as people of different races and backgrounds meeting together in workshops, and living together in peace and harmony. It was a place I was very reluctant to leave. I gained there the strength to persevere in my work for freedom, not just for blacks, but for all oppressed people" (cited in Horton, 1998: 150).

Two themes emerge from this story. First, Highlander was only one of many learning experiences in the lifelong preparation for Rosa Parks's unique form of protest. One particular learning experience in one setting (in this case, a workshop) is always in dialogue (complementing, expanding, contradicting, etc.) with previous learning experiences. Parks's experience in Highlander and her behavior in that bus cannot be explained without considering her history of increasing awareness about structures of discrimination and increasing involvement in a progressive social movement. Second, the workshop did not give her only (not even mainly) new information or new skills (e.g., about civil disobedience) but mainly a new sense of belonging, a renewed commitment to collective action for the common good. In other words, the main learning was at the emotional or attitudinal level. By experiencing firsthand the fact that blacks and whites could work together cooperatively, Rosa Parks gained a new perspective on and renewed certainty about the feasibility of a desegregated world.

From the perspective of an emancipatory education, the incremental and assimilative nature of Parks's learning is of great significance for two reasons. One is the importance of perseverance and that strength usually comes from a collective project. In the words of Paul Rogat Loeb (2000): "Parks's actual story conveys an empowering moral that is lost in her public myth. She began modestly by attending one meeting and then another. Hesitant at first, she gained confidence as she spoke out. She kept on despite a profoundly uncertain context as she and others acted as best they could to challenge deeply entrenched injustices, with little certainty of results. Had she and others given up after their 10th or 11th year of commitment, we might never have heard of Montgomery. Parks' journey suggests that social change is the product of deliberate, incremental action, whereby we join together to try to shape a better world" (7).

The second moral of the story, equally important for adult educators, is that it removes the godlike and individualistic perception of social leaders and illustrates that any ordinary citizen has the potential to influence social reality. I refer again to Paul Rogat Loeb (2000), who says it more eloquently and lucidly than I could:

> People like Parks shape our models of social commitment. Yet the conventional retelling of her story creates a standard so impossible to meet that it may actually make it harder for the rest of us to get involved. This portrayal suggests that social activists come out of nowhere to suddenly materialize to take dramatic stands. It implies that we act with the greatest impact when we act alone, or when we act alone initially. It reinforces a notion that anyone who takes a committed public stand—or at least an effective one—has to be a larger-than-life figure, someone with more time, energy, courage, vision or knowledge than any normal person could ever possess. . . . This in no way diminishes the power and historical importance of her refusal to give up her seat. But it does remind us that this tremendously consequential act might never have taken place without the humble and frustrating work that she and others did earlier on. It reminds us that her initial step of getting involved was just as courageous and critical as the fabled moment when she refused to move to the back of the bus. (7).

Likewise, in this long process of assimilative learning, sometimes we experience important transformations in the way we understand reality, but these experiences do not constitute "transformative learning" in the strictest sense because they do not imply a deep shift in perspective. This is the case, for instance, when we expand our horizons by meeting people whom we ordinarily do not meet or by traveling beyond our realm of ordinary experience. This could be conceptualized more as "expansive learning" than as transformative learning in its narrow definition, but it is nonetheless transformational, in the sense that the person moves, sometimes consciously, sometimes unconsciously, to a broader understanding of reality and a more comprehensive worldview. Parks Daloz (2000) describes a case of expansive learning when reviewing the life of Nelson Mandela and the process by which he started to think outside Thembuland (his village of origin) and began to think as a member of the African community. This transition from parochialism to a more continental or even cosmological approach takes time and encounters with "outsiders." In my own research in Porto Alegre, one of the most significant learning experiences that people mention as a result of participating in the budget process is that they slowly move from a concern about their own street or neighborhood to an understanding of the city as a whole. In other words, the participatory budget fosters the development of a new social consciousness by which people expand their concerns from narrow self-interests to collective needs and develop attitudes of respect and social solidarity.

In summary, I submit that important personal and social transformations can occur through learning processes that are more assimilative and expansive than "transformative" and that emotional learning experiences are as important as rational ones, especially in relation to collective social action. The early work of Mezirow (1978: 22) suggests that transformations tend to follow a learning cycle initiated by a disorienting dilemma and resulting in a reintegration into society on the basis of a new perspective. More research is needed to explore the possibility that transformations may occur incrementally, without necessarily experiencing a disorienting dilemma.

Pedagogy of the City: When Learning Sparks Are Darting

The Highlander workshop that Rosa Parks attended in 1955 was part of the Citizenship School program, which was directed by Septima Clark. By 1961, after a series of conversations between Myles Horton and Martin Luther King, the Highlander Institute transferred its Citizenship School program to the Southern Christian Leadership Conference.[5] On that occasion, Horton said a few words to the prospective teachers in his farewell speech.

> People learn faster and with more enjoyment when they are involved in a successful struggle for justice that has reached social movement proportions, one that is getting attention and support outside the movement, and it's socially big enough to go far beyond the individuals involved. It's a much bigger experience than anything you've had before as an individual. It's bigger than your organization, and it's qualitatively different, not just more of the same. I want the struggle for social and economic justice to get big and become so dynamic that the atmosphere in which you are working is so charged that sparks are darting around very fast, and they explode and create other sparks, and it's almost perpetual motion. Learning jumps from person to person with no visible explanation of how it happened. (1998:108)

Two insights in particular emerge from Horton's (1998) farewell speech. First, people learn faster and happier when they are not alone, when they are part of a social movement, and especially if the movement has an ethical stand, a utopia to chase, and some successes to celebrate. Second, learning in this context is dynamic, unpredictable, and often inexplicable, jumping like sparks from person to person. The stories of Rosa Parks and Nelson Mandela, as well as the stories of many ordinary citizens that I collected in my research on the Porto Alegre Participatory Budget, tend to confirm Horton's statement (Schugurensky, 2001).

The Participatory Budget is an interesting case in this regard because most local urban planners, city officials, community organizers, and participants do not perceive the pedagogical potential of participatory democracy. The majority of CROPS (Coordinadores Regionales do Orcamento Participativo: community advisors to the Participatory Budget) do not see their work as a pedagogical one and do not even see its pedagogical dimension (Moll and Fischer, 2000). However, several active participants (delegates and counselors) understand the Participatory Budget as an educational space. Some even labeled it a "citizenship school," and others equated it as "our university, the university of the people." It is also interesting to note that learning and changes occurred more quickly among ordinary citizens than among public servants, urban planners, and technical experts, whose attitudes and routines have been ingrained in an ethos of expertise and a centralized and authoritarian model of governing that has characterized Brazilian public bureacracy for centuries (Pontual, 2000). The pedagogical role of the participatory budget is impressive. By engaging actively in deliberation and decision making processes, individuals and communities learn and adopt basic democratic competencies and values. They learn to listen carefully and respectfully to others; to request the opportunity to speak and to wait for their turn; to argue and to persuade; to respect the time agreed on for individual interventions and for ending the meeting; and to advance the needs of their neighborhoods while understanding the complexity of the city as a whole. They also learn about the needs of other groups and to exercise solidarity with them. In summary, ordinary citizens develop their self-confidence, their political capital (the capacity to influence political decisions), their concern for the common good, and their democratic spirit (Alfonsisn, 1999; Abers 2000, 1998; Schugurensky, 2001).

Parks Daloz (2000) identifies four conditions under which engagement with others can lead to greater social responsibility: the presence of the other, reflective discourse, a mentoring community, and opportunities for committed action. For him, deep change takes time, strategic care, patience, the conviction that we are not working alone, and hope in a better future: "It calls us as teachers and citizens to seek out and encourage engagement with those different from ourselves, to foster critical reflection on the meaning of our differences, to create mentoring communities where socially responsible commitments can be formed and sustained, and to make available opportunities to practice these emerging and vital commitments. These are small steps, but each one can makes another possible" (121). The participatory budget of Porto Alegre, although it has clear problems and limitations, offers the potential for such a space among ordinary citizens, particularly among the most marginalized segments of the population. By encouraging open deliberation and decision making on public issues in a process controlled by the local communities and by creating solidarity networks, it promotes some of the conditions outlined by Parks Daloz.

As Freire (1970) has pointed out and as Belenky and Stanton (2000) remind us, developing the critical capacity for participating in democratic deliberation is particularly important for people who have been excluded, silenced, marginalized, and oppressed most of their lives. This capacity is more likely to develop when people have a space and an opportunity for making their voices heard, for questioning authority, for forging unity while recognizing and celebrating diversity, for engaging in collective action, and for transforming their own community, which in turn awakens their belief in their own potential to transform larger social realities. Indeed, people's belief in their own potential to influence reality (what is known as "personal efficacy" or "personal empowerment") is strongly connected to transformative learning. This brings us full circle back to transformative learning theory, which contends that transformative learning requires supportive relationships and a supportive environment that encourages a sense of personal efficacy. Given that growth in personal efficacy is dependent on broader relationships of power, significant changes are more likely to occur through a collective political-pedagogical process that aims simultaneously at increasing both critical reflection and social justice. A collective process is not only the precondition for social change; it is also a powerful catalyst for learning. As Ralph Peterson remarked, "when community exists, learning is strengthened: everyone is smarter, more ambitious, and productive" (1992: 2).

Notes

1. As in the case of andragogy, it is debatable whether transformative learning is a specific theory of learning applicable only to adults or to all learners, including children.

Although the literature suggests that both andragogical principles and transformative learning are exclusive provinces of adulthood or of adult education, it can be argued that they apply to any educational process along the whole life span (Kegan, 2000). Although Mezirow (2000) argues that autonomous agency, critical reflections on one's own assumptions (reflective judgment) and participation in critical discourse are features that are more likely to be observed in adults, this does not mean that they are totally absent among children and adolescents. Moreover, there is not enough evidence to claim that the four main ways of learning outlined by Mezirow (elaborating existing frames of reference, learning new frames of reference, transforming points of view, and transforming habits of mind) can occur only among adults.

2. Although Freire understood reflection and action dialectically, in his first works he tended to assume a causal relationship in which the first would lead to the second. In subsequent works he criticized his original view on this relationship as embedded by naïve psychologism. (See Schugurensky, 1998.)
3. Mezirow (1994) tends to agree with this characterization, as he points out that his theory is indeed a theory of adult learning and not an educational philosophy such as the one advanced by Freire (1970). In any case, as a theory of adult learning, transformative learning has plenty to contribute to an adult education agenda that fosters the development of a more democratic society.
4. Although the field is still in its infancy, the New Approaches to Lifelong Learning (NALL), a network of researchers that focus on informal learning in a variety of settings and is coordinated by David Livingston at OISE/UT, promises novel contributions and insights. See also the book by Griff Foley (1999).
5. By that time, the program had rapidly grown beyond the capacity of the Highlander Institute, and Luther King was precisely looking for an educational program for blacks in the south.

References

Abers, Rebecca. (1998). Do clientelismo à cooperaçao: Políticas participativas e organizaçao da sociedade civíl em Porto Alegre *Cadernos Ippur, 12* (1) (Janeiro-Julho): 47–78.

Abers, Rebecca. (2000). *Inventing local democracy: Grassroots politics in Brazil.* Boulder, Co.: Lynne Rienner Publishers.

Addams, Jane. (1930). *The second twenty years at Hull House.* New York: Macmillan.

Alfonsin, Betania. (1999) Elogio y critica del proceso de Porto Alegre. In Jose L. Coraggio (Ed.), *Municipio, democracia y desarrollo local: La experiencia de Porto Alegre* (pp. 27–32). Buenos Aires: Instituto del Conurbano, Univesidad Nacional de General Sarmiento.

Belenky, Mary, and Stanton, Ann. (2000). Inequality, development and connected knowing. In J. Mezirow and Associates (Eds.), *Learning as transformation: Critical perspectives on a theory in progress* (pp. 71–102) San Francisco: Jossey-Bass.

Berry, Jeffrey, Portney, K., and Thmpson, K. (1993). *The rebirth of urban democracy.* Washington, D.C.: The Brookings Institution.

Brookfield, S. (2000). Transformative learning as ideology critique. In J. Mezirow and Associates (Eds.), *Learning as transformation: Critical perspectives on a theory in progress* (pp. 125–150). San Francisco: Jossey-Bass.

Bush, R., and Folger, J. (1994). *The promise of meditation: responding to conflict through empowerment and recognition.* San Francisco: Jossey-Bass.

Clark, M. C. (1997). Transformational learning. In S. B. Merriam (Ed.), *An update of adult learning theory: New directions for adult andcontinuing education, No. 57,* (pp. 47–56). San Francisco: Jossey-Bass.

Collins, Michael. (1998). Critical returns: From andragogy to lifelong education. In Sue Scott, B. Spencer, and A. Thomas (Eds.), *Learning for life: Canadian readings in adult education* (pp. 46–58). Toronto: Thompson.

Courtenay, B. C., Merriam, S. B., and Reeves, P. (1998). The centrality of meaning-making in transformational learning: How HIV-Positive adults make sense of their lives. *Adult Education Quarterly, 48* (2): 63–82.

Cunningham, Phylllis. (1992). From Freire to feminism: The North American experience with critical pedagogy. *Adult Education Quarterly, 42* (3): 180–191.

Dewey, J. (1966 [1916]). *Democracy and education.* New York: Free Press.

Dewey, John. (1956 [1899]). *The School and society.* Chicago: University of Chicago Press.

Emler, Nicholas, and Frazer, Elizabeth. (1999) Politics: The education effect. *Oxford Review of Education, 25* (1,2) (Mar/Jun): 251–73.

Foley, Griff. (1999) *Learning in social action: A contribution to understanding informal education.* London: Zed Books.

Freire, Paulo. (1970). *Pedagogy of the oppressed.* New York: Seabury Press.

Giroux, Henry. (2001). *Public spaces/private lives: Beyond the culture of cynicism.* Lanham, M.D. Rowman and Littlefield.

Gramsci, A. (1971). *Selections from the prison notebooks.* (Qiton Hoae and Geoffry Nowell Smith, Eds. and Trans.). London: Lawrence and Wishart.

Green, Maxine. (1988). *The dialectic of freedom.* New York: Teachers College Press.

Habermas, J. (1984). *The theory of communicative action. Vol. 1: Reason and rationalization of society. Vol. 2: Lifeworld and system: a critique of functionalist reason.* (Thomas McCarthy, Trans.). Boston: Beacon.

Hanson, Ann. (1996). The search for a separate theory of adult learning: does anyone really need andragogy? In R. Edwards, A. Hanson, and P. Raggat (Eds.), *Boundaries of adult learning* (pp. 99–127). New York: Routledge.

Hart, M. (1990). Critical theory and beyond: further perspectives on emancipatory education. *Adult Education Quarterly, 40* (Spring): 126–138.

Horton, Myles. (1998). *The long haul: An autobiography.* New York: Teachers College, Columbia University Press.

Kegan, Robert. (2000). What 'form' transforms? A constructive-development approach to transformative learning. In J. Mezirow and Associates (Eds.), *Learning as transformation: Critical perspectives on a theory in progress* (pp. 35–70) San Francisco: Jossey-Bass.

Knowles, Malcolm. (1970). *The modern practice of adult education: Pedagogy versus andragogy.* New York: Association Press.

Lindeman, E. C. (1926). *The meaning of adult education.* New York: New Republic.

Loeb, Paul Rogat. (2000). Ordinary people produce extraordinary results. Heroism: Rather than mythologize those who act for justice, we can learn from what empowered them. *The Los Angeles Times,* January 14: 7

Lohman, Roger (1992). The commons. New perspectives on nonprofit organizations and voluntary action. San Francisco : Jossey-Bass.

Menard, L. (1997). Re-imagining liberal education. In R. Orril, (Ed.). *Education and democracy: Re-imagining liberal learning in America* (pp. 1–19). New York: College Board Publications.

Merriam, S. and Caffarella, R. (1999). *Learning in adulthood: A comprehensive guide.* Second edition. Jossey-Bass Publishers, San Francisco.

Mezirow, Jack. (1978). *Education for perspective transformation: Women's re-entry programs in community colleges.* New York: Teachers College Columbia University Press.

Mezirow, Jack. (1992). Transformation theory: Critique and confusion. *Adult Education Quarterly, 44* (4): 250–52.

Mezirow, Jack. (1994). Understanding transformation theory. *Adult Education Quarterly, 44* (4): 222–44.

Mezirow, Jack. (1995). "Transformation Theory of Adult Learning." In M. Welton (Ed.), *Defense of the lifeworld,* (pp. 39–70). New York: SUNY Press.

Mezirow, Jack. (1996). Contemporary paradigms of learning. *Adult Education Quarterly, 46* (3): 158–72.

Mezirow, Jack. (1998). On critical reflection. *Adult Education Quarterly, 48* (3): 185–98.

Mezirow, Jack. (2000). Learning to think like an adult: core concepts of transformation theory. In J. Mezirow and Associates (Eds.), *Learning as transformation: Critical perspectives on a theory in progress,* (pp. 3–34). San Francisco: Jossey-Bass

Moll, Jacqueline, and Fischer, Nilton Bueno. (2000) Pedagogias nos tempos do orcamento participativo en Porto Alegre: Possiveis implicaçoes educativas na ampliacao da esfera publica. Unpublished manuscript.

Newman, M. (1994). *Defining the enemy: Adult education in social action.* Sydney, Australia: Stewart Victor.

Parks Daloz, Laurent. (2000). Transformative learning for the comomon good. In J. Mezirow and Associates, (Eds.), *Learning as transformation: Critical perspectives on a theory in progress,* (pp. 103–25). San Francisco: Jossey-Bass.

Orril, Robert (Ed). (1997). *Education and democracy. Re-imagining liberal learning in America.* New York: College Board Publications.

O'Sullivan, Edmund. (1999). *Transformative learning: Educational vision for the 21st century.* Toronto: University of Toronto Press.

Pateman, Carole. (1970). *Participation and democratic theory.* London: Cambridge University Press.

Peterson, Ralph. (1992). *Life in a crowded place: Making a learning community.* Portsmouth, N.H.: Heinemann.

Pontual, Pedro de Carvalho. (2000). O processo educativo no orcamento participativo: Aprendizados dos atores de sociedade civil e do estado. Unpublished doctoral dissertation. Sao Paulo: Pontificia Universidade Catolica de Sao Paulo.

Schugurensky, D. (1998). The legacy of Paulo Freire: A critical review of his contributions. *Convergence, 31*(1,2): 17–29.

Schugurensky, D. (2001). Grassroots democracy: The participatory budget of Porto Alegre. *Canadian Dimension 35* (1): 30–32.

Scott, S. (1998). An overview of transformative theory in adult education. In S. Scott, B. Spencer, and A. Thomas (Eds.), *Learning for life: Canadian readings in adult education* (pp. 98–106). Toronto: Thompson.

Spencer, B. (1998). *The purpose of adult education. A guide for students.* Toronto: Thompson.

Tannen, D. (1998). *The argument culture.* New York: Random House.

Taylor, E. W. (1997). Building upon the theoretical debate: a critical review of the empirical studies of Mezirow's transformative learning theory. *Adult Education Quarterly, 48* (1): 34–59.

Taylor, Edward. (2000). Analyzing research on transformative learning. In J. Mezirow and Associates, (Eds.), *Learning as transformation: Critical perspectives on a theory in progress,* (pp. 285–328). San Francisco: Jossey-Bass.

Tennant, Mark. (1993). Perspective transformation and adult development. *Adult Education Quarterly 44* (1): 34–42.

Warren, M. (1992). Democratic theory and self-transformation. *American Political Science Review 86:* 8–23.

CHAPTER 7

The Signature of the Whole

Radical Interconnectedness and its Implications for Global and Environmental Education

David Selby

Down Green Lane

In the 1950s, as a boy between the ages of five to ten, I lived in the village of North Hykeham, a few miles outside the cathedral city of Lincoln, on the edge of the Lincolnshire fenlands, England. To reach the fens we had to walk from the village down a bridle track called Green Lane. To go down Green Lane for any child interested in nature was to enter a world of wonder. In wintertime, the fens would freeze over and it was possible to walk for miles over ice, slipping and sliding, looking for animal tracks, with one ear cocked for the sound of creaks and groans indicating that you literally were approaching thin ice and that it was time to retreat. Not that ice walking was life endangering—except for the drains and river ways, the water stood only about two feet deep for mile after mile. To fall through the ice was cold and unpleasant but also quite a thrill. In spring and summer the fens transformed into a vast wild garden of flowers with evocative names such as marsh marigold, lady's smock, ragged robin, red campion, monkey flower. I spent day after happy day searching for flowers and keeping an annual scrapbook of pressed flowers, noting the date of first seeing the flower in bloom each year. Each year we watched the coming of birds in the spring, their going in the fall. We knew the badger holes, the fox coverts, and the broken-down willows where the shrews nested.

In the early 1980s I took my children to see this place of wonder, the place where I had lived out some of the happiest times of my boyhood. Green Lane had become the principal road through a suburban housing estate. The fens had been drained in the name of agricultural development and efficiency (as understood by Strasbourg bureaucrats). The place where I lost my Wellington boots in a mire of mud one springtime—to be chided heavily by my mother in those impecunious times on returning home bootless—had been covered over with concrete. The sense of loss was palpable. Somehow part of me, a source of my identity, of my sense of self, had been taken away.

This personal story captures for me some of the life springs of modern environmentalism: the adult experience of loss of places of deep meaning in our lives through rampant development and urbanization; of increasing disconnection from nature and, through that estrangement, an existential crisis of identity (Tomashow, 1996). The tarmacking of Green Lane was both process and symbol of disconnectedness from Earth and erosion of identity.

The Global Environmental Crisis as Crisis of Worldview

In many of us this sense of loss has grown hand in glove with an uneasy sense that our window on the world—our worldview—is somehow distorted, deeply destructive in its impact, and quite insufficient either to understand what is happening to the planet or to do anything fundamentally about it.

A number of commentators have argued that mainstream western thinking has inherited a worldview from seventeenth- and eighteenth-century scientists and philosophers that is underpinned by notions of separation, otherness, and domination. That worldview, they argue, has been deeply influenced by Francis Bacon's view that the goal of science was to enslave nature (and in the process of which "to torture nature's secrets from her") as well as by Rene Descartes's division of the world into *res cogita* (things of the mind) and *res extensa* (mechanically extended substances or matter) and his consequent arrogation of mind and free will exclusively to humans. This has led to our creating a hierarchy within ourselves (mind above emotions and body) and to our locating ourselves outside and above nature (which has no mind). This also has correlated with our according only instrumental value to nature (nature as mindless resource); our denying ethical and moral status to other life-forms and to environments (in that they are mindless machines); and, thus, to our allowing ourselves virtually unfettered license to exploit (Bateson, 1973; Bohm, 1990; Capra, 1983, 1996; Evernden, 1985; Merchant, 1981). While fueling the hubris of uniqueness, it has fostered our modern sense of alienation and existential crisis. "We are distinct from everything around us and inexorably alone" (Zohar, 1990: 34).

The machine image and understanding of the world as put forward by the likes of Descartes, Bacon, and Sir Isaac Newton, the same commentators maintain, also has become deeply embedded in western thought. We try to understand how something works by dividing it into what are held to be its discrete component parts. If an identified part malfunctions, we tend to it without reference to the whole. Understanding—and control—is achieved through compartmentalizing, pigeonholing, and analyzing, through atomism or reductionism. Separation is the name of the game (Capra, 1983; Callicott, 1986; Merchant, 1981). The general practitioner or specialist tends to a pain in a part of the body without reference to the rest of the body, to the patient's psyche, to social and environmental relationships (the specialist often particularly so in that the greater the degree of speciality the more frequent the occurence of speciality-myopia). The corporate executive toasts a hefty credit over debit account without factoring in the environmental, cultural, social, and psychological costs of gathering raw material, processing, and distributing the firm's product. The science teacher teaches the flower by having the child name the parts but misses the essence of the flower (more than the sum of its parts) and of the flower in its context. A reductionist mentality also tends to wed its adherents to a deterministic outlook. Just as in a machine process, in which nearby components react one upon the other, events in the world are viewed as happening in an inexorably linear fashion, while instability and chance are seen as shortcomings in our present capacity to control—as "physical problems awaiting mechanical solutions" (Callicott, 1986: 303). Within reductionism, "all causal relationships are reducible to the motion or translation from point to point of simple bodies or the composite bodies made up of them. The mysterious causal efficacy of fire, disease, light, or anything else is explicable, in the last analysis, as the motion, bump, and grind of the implacable particles" (Callicott, 1986: 303). Only our minds, Cartesianism holds, are free to range as and where they want (Cottingham, 1986).

The dualisms spawned by Cartesian thought (e.g., human-animal, mind-body, masculine-feminine, us-them, inner-outer, subject-object; reason-emotion, spirit-matter, culture-nature, teacher-learner) and the hegemonic thinking they inspire also have become ingrained in the western mind-set. Overlay one or more dualisms, as mainstream western culture has done and continues to do, and we create the hegemonic attitudes and structures that liberationist and transformative educators are now called upon to confront. Masculine-feminine, mind-body, reason-emotion, subject-object, for instance, superimposed, yield the mental and social scaffolding of patriarchy (Plumwood, 1993: 43). Of these dualisms, more later.

Global and Environmental Education as Responses to the Mechanistic Mind-set

In their most transformative expressions, global and environmental education can be viewed as educational countercultures to mechanism and reductionism as they have colonized education and as educational expressions of a holistic paradigm (Selby, 1999, 2000a). This is often expressed symbolically using the billiard ball and web models (figures 7.1 and 7.2). The billiard ball model—depicting a cluster of billiard balls on a billiard table—has been employed to indicate separateness, discreteness, and forms of external relationship between things where the relationship has no effect on their internal structure and dynamics (Zohar, 1990: 81). In education, the model finds expression in the division of arts and sciences, separate subject disciplines, grade apartheid, individualized learning, the strict delineation of who is the teacher and who the learner, and the arm's-length relationship between school and community (Greig, Pike, and Selby, 1989: 18–24).

Transformative global and environmental educators have countered the billiard ball model with the web model (understood dynamically). The latter has seemed to convincingly capture understandings drawn from ecological and quantum (subatomic) science that: everything is dynamically connected and related to everything else; nothing can be completely understood save in relationship to everything else; identity is multifaceted and includes a significant near-and-far contextual element; what happens somewhere will impact to a greater or lesser extent elsewhere, even everywhere (captured to some extent in the environmentalist's saying "You are always downstream of someone"); what happens locally is also a global phenomenon (a part of the whole, itself acting to inform the whole) and the signature of global events will

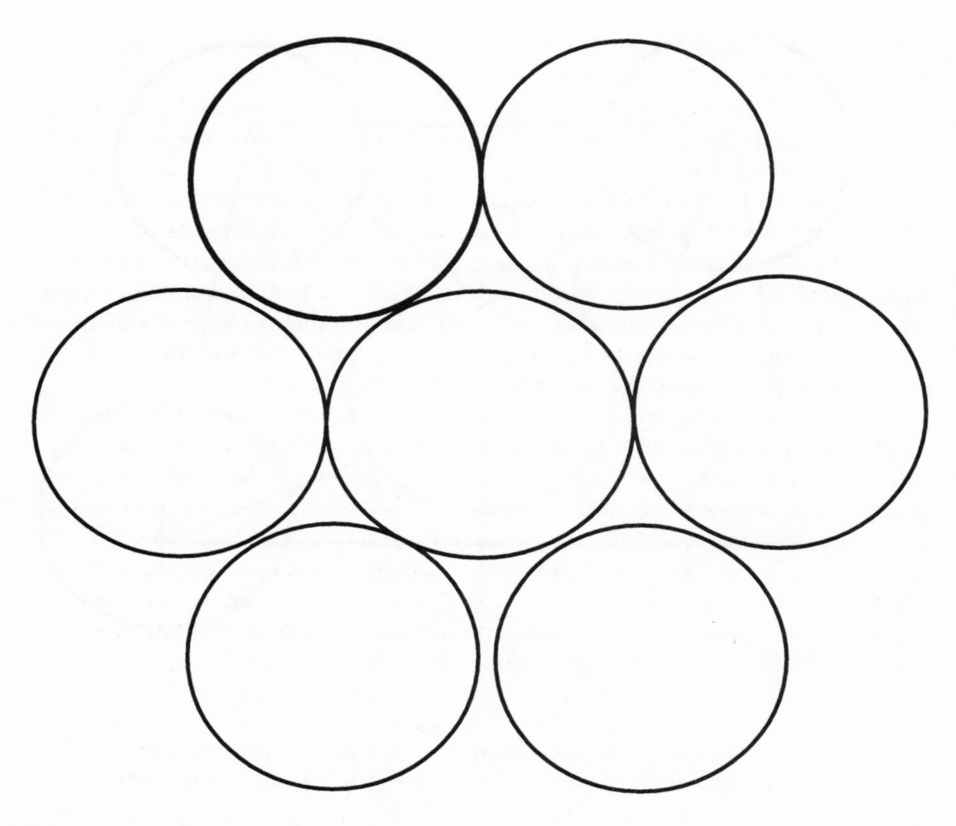

Figure 7.1 Billard Ball/Building Clock/Clock

be manifest locally; different global issues—such as environment, development, health, peace, rights—are interconnected; past, present, and future are interwoven, coevolving and cocreating elements of time. Using such insights as captured in the web metaphor, educators have developed curricula, teaching materials, and learning activities built on the concepts of interconnectedness, interdependence, and interrelationship. (See, e.g., Fountain, 1995; Pike and Selby, 1999, 2000; Townsend and Otero, 1999). My question is: Have transformative global and environmental educators gone far enough in responding to mechanism and reductionism?

Metaphors tell their tale. The metaphor of the web is instructive. It can signify frailty—think of the delicate filaments of a spider's web caught in the dew of the early fall. As such, the web suggests the delicacy of the strands of any ecosystem—so easily disrupted by human interference. It can signify strength and security. Think of the strength and resilience of ecosystems through the ages. Think of the net that holds the falling acrobat. It also can signify entrapment. Think of the spider's web from the perspective of the fly. Think of the marginalized caught in the web of economic globalization. So we need to be clear what our web metaphor is suggesting. Webs can be liberating; they also can be constrictive. There are positive and shadow sides to any metaphor (Heshusius, 1991).

Figure 7.2 Web Model

RADICAL INTERCONNECTEDNESS: A THIRD LEVEL OF PRESENCE

My proposal is not that an emphasis on interconnectedness, interdependence, and interrelatedness, as captured in the web metaphor, is misconceived—far from it—but rather that it overlooks an important element of reality or third level of presence that is profoundly important to a holistic or ecological worldview. Long-standing models and metaphors retain some usefulness. The billiard ball and the accompanying classic metaphors of mechanism, the building block and the clock, continue to have a rightful but limited place in our scheme of things. They represent one level of presence. I need to know a chair is a chair and that my car engine will work and can be put right if a part becomes faulty. But they fall far short of enabling us to understand human-human or human-Earth relationships. Hence, the web has been proposed as a persuasive metaphor—of wider sway and significance—for understanding a second level of presence—the dynamic and interconnected nature of our world. This, I suggest, also has its limitations for evoking transformative Earth consciousness and behaviors. I would like to propose dance (of the free-form variety) as a metaphor for the way we need to conceive of the world at a deeper and third level of presence.

Leading-edge ecologists and quantum physicists have suggested to us that there is a world of "unbroken wholeness" or "holomovement" (Bohm, 1983, 1990) underlying the world of separate things and the world of interconnections. Inspired by the web metaphor, global and environmental educators have sought to depict in curriculum programs and learning modules the complex interactions between elements of ecosystems (plants, birds, insects, humans, fungi, etc.) and between entities in the human world (individuals, communities, nation, states, nonstate actors). But the overt and covert agenda of these curricular offerings still depicts the entities as primary, solid, and separate (even though interconnected).

At the third level, the entities are not primary, solid, or separate. The relationship becomes primary, and the entity is itself a secondary manifestation.

Physicist David Bohm (1983, 1990) has described the subatomic world of relatively separate things (neutrons, electrons, etc.) as the "explicate order" behind which is an "implicate order," in which everything is enfolded within everything else. Bohm extrapolates from the subatomic world to suggest we would be wise to countenance the implicate order in our understanding of our macroworld—to see that in a profound and very real way, everything is embedded in everything else and that things or objects are ontologically subordinate to flows and patterns. Everything is thus the signature of the whole. Ecologist Paul Shepard (1959) and biophysicist Harold Morowitz (1972) have said very much the same thing. From a modern perspective, says Shepard, "nature is epitomized by living objects rather than complex flow patterns of which objects are temporary formations. The landscape is a room-like collection of animated furniture . . . but it should be noted that it is best described in terms of events which constitute a field pattern" (1959: 505–6). Or as Morowitz puts it: "Viewed from the point of view of modern ecology . . . the reality of individuals is problematic because they do not exist per se but only as local perturbations in [the] universal energy flow. . . . Consider a vortex in a stream of flowing water. The vortex is a structure made of an ever-changing group of water molecules. It does not exist as an entity in the classical Western sense; it exists only because of the flow of water through the stream. If the flow ceases the vortex disappears. In the same sense structures out of which biological entities are made are transient, unstable entities with constantly changing molecules dependent on a constant flow of energy to maintain form and structure" (1972: 156).

Here are environmental philosopher Holmes Rolston III's reflections as he stood on the shoreline of a Rocky Mountain wilderness lake: "Does not my skin resemble this lake surface? Neither lake nor self has independent being. . . . Inlet waters have crossed this interface and are now embodied within me. . . . The waters of North Inlet are part of my circulatory system; and the more literally we take this truth the more nearly we understand it. I incarnate the solar energies that flow through this lake. No one is free-living. . . . *Bíos* is intrinsically symbiosis" (1975: 122).

Entities—including ourselves—according to new physics and new ecology are momentary configurations of energy, local perturbations in a total energy field or holomovement. We emerge into the explicate, become manifest, only to resubmerge into the implicate order of being (which at one level of presence we never left). We are ephemeral manifestations of a fertile no-thing-ness from which all things emerge and to which all return (Zimmerman, 1988, 22). David Bohm (1990) sees this conception of reality as closing the Cartesian bridge between mind and matter. Just as all things emerge from the holomovement and their "existence is sustained in a constant process of unfoldment and re-enfoldment," giving rise to "their relatively stable and independent forms in the explicate order" prior to their resubmergence, so behaves mind "with its constant flow of evanescent thoughts, feelings, desires, and impulses, which flow into and out of each other, and which, in a certain sense, enfold each other" (273).

Thus, at the deeper third level of presence, where the web metaphor becomes unsatisfactory, we need to consider things as expressions of the dynamic unfolding, the being and becoming, of the whole. We need to see entities—ourselves, nonhuman animals, rocks, nation-states, political groupings—not first and foremost as objects but primarily as processes or dances. Phenomena (people, other-than-human lifeforms, places, countries) at this level are coevolving manifestations of a multileveled and multidimensional dance of internal and external relationships. Global and environmental educators need to embrace the metaphor of dance and the level of presence of unbroken wholeness it represents in theory and practice while continuing to work with the metaphor of the web of relationship.

ON INDIVIDUALISM, SELF, ALTRUISM, AND NARCISSISM

Such a concept of radical interconnectedness helps us recall that the word "individual" has become distorted in modern times to denote she or he who is separate. It originally meant "a person undivided from the whole," a meaning arising from an intuitive and spiritual understanding that richness and uniqueness emerge from deep connectivity—that the more profoundly connected we are with Earth and with each other then, paradoxically, the more we become special and distinct. As David Steindl-Rast puts it: "The more you know a friend, the more you know that friend as unknowable" (Capra and Steind-Rast, 1992: 102). Deeper connectedness, deeper mystery. Deeper connectedness, deeper awe.

If we embrace the third level of presence and its metaphor of dance—that we are processes, not objects, expressions of a perpetual dance of inner and outer relations—we also move to a radically different conception of self. The dance suggests that the world is within us and that, in some mysterious way, we are in the world. This is what Australian environmentalist Paul Seed (Seed, Macy, Fleming, and Naess, 1988: 6) meant in part when he exclaimed "we are the rocks dancing!" and when he could assert that the destruction of the rain forests had become for him as painful as losing a finger. When Seed began to campaign actively to protect the remaining rain forests of New South Wales, he wrote: "I knew then I was no longer acting on behalf of myself or my human ideas but on behalf of the Earth . . . on behalf of my larger self, that I was literally part of the rain forest defending herself" (37). I am the Green Lane; the Green Lane is me.

In embracing radical interconnectedness, the unending debate between working to save Earth for reasons of self-interest (e.g., arguments in favor of preserving the rain forest because unknown plants may provide cures for human disease) or for pure ecoaltruism becomes rather redundant. If at a deep and equally real level, self has no boundaries, saving the rain forest is the highest self-interest in that in a profound way you know you are the rain forest or know the rain forest is within you and making you what you are. In the same way we give new meaning to narcissism (excessive love of/interest in self). The "Song of Self" becomes the "Song of Earth," a fierce awareness of our short period of emergence from but still unbroken deep connection to, the whole (Roszak, 1992: 264). Put another way, we need to reconsider ongoing discussions concerning intrinsic value. Among global and environmental educators there has been a general (but not complete) embrace of the notion of self as intrinsically valuable. The argument has turned over whether we need to care for environments and other-than-human life-forms because of their extrinsic (instrumental) value or because they have value in, of, and to themselves. If, as quantum and ecological theory suggest, there is a continuity of self and nature, and if self is intrinsically valuable, then nature is intrinsically valuable. There is axiological complementarity (Callicott, 1985). "If it is rational for me to act in my own best interest, and I and nature are one, then it is rational for me to act in the best interests of nature" (275). The conventional separation of self and world—myself and Green Lane—cannot easily withstand the implications of a quantum/ecological worldview.

Val Plumwood (1993: 176–81) critiques deep ecological notions of expanded or oceanic self, arguing that denial of boundary demeans the independence and devalues and disrespects difference and particularity in the other-than-human while enlarging and extending egoism. Such criticisms clearly require engagement but do seem to assume that egoism remains a constant, a "conventional, constricted ego" (Callicott, 1985: 275), thus denying the leaven of axiological complementarity once the individual has consciously and mindfully embraced reality at different levels of presence, while they also overlook the dynamic and tensile interplay among the three levels within the mindful individual.

On Embracing Instability, Uncertainty, Awe, and Wonder

The quantum and ecological worldview show us that we can never know anything for sure. While the mechanistic worldview (and its educational manifestations) trades in certainties and stable understandings, a holistic worldview espouses instability and uncertainty. In a world in which, at one level of presence, everything relates to everything else and, at a deeper level, everything is embedded in everything else, we have to acknowledge that flow, movement, and complexity, allied to our limited vision and inability to comprehend and entertain all the questions to ask, make for, at best, provisional knowing.

Werner Heisenberg looked into the atom and found that subatomic entities are unknowable in any comprehensive way. Look for the momentum of a particle and you cannot know its position; establish its position and you cannot be certain about its momentum (Zohar, 1990: 10–11). Allied to that, entities within the atom simultaneously manifest themselves as particles and waves but if you measure one, you cannot see and measure the other. "Nothing," writes Danah Zohar (1990: 11), is fixed or fully measurable, everything remains indeterminate, somewhat ghostly, and just beyond our grasp." What we observe is not nature itself but nature exposed to the nature and limitations of our questioning. Elusiveness is a quality of world as dance.

Ilya Prigogine (1989) asks us to think of a pendulum. If we agitate a pendulum, we can predict that it will move inexorably toward minimal then no swing with its center of gravity as low as possible. We can be certain what will happen. But what, he asks, if we turn the pendulum on its head? It is difficult to predict what will follow. Fluctuating forces may make it fall to left or right. It is difficult to control. The notion of the upturned pendulum, Prigogine avers, has been "ideologically suppressed" (396) in that its message of instability is inconvenient for a culture that seeks to dominate and exploit nature. "In a deterministic world nature is controllable, it is an inert object susceptible to our will. If nature contains instability as an essential element, we must respect it, for we cannot predict what may happen" (397).

Mainstream western thinking has viewed—and still largely views—nature as deterministic; nature as swinging pendulum. (There is still determinism, albeit complex, in the web.) But what if we allow that nature is nondeterministic and unstable? First, we bring the internal world of the human mind (seen by the dominant western worldview as free, nondeterministic, and outside nature) and the external world of nature (seen, by that same worldview, as machinelike and deterministic) together. We subvert the mind-nature divide of Descartes.

Second, in denying certainty and recognizing our inability to control or predict, we are better able to accord respect, awe, wonder, and reverence to nature. There is a close connection between embracing instability and cultivating a sense of wonder and reverence. Calling for "respect, not control," Prigogine writes: "We need to be aware that our knowledge is still a limited window on the universe; because of instability we must abandon the dream of total knowledge of the universe" (1989: 399).

Instability and radical interconnectedness are themselves in perpetual dance. Whenever nature, the world, a particular environmental or social situation moves closer to disequilibrium, the wider and more coherent the range of forces necessary to bring the situation to a new level of (complexified) equilibrium (Capra, 1996: 181). Whenever a situation becomes static and moribund, the web and dance are there to restore dynamism. We can speculate that we would have a much less connected world if everything were in constant balance, just as we would have less exciting human minds and psyches in a world lacking natural and cultural diversity.

All this, I suggest, makes me worry about global and environmental educators who continue to genuflect at the altar of "balance." What kind of balance do they have in mind between forces that are profoundly unequal and voices that are unequally heard in a world that is out of kilter? Is balance an appropriate objective if the overall goal is transformation? Or should we encourage tilt toward the disequilibrium that will effect radical change leading to new, more complex, configurations within a new equilibrium? Disequilibrium is probably a prerequisite of holistic, global, and transformative perception. "Coherence far from a state of equilibrium acquires huge dimensions in comparison with what happens in a state of equilibrium. In equilibrium each molecule can only see its immediate neighbours. Out of equilibrium the system can see the totality of the system. One could almost say that matter in equilibrium is blind, and out of equilibrium starts to see" (Prigogine, 1989: 399).

Educational Implications of Radical Interconnectedness

So what does the radical interconnectedness of the dance have to say to global and environmental educators? I believe it suggests five things.

First, it suggests that we take a both/and rather than an either/or approach to the ongoing debate between those who think our environmental education should be locally/bioregionally focused and those calling for a global focus.

David Orr (1992), Madhu Prakash (1994), and others have called for place-based environmental education and have discounted global environmental education as outside our experience and beyond our knowledge. Prakash has argued that we cannot "know" the globe except by reducing the whole to statistics, as it is too big (51). Orr argues that you can only know and appreciate what is really close but concedes that place-oriented environmental education can be "inherently parochial and narrowing" (131). These thinkers not only seem to harbor questionable mechanistic understandings regarding what it means to "know" but also posit local and global as dichotomies—false dichotomies. Local and global are embedded or nested in each other. Both web and dance are everywhere. My Green Lane experience was and remains both a local and global (a glocal) experience. As global and environmental educators, we need to allow both web and dance to inform our conceptual frameworks as well as the learning programs and learning experiences we offer.

Second, we should help students move beyond the mechanistic sense of the individual that mainstream western culture propagates. Too much environmental and global education has been outer-directed (looking out on the world) and has denied interiority (inner journeying). The hidden agenda of this tendency is to collude with mechanism by implying that our inner self is outside the universe. (The English word "environment" is itself problematic here—that which surrounds but, by implication, is not, and does not include, us!)

Through appropriate topics and methodologies, we need to help those in our learning communities know and experience at one and the same time the discrete self, the relational self, and the dancing self. (Ways of doing this are summarized in

Figure 7.3 The Billard Ball, Web, Dance Models and Some Eductional Implications

Metaphors	*Underlying Concepts*	*Curriculum*	*Process*
Billard Ball/Building Block/Clock	Separateness Fragmentation Compartmental-ism Linear connection	Subjects Disciplines Arts/Sciences	Individualised, competitive learning Machine-image education (input-output) Fast learning
Web	Interconnection Interdependence Interrelation-Ship	Intergration Interdisciplinary	Co-operative, interactive learning Children (not child) centered Mixed-paced learning
Dance	Embeddedness Enfoldment Interpenatra-tion	Other-than-disciplinary Experience	Empathetic, embodied learning Spiritual learnng Slow learning

figure 7.3). In western education, we are very good at the first, weak at the second (despite the efforts of global, environmental, holistic, and transformative educators), and usually blind to the third. This need to experience the relational self and the dancing self speaks, for example, to working with relational modes of knowing that would help us to recognize our inner connectivities (the embedded nature of body, mind, emotions, and spirit) and our deep connectivities with each other and with nature. It also would mean introducing new modalities, enabling students to explore their inner ecology, to cultivate their attunement to their senses and body rhythms, and, thus, to develop an embodied relationship to nature—contemplative and therapeutic art, artful self-inquiry, dance, deep-breathing exercises, yoga, meditation, relaxation, and peer reflexology would all become valued features of a truly global learning process (Houston, 1982; Liebmann, 1986; Lipsett, 2001; Macy, 1991; Miller, 2000; Nakagawa, 2000; Nhat Hanh, 1990, 1992; Selby, 1996).

These modalities of inner journeying clear the clutter of explicate reality; limit or stop thought, bring together the physical, mental, and emotional aspects of our being; and can create an awareness of the oneness of everything. They are ways to meet the dancing self. Many of us have experienced that occasional sense of self as oceanic—from the thrill of climbing a mountain, of weaving the waters of a difficult river in a canoe to other manifestations of what Abraham Maslow (1985) calls "peak experiences," such as T. S. Elliot's "moment in a rose garden." But here the suggestion is that we cultivate inner journeying within our formal learning programs. Beginning will be difficult but this is a kind, not all-or-nothing, philosophy. We can feel good about small beginnings—because what we are doing is difficult and countercultural—knowing that the ripples will go where they will and remembering that what happens somewhere is, in a strange way, happening everywhere.

The inner journeying modalities also speak to mindful, still, and slow learning as a counterbalance to the packaged rush and treadmill of transmissional/mechanistic learning *and the* swift-paced quality of much learner-centered learning. Slow learning is also an attunement to the pace of nature. "The natural world is really slow. Save for the waving of trees in the wind, or the occasional animal movement, things barely happen at all. To experience nature, to feel its subtleties, requires human perceptual ability that is capable of slowness. It requires that human beings approach experience with patience and calm" (Mander, 1991: 86). As Krishna says in the *Bhagavad Gita:*

> When all desires are peace and the mind, withdrawing within, gathers the multitudinous straying senses into harmony of recollection,
>
> Then, with reason armed with resolution, let the seeker quietly lead the mind into spirit, and let all his [*sic*] thoughts be silence. . . .
>
> He sees himself in the heart of all beings and he sees all beings in his heart [*sic*].
> (1982: 6: 24–31)

If this sounds like spirituality in the curriculum, that would be an appropriate conclusion. It is unlikely that environmental and global education can ever impact our culture unless we embrace a radical interconnectedness that revives mystery, a sense of the ineffable, the unknowable. A common deep ecological reading is that spirituality is a recognizing of deeper levels of connection within ourselves and between ourselves and the world. Theodore Roszak (1992: 45, 63) suggests that there is no likely way to return to planetary and societal health unless we heal the dichotomy between psyche and nature born of industrialism and seventeenth-century mechanistic science. He adds: "The great changes our runaway industrial civilization must make if we are to keep the planet healthy will not come by the force of reason alone or the influence of fact. Rather, they will come by way of psychological transformation. What the earth requires will have to make itself felt within us as if it were our own private desire. Facts and figures, reason and logic can show us the errors of our present ways; they can delineate the risks we run. But they cannot motivate, they cannot teach a better way to live. That must be born from inside our own convictions. And that birth may have to be a painful one" (47).

As an afterword on the nature of self, I would like to make the likely controversial suggestion that we bring death into the curriculum. Death denial is, perhaps, a central aspect of our planetary crisis. We buy and consume and rush for seeming immortality. As Susan Griffin (1995) puts it: "Fragmentation creates a temporary reprieve from the fear of death and loss. But it also creates its own grievous sense of death and loss. . . . In dividing itself from mortality, the European psyche dulls its own experience of the world" (51–52). If we wish school-age or adult learners to see themselves at one level of presence as processes or perturbances in the energy field, then the return to the implicate order is something we need to talk about and reflect upon. The cycles of birth and death are central to an ecological perspective. How to do this—within a dominant death-denial culture—is something we need to address (and in multiple and complex ways given environmental, socioeconomic, and cultural diversity).

Third is radical interconnectedness calls for multidimensional ways of knowing. Transformative global, environmental (especially ecofeminist), holistic educators have been at the forefront in trying to move learning away from an overemphasis on reason, thought, analysis, and objectivity (Russell and Bell, 1996; Selby, 1996). Inspired by the metaphor of the web, they have called for intuition (the ability to be immediately sensitive to the whole), synthesis, the sharing of subjectivities, and relational sensibility to be accepted as equally valid ways of knowing. But, perhaps in deference to prevailing culture, we have not pushed these ideas with the conviction we might have brought to bear. The dance metaphor calls for a thorough reclaiming of emotion, subjectivity, bodily sensibility, intuition, empathy, caring and compassion, love, and relational and spiritual sensibility as means of knowing (Russell and Bell, 1996; Miller, 1993, 2000; Selby, 1996).

In seeking multidimensional ways of knowing, a caution against computers. Computers, we are often told with almost hysterical fervor, can connect us to the world. As one advocate (cited in Maxwell, 1999) exhorts: "Let's put a computer in every home and every classroom. . . . Let's connect Canadians of every age, race, and gender to each other and to the rest of the planet." Yet it is important to recognize that computers offer a disembodied form of connectivity that denies physicality, compresses emotions through a cognitive prism, cushions us from direct experience of others and the outdoors, and ignores spirituality (Maxwell, 1999). A radical rendition of interconnectedness would resist the onward rush to dot-com the learning community. While computers have their uses, we should recognize that they are among the latest technical phenomena in the process of disconnecting humans from nature. We should see them for what they are—machines that have their uses. They are no substitute for lived and embodied connectedness with nature and people. It is significant that, while we understand "media literacy" as the ability to critically deconstruct and decode media of various kinds, we are becoming conditioned to interpret "computer literacy" as the ability to use computers efficiently (while relating to them uncritically).

Fourth, radical interconnectedness suggests that more environmental educators recognize that they are part of a wider community of countercultural and liberationist educators and that new coalitions and alliances are necessary.

Although alliance building among environmental, development, health, humane, human rights, and peace educators as well as educators working against discrimination has been central to global education (Goldstein and Selby, 2000), the majority of environmental educators have been shy of actively and concretely recognizing that they are part of a community of educators seeking environmental and social justice and nonviolent change. For instance, peace education has long identified "environmental damage" as a problem of peace and "ecological balance" as a value underlying peace (Hicks, 1988; Smith and Carson, 1998) but the concepts, models, and theories of peace education have found little space within environmental education discourse. With some notable ecofeminist exceptions (e.g., Bell and Russell, 1999; Donovan, 1993), environmental education has shied away from creative engagement with humane education with its emphasis on animal-related issues, challenging anthropocentrism, and exploring the correlation of human and nonhuman oppressions. Soul sisters have barely talked, and it is environmental education that has fought shy of engagement (Selby, 1995). Also, most environmental educators have not seriously engaged with multicultural education by recognizing the interplay of different cultural perspectives around understandings of environment and environmental issues (Running Grass, 1996). Biodiversity and cultural diversity have not danced together. Preservation of biosphere and preservation of ethnosphere (Davis, 2000: A15) have not coalesced in environmental education's learning and teaching. Finally, save at the cutting edge (Russell and Bell, 1996; Russell, Bell and Fawcett, 2000), environmental education has not combined with antidiscriminatory education to any extent. Environmental issues are very much social justice issues if people of a different gender and/or belonging to different racial and ethnic groups contribute to, or feel the effects of, environmental despoilation differentially (Lousley, 1998, 27).

The radical interconnectedness of the dance suggests that the respective fields are mutually enfolded. We need to see each field as one among a "network of pearls" as in this passage from the *Avatamsaka Sutra:* "In the heaven of Indra, there is said to be a network of pearls so arranged that if you look at one you see all the others reflected in it. In the same way each object in the world is not merely itself but involves every other object, and in fact IS everything else" (cited in Pike and Selby, 1995: 13).

As a basis for broadening the community of liberationist and countercultural educators, it is important that we recognize, as ecofeminist educators and transformative humane educators have (Russell and Bell, 1996; Selby, 1995), that not only are oppressions mutually reinforcing but their dynamics are similar—whether the oppression is of women, ethnic or sexual minorities, environments, or animals. The oppressors treat the object of the oppression as "other" and proceed (Plumwood, 1993, 1996) to:

- Radically exclude—creating sharp boundaries and maximum separation of identity between themselves and the "other" as seeming justification and reconfirmation of superiority
- Homogenize or stereotype—hence disregarding or denying difference and diversity in characteristics, motivations, tendencies, and perspectives among the "other"
- Inessentialize—denying dependency on, and backgrounding, the "other"
- Incorporate—defining only in relationship to themselves and denying the intrinsic needs and independent agency, creation of value, and motivations of the "other" ("Humanity is male and man defines woman not in herself but as relative to him"—Simone de Beauvoir, The Second Sex, cited in Plumwood, 1996);
- Instrumentalize—denying any value in the "other" beyond the useful.

Fifth, radical interconnectedness suggests that we need to rethink how we try to bring about educational change.

Our approaches to change have been wedded to mechanism. We have opted for restricted change focuses (e.g., developing a global or environmental pack or program for a specific grade and school subject; reduce, reuse, and recycle programs, schoolyard naturalization) when our ecological understanding tells us that change is about strength/resilience through diverse yet connected initiatives, coalitions, and partnerships and dynamic and synergistic interplays between different change initiatives (Selby, 2000b). Change, in short, has to be holistic to be effective. A challenge we face, given the marginality of the fields of environmental and global education, is how to mount the kind of holistic, multifaceted change initiatives our hearts and minds tell us are essential if we are to have sustained impact on educational institutions and systems and if we are to remain faithful to ecological principles and processes of change.

ENDWORD

"Radical" means going to the roots of things. We have to ask ourselves deeply whether we are about reform (which may simply buttress attitudes and structures that are at the root of the ecological crisis) or transformation. We have to ask whether our aim is to tamper with or turn around. In a more sophisticated and contemporary version of the story of the emperor's new clothes, Douglas Adams wrote in *The Hitchhiker's Guide to the Galaxy*: "It is an important popular fact that things are not always what they seem. For instance, on the planet Earth, man had always assumed that he was more intelligent than dolphins, because he had achieved so much—the wheel, New York, war, and so on—whilst all the dolphins had ever done was muck about in the water and have a good time. But conversely, the dolphins had always believed that they were more intelligent than man—for precisely the same reasons" (1979: 119).

NOTES

In its original form, this chapter was given as a keynote presentation at the May 2000 annual conference of the Global, Environmental & Outdoor Education Council (GEOEC) of the Alberta Teachers' Association. I would like to thank Connie Russell, Faculty of Education, Lakehead University, for her invaluable comments on an earlier draft. The standard rider obtains, however, that the opinions expressed here remain my responsibility.

REFERENCES

Adams, D. (1979). *The hitchhiker's guide to the galaxy.* London: Pan.

Bateson, G. (1973). *Steps to an ecology of mind.* London: Granada.

Bell, A. C., and Russell, C. L. (1999). Life ties: Disrupting anthropocentrism in language arts education. In J. Roberstson, (Ed.), *Teaching for a tolerant world,* (pp. 68–89). Urbana, Ill.: National Council of Teachers of English.

Bhagavad gita. (1982). (J. Mascaro, Trans.). Harmondsworth: Penguin.

Bohm, D. (1983). *Wholeness and the implicate order.* New York: Ark.

Bohm, D. (1990). A new theory of the relationship of mind and matter. *Journal of Philosophical Psychology, 3*(2), 271–86.

Callicott, J. B. (1985). Intrinsic value, quantum theory, and environmental ethics. *Environmental Ethics, 7* (Fall): 257–75.

Callicott, J. B. (1986). The metaphysical implications of ecology. *Environmental Ethics, 8* (Winter): 301–15.

Capra, F. (1983). *The turning point: Science, society and the rising culture.* London: Flamingo.

Capra, F. (1996). *The web of life: A new scientific understanding of living systems.* New York: Anchor.

Capra, F., and Steindl-Rast, D. (1992). *Belonging to the universe: Explorations on the frontiers of science and spirituality.* San Francisco: Harper Collins.

Cottingham, J. (1986). *Descartes.* Oxford: Basil Blackwell.

Davis, W. (2000). A dead end for humanity. *The Globe and Mail,* December28: A15.

Donovan, J. (1993). Animal rights and feminist theory. In G. Gaard (Ed.), *Ecofeminism: Women, nature, animals,* (pp. 167–94). Philadelphia: Temple University Press.

Evernden, N. (1985). *The natural alien: Humankind and environment.* Toronto: University of Toronto Press.

Fountain, S. (1995). *Education for development: A teacher's resource book for global learning.* London: Hodder and Stoughton.

Goldstein, T., and Selby, D. (Eds.). (2000). *Weaving connections. Educating for peace, social and environmental justice.* Toronto: Sumach.

Greig, S., Pike, G., and Selby, D. (1989). *Greenprints for changing schools.* London: World Wide Fund for Nature/Kogan Page.

Griffin, S. (1995). *The Eros of everyday life: Essays on ecology, gender and society.* New York: Doubleday.

Heshusius, L. (1991). On paradigms, metaphors, and holism. *Holistic Education Review* (Winter), 38–43.

Hicks, D. (1988). *Education for peace: Issues, principles, and practice in the classroom.* New York: Routledge.

Houston, J. (1982). *The possible human: A course in enhancing your physical, mental and creative abilities.* Los Angeles: J. P. Tarcher.

Liebmann, M. (1986). *Art therapy in groups: A handbook of themes, games and exercises.* London: Croom Helm.

Lipsett, L. (2001). On speaking terms again: Transformative experiences of artful earth *connection.* Unpublished dissertation, Department of Adult Education, Ontario Institute for Studies in Education of the University of Toronto.

Lousley, C. (1998). (De)politicizing the environment club: Environmental discourses and the culture of schooling. Unpublished M.A. thesis, Department of Currriculum, Teaching and Learning, Ontario Institute for Studies of Education of the University of Toronto.

Macy, J. (1991). *World as lover, world as self.* Berkeley, Calif.: Parallax.

Mander, J. (1991). *In the absence of the sacred: The failure of technology and the survival of the Indian nations.* San Francisco: Sierra Club.

Maslow, A. (1985). *The farther reaches of human nature.* New York: Penguin.

Maxwell, M. (1999). *Toward an ecology of being: Earth, spirit and education.* Unpublished. dissertation, Department of Curriculum, Teaching and Learning, Ontario Institute for Studies in Education of the University of Toronto.

Merchant, C. (1981). *The death of nature: Women, ecology and the scientific revolution.* San Francisco: Harper and Row.

Miller, J. P. (1993). *The holistic teacher.* Toronto: OISE Press.

Miller, J. P. (2000). *Education and the soul: Toward a spiritual curriculum.* Albany NY: State University of New York Press.

Morowitz, H. J. (1972). Biology as a cosmological science. *Main Currents in Modern Thought, 28:* 156.

Nakagawa, Y. (2000). *Education for awakening: An eastern approach to holistic education.* Brandon, Vermont: Foundation for Educational Renewal.

Nhat Hanh, T. (1990). *Breathe! You are alive: Sutra on the full awareness of breathing.* Berkeley, Calif.: Parallax.

Nhat Hanh, T. (1992). *Touching peace: Practising the art of mindful living.* Berkeley, Calif.: Parallax.

Orr, D. (1992). *Ecological literacy: Education and the transition to a postmodern world.* Albany: State University of New York Press.

Pike, G., and Selby, D. (1995). *Reconnecting: From national to global curriculum.* Godalming, UK: World Wide Fund for Nature UK.

Pike, G., and Selby, D. (1999). *In the global classroom 1.* Toronto: Pippin.

Pike, G., and Selby, D. (2000). *In the global classroom 2.* Toronto: Pippin.

Plumwood, V. (1993). *Feminism and the mastery of nature.* New York: Routledge.

Plumwood, V. (1996). Androcentrism and anthrocentrism: Parallels and politics. *Ethics and the Environment, 1* (2) (Fall): 119–32.

Prakash, M. S. (1994). From global thinking to local thinking: Reasons to go beyond globalization toward localization. *Holistic Education Review, 7* (4): 50–56.

Prigogine, I. (1989). The philosophy of instability. *Futures, 21* (4): 396–400.

Rolston, Holmes III. (1975). Lake Solitude: The individual in wildness. *Main Currents in Modern Thought, 31* (4): 121–26).

Roszak, T. (1992). *The voice of the earth.* New York: Simon and Schuster.

Running Grass. (Ed.) (1996). Multicultural environmental education. *Race, Poverty and Environment, 1* (2/3).

Russell, C. L., and Bell, A. C. (1996). A politicized ethic of care: Environmental education from an ecofeminist perspective. In K.Warren (Ed.), *Women's voices in experiential education,* (pp. 172–81). Dubuque, Iowa: Kendall/Hunt.

Russell, C. L., Bell, A. C., and Fawcett, L. (2000). *Navigating the waters of Canadian environmental education.* In T. Goldstein and D. Selby (Eds.), *Weaving connections: Educating for peace, social and environmental justice,* (pp. 196–217).Toronto: Sumach.

Seed, J., Macy, J., Flerming, P., and Naess, A. (1988). *Thinking like a mountain: Towards a council of all beings.* London: Heretic.

Selby, D. (1995). *Earthkind: A teacher's handbook on humane education.* Stoke-on-Trent: Trentham.

Selby, D. (1996). Relational modes of knowing: Learning process implications of a humane and environmental ethic. In B. Jickling (Ed.), *A colloquium on environment, ethics and education,* (pp.49–60). Whitehorse, Yukon: Yukon College.

Selby, D. (1999). Global education: Towards a quantum model of environmental education. *Canadian Journal of Environmental Education, 4:* 125–141.

Selby, D. (2000a). Global education as transformative education. *Zeitschrift for Internationale Bildungforschung und Entwicklungspadagogik, 23* (3): 2–10.

Selby, D. (2000b). A darker shade of green: The importance of ecological thinking in global education and school reform. *Theory into Practice, 39* (2): 88–96.

Shepard, P. (1959). A theory of the value of hunting. *Transactions of the Twenty-Fourth North American Wildlife Conference,* March 2–4, (pp. 504–512). Washington, D.C.: Wildlife Management Institute.

Smith, D, and Carson, T. (1998). *Educating for a peaceful future.* Toronto: Kagan and Woo.

Tomashow, M. (1996). *Ecological identity: Becoming a reflective environmentalist.* Cambridge, Mass.: MIT Press.

Townsend, T., and Otero, G. (1999). *The global classroom.* Cheltenham, Victoria, Australia: Hawker Brownlow.

Zimmerman, M. E. (1988). Quantum theory, intrinsic value, and pantheism. *Environmental Ethics, 10:* 3–30.

Zohar, D. (1990). *The quantum self: A revolutionary view of human nature and consciousness rooted in the new physics.* London: Bloomsbury.

CHAPTER 8

Learning from a Spiritual Perspective

JOHN (JACK) P. MILLER

In the last few years there has been an increasing interest in the whole area of spirituality in education (Kessler, 2000; Miller, 1999; Palmer, 1998). Part of this interest stems from the perception that spirituality is an inherent part of human existence and cannot be ignored in educational settings. Some would suggest too that the overemphasis on a materialist conception of life has led to a host of problems in our society. Consider the words of a Columbine student after the murder of fellow students in April 1998:

> The paradox of our time in history is that we have taller buildings but shorter tempers; wider freeways but narrower viewpoints; we spend more but have less; we buy more but enjoy it less. We have bigger houses and smaller families; more conveniences but less time; we have more degrees but less sense; more knowledge but less judgment; more experts but more problems; more medicine but less wellness. We have multiplied our possessions but reduced our values. We talk too much, love too seldom, and hate too often. We've learned how to make a living but not a life; we've added years to life, not life to years. We've been all way to the moon and back but have trouble crossing the street to meet the new neighbor. We've conquered outer space but not inner space; we've cleaned up the air but polluted the soul; we've split the atom but not our prejudice. We have higher incomes but lower morals; we've become long on quantity but short on quality. These are times of world peace but domestic warfare; more leisure but less fun; more kinds of food but less nutrition. These are days of two incomes but more divorce; of fancier houses but broken homes. It is a time when there is much in the show window and nothing in the stockroom; a time when technology can bring this letter to you, and a time when you can choose either to make a difference . . . or just hit delete. (Miller, 2001: 2)

As a result of the inadequacies of modern life, which also include the problems of poverty, homelessness, and racism, many people are looking for a different way of approaching education that includes a spiritual perspective (Lerner, 2000). But what does it mean to learn from a spiritual perspective? This chapter explores this question, since learning from a spiritual perspective is radically different from present approaches to education. When we view life from a spiritual perspective, we see ourselves connected to something larger than ourselves. This "something" has a mysterious quality that can give rise to a sense of awe and wonder.

LEARNING INVOLVES LETTING GO

Learning from a spiritual perspective does not consist of the continual accumulation of knowledge. From a spiritual viewpoint, letting go is as important as acquiring. Our heads can become stuffed with irrelevant facts that can prevent us from seeing things as they are. Spiritual teachers have talked about the importance of "emptying the mind." The *Tao te Ching* states that "In the pursuit of knowledge, everyday something is added. In the practice of the Tao, everyday something is dropped" (Lao Tzu, 1988: 48).

Thoreau also made this point: "It is only when we forget our learning that we begin to know. I do not get nearer by a hair's breadth to any natural object so long as I presume that I have an introduction to it from some learned man. To conceive of it with a total apprehension I must for the thousandth time approach it as something totally strange. If you would make acquaintance with the ferns you must forget your botany" (cited in Bickman, 1999: 2).

THE IMPORTANCE OF SPIRITUAL PRACTICE

In order to let go, or unlearn, the student usually engages in some form of spiritual practice. Most spiritual practices focus on releasing hold of the ego. One form of spiritual practice is meditation. Meditation, as well as other forms of spiritual practice, encourages the student to quiet down so that he or she can begin to see with some clarity. When the mind is filled with thoughts, some of which are compulsive and unexamined, it is very difficult to let go. Through meditation the person begins to quiet down and witness the flow of thoughts. By witnessing the thoughts, the student can realize that he or she is not the thought but merely the observer of thoughts. For example, if I have a negative thought about another person, I can see that as merely a thought floating by and not taking hold. This realization can greatly lessen the impact of the negative thought on myself and on my behavior. Through this process I can learn to let go of past conditioning and learnings.

ATTENTION

The key to spiritual practice is attention. By attending to what is happening in the here and now, I became more present. Attending to what is happening in the moment means that we are not caught up in fantasizing about the future or going over the past. When we cannot hold our attention, our mind wanders and loses focus so that we cannot function well.

One of the ways that we can learn attention in daily life is through mindfulness. Bhante Gunaratana (1999) has identified some of the main characteristics of mindfulness. First, mindfulness is *nonjudgmental observation.* We witness what is happening without criticizing or praising what is going on. Thus mindfulness is also *an impartial watchfulness.* Mindfulness does not classify thoughts or events as good or bad but attempts to see them as they are. Thus, there is no attempt to repress thoughts.

Mindfulness is not thinking but rather *nonconceptual awareness.* Krishnamurti called it "choiceless awareness," which avoids labeling and analysis. Gunaratana also emphasizes that mindfulness is *present-time awareness.* If you slip into remembering past events, you are no longer mindful. Mindfulness is *awareness without ego or self.* It does not see events or things in terms of "me" or "mine." Gunaratna notes: "For example suppose there is a pain in your left leg. Ordinary consciousness would say, 'I have pain.' Using mindfulness, one would simple note the sensation" (1999: 136).

Mindfulness is *awareness of change.* It does not try to cling to experience but simply attempts to witness the flow of events, both external and internal. Of particular importance is the awareness of internal change and the arising and passing away of thoughts and feelings. Finally, Gunaratana suggests that mindfulness is *participatory observation,* where we are both the participant and observer at the same time. Mindfulness is not a cold detachment from life but "an alert participation in the ongoing process of giving" (1999: 137).

LEARNING IS HOLISTIC

Education has tended to focus on the head and to ignore the rest of our being. From a spiritual perspective, learning does not just involve the intellect; instead, it includes every aspect of our being including the physical, emotional, aesthetic, and spiritual. These aspects also are interconnected; we cannot compartmentalize learning. For example, if we are studying a poem, we maybe moved both emotionally and spiritually. Some researchers attempt to reduce all learning to the brain. This is simply too reductionistic from a spiritual perspective. Of course, there may be links between learning and the brain, but learning and development also have a mysterious and spontaneous element.

Gandhi (1980) argued for a holistic approach to learning:

> I hold that true education of the intellect can only come through a proper exercise and training of the bodily organs, e.g., hands, feet, eyes, ears, nose, etc. In other words an intelligent use of the bodily organs in a child provides the best and quickest way of developing his intellect. But unless the development of the mind and body goes hand in hand with a corresponding awakening of the soul, the former alone would prove to be a poor lopsided affair. By spiritual training I mean education of the heart. A proper and all-round development of the mind, therefore, can take place when it proceeds *pari passu* with the education of the physical and spiritual faculties of the child. They constitute an indivisible whole. According to this theory, therefore, it would be a gross fallacy to suppose that they can be developed piecemeal or independently of one another. (138)

LEARNING THROUGH CONTEMPLATION

Education has tended to limit learning to the recall of information and conceptual knowing. From a spiritual perspective, knowing can occur in a variety of ways.

Ken Wilber (1983) has described three levels of knowing. Drawing on the thought of St. Bonaventure, a favorite philosopher of western mystics, he cites three levels of knowing: technical rationality, reflection, and being. Bonaventure describes three modes of knowing or three "eyes." The first eye is of the flesh, where we perceive the external world of space, time, and objects. The second eye is reason, where knowing comes through philosophy, logic, and reflection. The third eye is that of contemplation, where we gain knowledge of transcendent realities. Wilber notes: "Further, said St. Bonaventure, all knowledge is a type of illumination. There is exterior and inferior *illumination* (*lumen exterius* and *lumen inferius*), which lights the eye of flesh and gives us knowledge of sense objects. There is *lumen interius,* which lights the eye of reason and gives us knowledge of philosophical truths. And there is *lumen superius,* the light of transcendent Being which illumines the eye of contemplation and reveals salutary truth, 'truth which unto liberation'" (1983: 3).

Wilber (1983) also states that Bonaventure's three levels correspond with the ideas of Hugh of St. Victor, who distinguished among *cogitatio, meditatio,* and *contemplatio* (cognitive, meditative, contemplative). *Cogitatio* is empiricism and thus is based on knowing the facts of the external world. *Meditatio* involves internal reflection and seeing the truths of the mind. *Contemplatio,* again, is beyond duality "whereby the psyche or soul is united instantly with Godhead in transcendent insight (revealed by the eye of contemplation)" (3). Emerson has described contemplation thus:

> We live on different planes or platforms. There is an external life, which is educated at school, taught to read, write, cipher and trade; taught to grasp all the boy can get, urging him to put himself forward, to make himself useful and agreeable in the world, to ride, run, argue and contend, unfold his talents, shine, conquer and possess.
>
> But the inner life sits at home, and does not learn to do things nor values these feats at all. 'Tis quiet, wise perception. It loves truth, because it is itself real; it loves right, it knows nothing else; but it makes no progress; was as wise in our first memory of it as now; is just the same now in maturity and hereafter in age, as it was in youth. We have grown to manhood and womanhood; we have powers, connection, children, reputations, professions: this makes no account of them all. It lives in the great present; it makes the present great. This tranquil, well founded, wide-seeing soul is no express-rider, no attorney, no magistrate: it lies in the sun and broods on the world. (cited in Geldard, 1993: 172)

Contemplative knowing is essential to learning from a spiritual perspective. As noted, it involves a direct form of knowing where the barrier between the knower and the known disappears. Young children learn this way through direct experience with the world. They do not experience themselves as separate from the world but as part of everything that they encounter. One of the best descriptions of contemplation comes from Jacques Lusseryan (1987):

> Being blind I thought I should have to go out to meet things but I found that they came to meet me instead. I have never had to go more than halfway, and the universe became the accomplice of all my wishes. . . .
>
> If my fingers pressed the roundness of an apple, each one with a different weight, very soon I could not tell whether it was the apple or my fingers which were heavy. I didn't even know whether I was touching it or it was touching me. As I became part of the apple, the apple became part of me. And that was how I came to understand the existence of things.
>
> Touching the tomatoes in the garden, and really touching them, touching the walls of the house, the materials of the curtains or a clod of earth is surely seeing them as the eyes can see. But it is more than seeing them, it is tuning in on them and allowing the current they hold to connect with one's own, like electricity. To put it differently, this means an end of living in front of things and the beginning of living with. Never mind if the word sounds shocking for this is love. (27–28)

Traditionally schools have not supported contemplative knowing. We need to nurture contemplation through activities like meditation, visualization, and creative thinking in classrooms. I have described these approaches in a variety of other contexts (Miller, 1996, 1999).

AWE AND WONDER

Learning from a spiritual perspective is often accompanied by the sense of awe and wonder that Lusseryan (1987) describes. When we observe almost any aspect of the cosmos, we can experience this sense of reverence. Michael Lerner (2000), in writing about education, says: "Let awe and wonder be the first goals of education" (243). Emerson believed that all human beings have an inherent right to an "original relationship to the universe" (cited in Geldard, 1993: 63). This is an unmediated relationship where we experience an organic connection with life itself. This connection gives rise to a sense of reverence and wonder that can be nurtured in students by letting them be in nature and the world. Again, Thoreau put it well: "I wish to live ever as to derive my satisfactions and inspirations from the commonest events, every-day phenomena, so that what my senses hourly perceive, my daily walk, the conversation of my neighbors, may inspire me, and I may dream of no heaven but that which lies about me. . . . The sight of a marsh hawk in Concord meadows is worth more to me than the entry of the allies into Paris" (cited in Bickman, 1999: 42).

Viewing the world in this way requires what the Buddhists call the beginner's mind, or the mind that sees things freshly, as if for the first time. Through conditioning we can no longer see what is actually around us, so letting go and unlearning become important in restoring a sense of awe and wonder.

COMPASSION

Learning from a spiritual perspective requires compassion for all beings including oneself. We might call this compassionate knowing, which arises from the recognition that we are part of an interconnected universe. Like Lusseryan (1987), we see that we are part of everything and that everything is part of us. Since we are interconnected in this way, a natural compassion for all beings arises. Compassion can let us encounter the world in a more authentic way. Rubin (1975) comments: "In establishing an increasingly realistic frame of reference, compassion has the effect of thrusting us into the very middle of ourselves and into the middle of life generally" (195–96). Rubin also suggests that compassion must include self-acceptance: "I treat myself as I treat a child I love. In respecting him, I dignify all aspects of the human condition. In observing him, I eagerly expect him to demonstrate much that is human. . . . I bring no harsh judgment to him. In accepting all that he is, he need not fear me. I love him because he is who he is and I will not and indeed cannot hurt him. Thus, we exist in a state of grace, in a state relatively free of tensions and fears born of chronic impending destructive judgment, criticism and castigation" (165).

It is important that teachers experience a compassion where they see themselves in their students: "You do not feel set off against them or competitive with them. You see yourself in students and them in you. You move easily, are more relaxed, and seem less threatening to students. You are less compulsive, less rigid in your thoughts and actions. You are not so tense. You do not seem to be in a grim win-or-lose contest when teaching" (Griffin, 1977: 79).

In presence of a compassionate teacher, the student feels "psychologically safe" and thus is able to take risks and learn. The student feels accepted at a deeper level and thus can go beyond learning that is merely performance that tries to impress the teacher. Through the presence of a compassionate teacher, the natural compassion of the student is also supported. The student does not see himself or herself as a separate ego competing with other students but as someone who is connected to others.

Joy

Joy has long been seen as an integral part of a spiritual perspective. Montaigne once wrote that "The most evident token and apparent sign of true wisdom is a constant and unconstrained rejoicing" (cited in *Secrets of Joy*, 1995: 125). Learning from a spiritual perspective is also joyful. This does not mean that there are not ups and downs in the learning process. However, it means that learning is seen basically as a joyous act. A classroom rooted in spirituality will have sense of joy and purpose that pervades almost everything that happens there. I would like to quote one teacher who was practicing mindfulness in her teaching. "As a teacher, I have become more aware of my students and their feelings in the class. Instead of rushing through the day's events, I take the time to enjoy our day's experiences and opportune moments. The students have commented that I seem happier. I do tend to laugh more and I think it is because I am more aware, alert and 'present,' instead of thinking about I still need to do" (cited in Miller, 1995: 22).

The school and classroom rooted in spirituality are places that children and teachers want to be because they feel affirmed there. They are also experiencing authentic learning that is not focused on what will be on the next test; instead it is learning that leads to a deep sense of joy and fulfillment.

Spiritual Learning and Transformative Learning

How is spiritual learning transformative? It is transformative in that it allows us to see the world anew. We begin to see the interconnectedness of life at every level of the cosmos. This leads to a natural compassion that I discussed above. Gandhi (1980) summarizes this idea:

> Man's ultimate aim is the realization of God, and all his activities, political, social and religious, have to be guided by the ultimate aim of the vision of God. The immediate service of all human beings becomes a necessary part of the endeavor simply because the only way to find God is to see Him in his Creation and be one with it. This can only be done by service of all. And this cannot be done except through one's country. I am part and parcel of the whole, and I cannot find Him apart from the rest of the humanity. My countrymen are my nearest neighbours. They have become so helpless, so resourceless, so inert that I must concentrate on serving them. If I could persuade myself that I should find Him in a Himalayan cave I would proceed there immediately. But I know that I cannot find Him apart from humanity. (57)

Michael Dallaire (2001) has explored the relationship between contemplation and social action. He states: "Concurrent with the growing awareness of the need to free the inner person of internalized dualism there is also a growing awareness that our world is more interdependent that we ever imagined. As this consciousness grows we are coming to see that 'the particular "I" cannot have justice unless the other "I" has justice'"(111).

Social action from a spiritual perspective comes from the heart rather than the ego. If we attempt to bring about change through our egos, we usually end up creating more suffering. One of the best examples of social action from a spiritual perspective was the salt march by Gandhi. Gandhi was under intense pressure to respond to the British tax on salt. He meditated and finally had a dream of walking to the ocean and taking salt out of ocean. This led to the salt march, where tens of thousands of Indians were arrested for following Gandhi into the ocean and taking water and letting it evaporate. This simple act was one of the most important in the struggle for Indian independence.

CONCLUSION

Learning from a spiritual perspective also requires humility. Traditional forms of learning emphasize mastery and achievement. Mastery and achievement, of course, have their role in education, but from a spiritual perspective we also realize that there is much that we cannot control and that this realization leads to humility. Jack Kornfield (2000) tells the following story told by one his translators in Indonesia: "My uncle was a rice farmer who learned to heal by meditating and going into a trance. From the first day he started healing people, the energy of the gods would come to help him see the illness in his patients; they would show him which herbs to use and where to touch. For twenty years the gods came but then one day the gods stopped appearing. So my uncle told people he could longer heal and went back to being farmer" (153).

To teach and learn from a spiritual perspective, we must have the humility of this farmer. In a sense we can view ourselves as vessels attempting to assist others in awakening; yet we can never ultimately know the energies that are both within and surrounding us and what effects these energies will have on others. Teaching and learning from spiritual perspective requires us to trust ourselves and the larger forces in the universe that we are connected to in deep and mysterious ways. Learning to trust ourselves, and these larger forces, allows us to teach in a way that does not succumb to the pressures of modern society.

REFERENCES

Bickman, M. (Ed.). (1999). *Uncommon Learning: Thoreau on education.* New York: Houghton Mifflin.

Dallaire, M. (2001). *Contemplation in liberation: A method for spiritual education in the schools.* Queenston, Ont.: The Edwin Mellen Press.

Gandhi, M. (1980). *All men are brothers: Autobiographical reflections.* (Krishna Kripalani, Ed.) New York: Continuum.

Geldard, R. (1993). *The Esoteric Emerson: The spiritual teachings of Ralph Waldo Emerson.* Hudson, N.Y.: Lindisfarne Press.

Griffin, R. (1977). Discipline: What's it taking out of you? *Learning* (February): 77–80.

Gunaratana, B. (1999). Mindfulness. In S. Salzberg (Ed.), *Voices of insight* (pp. 133–142). Boston: Shambhala.

Kessler, R. (2000). *The soul of education: Helping students find connection, compassion and character at school.* Alexandria, Va.: Association for Supervision and Curriculum Development (ASCD).

Kornfield, J. (2000). *After the ecstasy, the laundry: How the hear grows wise on the spiritual path.* New York: Bantam.

Lao Tzu. (1988). *Tao te ching.* (S. Mitchell, Trans.). New York: Harper & Row.

Lerner, M. (2000). *Spirit matters.* Charlottesville, Mass.: Hampton Roads Publishing.

Lusseryan J. (1987). *And there was light.* New York: Parabola.

Miller, J. (2001). Letting the soul sing. *Living Lightly* (Spring): 2.

Miller, J. (1999). *Education and the soul: Toward a spiritual curriculum.* Albany, N.Y.: SUNY Press.

Miller, J. (1996). *The holistic curriculum.* Toronto: OISE Press.

Miller, J. (1995). Meditating teachers. *Inquiring Mind, 12*: 19–22.

Palmer, Parker. (1998). *The courage to teach: Exploring the inner landscape of a teacher's life.* San Francisco: Jossey-Bass.

Rubin, T. (1975). *Compassion and self-hate: An alternative to despair.* New York: Ballantine.

Secrets of Joy. (1995). Philadelphia: Running Press.

Wilber, K. (1983). *Eye to eye: The quest for a new paradigm.* Garden City, N.Y.: Anchor Press/Doubleday.

CHAPTER 9

The Labyrinth

Site and Symbol of Transformation

Vanessa Compton

In these extraordinary times, we stand in wonder and bafflement as witnesses to a historical moment of evolutionary change, challenged to participate in this new stage of existence unfurling on so vast a scale that cultural historian and ecotheologian Thomas Berry says we are living "not in a cosmos but a cosmogenesis, a universe ever coming into being through an irreversible sequence of transformations moving . . . from a lesser to a greater order of complexity and from a lesser to great consciousness" (1999: 26). The human role in the transition from the "terminal Cenozoic" period of Earth history to the "emergent Ecozoic," divisions marked by the impact of the human presence on Earth's processes (O'Sullivan, 1999: 17), is pivotal. "There is liable not to be a blade of grass unless it is accepted, protected, and fostered by the human" (Swimme and Berry, 1992: 20).

Such sensitive responsibility requires our own personal evolutions, out of the late Cenozoic modernist mind-set that has brought us to this pass and into alignment with the emerging "great consciousness." There is no shortage of evidence that the project of modernity, in which we are as deeply enmeshed as its patterns of thought are embedded in us, has reached its limit as a useful cultural paradigm. George Grant (1998) has observed that, where "technology is the ontology of the age," the faculty we rely on, reason itself, is reduced to the instrumental in the service of the myopic: "Indeed, to think 'reasonably' about the modern account of reason is of such difficulty because that account has structured our very thinking in the last centuries. . . . The very idea that 'reason' is that reason which allows us to conquer objective human and non-human nature controls our thinking about everything" (Grant, 1998: 427).

With what faculties then can we engage in this evolutionary process? How do we get out of our own way?

From the perspective of cosmogenesis, three principles govern the formation of this emergent consciousness, and every other being in the universe as well: these are differentiation, subjectivity, and communion (Berry, 1999: 162). None figures in the mantra of the competitive global market. It is time to generate a more inclusive account, with all the parts that up to now have been imperceptible to the "single vision": a more heterogeneous understanding, less belligerent, conformist, and fearful than our contemporary myths of globalism and progress; a new story of the world where all subjectivities find a place. This is not impossible. Our species always has told itself stories, and we are especially good at creation myths. In these, we "remember" the way our ancestors came from mud and were taught the arts and skills, the *techne,* necessary for life as well as the limits to our powers, the consequences of our hubris, and our place in the universe. These stories, in the form of archetypal figures and situations, are so prevalent cross-culturally (including, I will argue, in the nonverbal forms) that one could say we are hard-wired to do this: It is a faculty we can rely on, whether or not we are aware of it.

There is a sense of urgency about bringing to collective consciousness a compelling alternate vision to a blindly destructive socioeconomic juggernaut. I struggle with my own impatient tendency toward a motherly but coercive "ecototalitarianism" in the face of resignation and a sense of futility, all signs of an embattled imagination: "To free oneself from old ways of seeing requires imagination. Every important social movement reconfigures the world in the imagination. What was obscure comes forward, lies are revealed, memory shaken, new delineations drawn on old maps: it is from this new way of seeing the present that hope for the future emerges. One fears that solitary and unique voices will be silenced by mass tyranny, yet this is not the only danger. In an atomized and alienated society, the imagination is endangered by the limitations of the ego . . . of being restricted to a canvas that is too small" (Griffin, 1996: 45–46).[1] As a practicing artist and researcher in art-based holistic education, I deplore the systematic cutback (and simultaneous corporatization and commodification[2] of public access to opportunities for encouraging and honing the creative imagination: it constitutes a mass deprivation and "internal exile" on the basis of class, in the service of profit and social control. But this sorry situation is by no means irreparable; the imagination was honored in premodern cultures, whose tools and practices remain for us now to rediscover and take up.

Imagination refers to several related functions and capacities. As the source for the language of the soul, which speaks in metaphor and "never thinks without an image" (Aristotle cited in Ackrill, 1987: 199), the imagination accesses deep, resonant psychological strata. It is also the mechanism for juggling the external, imported images ranging from design specifications to the distractions characteristic of "monkey mind." The difference between these various states is particularly noticeable during the ancient practices of meditation, which quiets and focuses the mind, and of ritual, which awakens the "old ones" inside of us, those parts of the brain that are usually ignored, the parts that do not speak in English but in images such as candlelight and color. Discernment among these states of mind is a kind of interior mapping. As we set out to restore our fundamental connections within the ecological matrix, road maps of our underlying structures and patterns of development can temper the urgency that so easily slides into the arrogance of progress, that deafening willfulness of which Lyotard warns: "The modern is all too easily snapped up by the future, by all its values of pro-motion, pro-gram, pro-gress . . . dominated by a very strong emphasis on wilful activism. Whereas the postmodern implies, in its very movement (going further than modernity in order to retrieve it in a kind of 'twist' or 'loop') . . . a capacity to listen openly to what is hidden within the happenings of today" (cited in Kearney, 1988: 27).

One intriguing road map–*techne,* developed expressly for going further and returning with an increased capacity to listen for the hidden, is the cathedral pavement labyrinth. A medieval design based in the holistic principles of Sacred Geometry, and first constructed in honor of Mary at Notre Dame de Chartres Cathedral in the twelfth century (itself built over an ancient Black Madonna site), this unicursal path is a metaphorical container for spiritual journeying, mirroring, as it winds back and forth through the eleven layers of concentric circles, the processes of human thought. While labyrinths have always represented the human quandary of how to proceed through multiplicity and confusion,[3] the reports from contemporary participants in reflexive walking meditation in the labyrinth are surprisingly consistent in describing the occurrence of physiological, emotional, spiritual, and cognitive events such as release of tension; an increased sense of clarity, well being, and communion; triggering of early memories and related insights; and coming to terms with difficult decisions or problems. Most often these events are revealed through symbolic imagery, auditory, sensate, and interioceptive as well as visual. Long after the walk, incubation continues in the imagination, illuminating situations and allowing "what was obscure to come forward" into consciousness.

A restoration of access to the imagination would accomplish many things: an emancipation and bringing home from exile, a healing and making whole of self and community, a sense of spaciousness and opening up to possibility, a sense of our own rhythm. When I see all these and more happen for all kinds of people walking the labyrinth pilgrimage, I feel privileged to be part of a worldwide movement to return the cathedral labyrinth to public use. Beyond the altruistic motivation, what has intrigued me is the question of how it is that this can be at all. As I review the literature and reflect on my years with the labyrinth, a list forms: labyrinth as archaeology, architecture, medieval software, liminal site, geometrical figure, public sacred art, transformative crucible, spiritual container, developmental scaffolding, spatial mnemonic, search algorithm, Zen koan, Goddess temple, western medicine wheel, dance notation, premodern social technology, *omphalos* and *temenos.* The multiple layers of meaning to be found beneath the labyrinth's deceptive simplicity suggest to me an access route, through our contemporary imagination and that of the medieval designer-genius, to an individual and collective embodiment of that "greater order of complexity . . . and consciousness" where we need to learn to make our home. I invite you to trace with me some of the entry points into this mysterious and powerful site.

MEETING THE MINOTAUR

I came to the labyrinth by way of the Minotaur, in an epiphanic moment with the catalogue from Picasso's *Vollard Suite* exhibition of 1977. Picasso had given the Minotaur a life: girlfriends, parties at the sculptor's studio, a tender gaze upon a sleeping lover, but tragic too, suffering violence but never inflicting it; now he is grievously wounded, now pathetically blinded, led around by a little girl holding flowers or a dove, past sailors with a boat. What was going on here? These were not illustrations of a quaint archaic story cycle for a children's picture book! The way Picasso had drawn the Minotaur, he appeared to be a regular character in the neighborhood.

There was something reassuring about the scribbly quality in the drawings and etchings, like being in a foreign country with enough knowledge of the language to understand what people are saying. The quick contouring marks Picasso had used to

Figure 9.1 The Chartres Cathedral Labyrinth

capture the solidity of a moving figure was a technique so familiar that, as I traced the gestures in the rhythm of the lines, I could practically feel the stiff curly mane and stocky body he had been drawing. It was easy to see that the Minotaur's presence had felt "real" to Picasso. The bull's head was a closely observed study—Picasso was Spanish, he went to bullfights—and the man's body was drawn from the inside out, by someone who knew how living in it felt; but the pathos and desire in the creature's character took the drawings out of the genre of documentary life drawing into the realm of psychological, maybe even autobiographical, portraiture. I remember being intrigued. What experience did these marks on paper *signify*? Were they not the traces left behind sometimes when something important—a Minotaur, maybe, or the curve of your child's upper lip—has come toward you through the veil? I had spent many hours at life-drawing class, enough to know the difference between practicing and performance. Practicing was that regular, faithful exercise to hone skills. But what was being "performed" in these portraits of Asterion?

I had a background in the classics; I knew the Olympians hung around mountaintops, particular groves of trees, and underground in certain caves, and that temples and cathedrals sometimes were built in places that were known to be particularly suitable in ways that had nothing to do with architecture or town planning or transportation or drainage. From Jung (1971) and Campbell (1968) I learned how myth and archetype functioned within the human psyche and in society, translated through iconography and architecture into forms accessible to anyone who paid attention. I had visited temples and cathedrals in Europe, guidebook in hand. These places were magnificent and beautiful but my response was more one

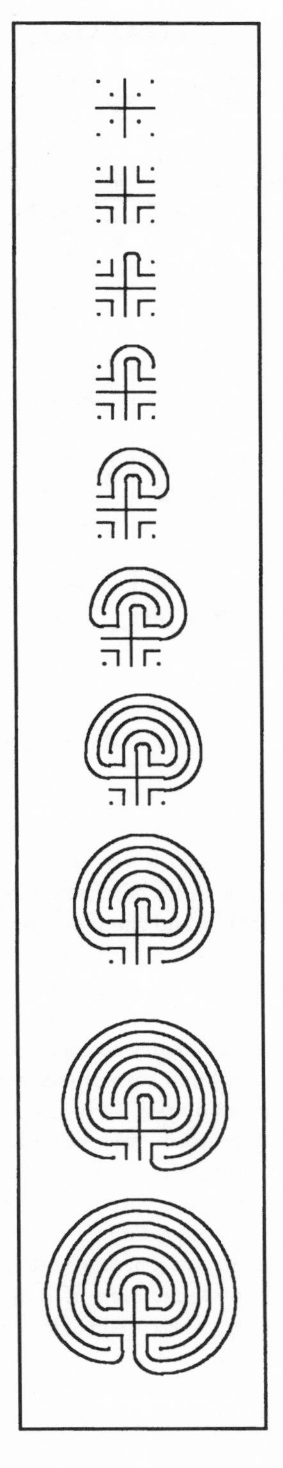

Figure 9.2 Drawing the Seven Curcuit Classical Labyrinth

of appreciation for the skill and devotion of the builders than a sense of transcendent awe. I recognized and admired their mastery and perseverance but I could not share their apparent certitude in the holiness of the space they inhabited and manipulated. It was as if I were on the outside looking in, nose pressed against the glass. Gallery space, although similarly remote, is necessarily secular as part of the culture industry, no matter what the artist in the studio may think of what s/he is doing or for whom. The division between sacred intent and secular context was becoming a problem of authenticity for me.

All this changed dramatically when, with no warning, during a seminar presentation for an undergraduate religious studies course in 1997, the power of created sacred space to provoke and effect profound personal transformation was made unmistakeably clear to me. I had left art practice and returned to the university full time, wanting to find out what was causing the students in my foundry workshops to transform before my very eyes. I was passionately interested in this question but the impervious rationality of the school administration and the strain of having to learn to drive the computer were, together, overwhelming. I was so tired that I started "seeing" things out of the corners of my eyes in shadowy stairwells and corridors. My apprehension about it all resolved itself into a Minotaur lurching out of the archaic darkness. I made a large "found object" Minotaur sculpture out of scrap metal and a cow skull, which expressed for me the brutal clash of human and technology, and represented also the hidden unlovable parts of ourselves ("I'll never be able to do this! What was I thinking?") that become huge and fearsome in the shadow. I took the sculpture into the seminar, along with handouts and a method I had found for the class to create a temporary masking-tape version of the unicursal seven-circuit classical labyrinth (more archaic and simpler than the Chartres labyrinth, I had discovered) on the cafeteria floor. This we did, making it large enough for everyone to walk through and double back out again in one long continuous line.

Joan, one of the older students, felt very uncomfortable being in such close proximity to others. This surprised me, as no one else had commented one way or the other on that part of it. At class the following week, however, Joan reported a remarkable incident. Her husband suffered from Alzheimer's disease, and over the weekend, he had fallen and hit his head, then blamed it on her in a violent angry outburst. Joan had retreated to her customary pattern of thirty years of marriage, apologizing and trying to "fix" it. Suddenly she realized she was not responsible for his mishap, and even if she were, he had no right to abuse her that way. This awareness illuminated the rest of their relationship. Joan said, "I saw that our marriage was the labyrinth, and he was the Minotaur at the center of it." She described him as being such a bad-tempered individual that the onset of Alzheimer's had been masked by his normally unpleasant and belligerent manner. Over the weekend, she had decided to take her grown children's advice and put her husband in a nursing home.

Joan was eager to talk, as if the experience of walking the labyrinth had broken open years of dammed-up feelings. I was astonished both at the content of the story and that she would share it with me. I knew I was witnessing an extraordinary transformation. Trying to make sense of it later, I wrote: "The power of the ancient pattern, activated by [Joan's] physical engagement, provided her with a clear symbol with which to see and interpret her real-time situation, and she felt herself liberated—truly a coming together of the recurring motifs of maiden and monster captive in the centre. What is remarkable is that she became her own liberator, in that her detachment from the oppressive power dynamics of a dysfunctional patriarchal arrangement was of her own creation through the recuperation and identification on her terms of her strengths, which she had previously projected onto the abusive husband. In a sense, the labyrinth site/process provided the internal 'communal' support for her shamanism on behalf of her liminal self" (Compton, 2001).

We had unwittingly re-created a very powerful device, with life-changing potential. I had an obligation to treat it with great respect or else stay away from it. I was full of questions: How could masking-tape on a lunchroom floor have such power? All we had been was curious; was a degree of belief on our part not necessary? What was the operative factor anyway: the group? the pattern? the walking itself? Who had invented it in the first place? How had they used it originally? Had only Joan undergone a transformation, or had everyone? How come they were not talking about it? How could I find out? How come nothing like that had happened to me? In retrospect, I could see I had been too preoccupied making sure everything was taken care of to be able to appreciate the actual experience. Then it dawned on me that this was not unusual, that I was often overly controlling, if well intentioned, about social situations and detached from the immediacy of experience as a result. It took months for this insight to come to consciousness. I realized that the labyrinth experience could be slow and subtle as well as dramatic and immediate as Joan's had been.

When I discovered that the eleven-circuit cathedral labyrinth was connected with the medieval neo-Platonic School of Chartres, my questions proliferated with the entry points. Was this the technology of a premodern holistic curriculum? Was it relevant for us in education now? Was the current interest in labyrinths another sign that were we in a transitional state similar to the one the medieval Europeans had experienced? I was reading about implicate order in the holographic universe and quantum consciousness, while in another part of the academic maze, devised with the help of my geometry professor, I was deciphering the mathematics of the labyrinth in both its twelfth-century cathedral setting and late-twentieth-century topological algebra; it was an inflammatory combination.[4] I had a clear intuition that the labyrinth pattern was somehow an analogue of synthetic thought process and that I would need to trace an equally meandering path to see its relevance to education. Being not mathematician, psychologist, medievalist, nor even very far along in the study of education, I wondered about the wisdom of undertaking academic wild goose chases. But it was impossible to turn away from Joan's story. I needed to find a teacher. Synchronicity took over, and within six months of conjuring up the Minotaur, I was in a labyrinth pilgrimage and facilitator-training session at Grace Cathedral in San Francisco.

A CULTURE IN DENIAL

The rupture in the fundamental continuity between humans and the universe—the feeling of being on the outside looking in, nose pressed to the glass—has developed out of a multitude of contributing factors but in common is a fragmentation, discernible at every level of social organization, originating within ourselves. The icon is Descartes bifurcating human being into a reasoning intellect residing in the brain, transported by, and distinct from, a mechanistic, experiencing body.[5] Farther out in the countryside around the same time, the forced depopulation and the infamous enclosures of common lands utterly disrupted traditional geographically based kinship patterns, relations between humans and the cycles of nature, and Christian ethics of stewardship and natural law, all of which, up until then, had contributed to maintaining balance and constraint in land use and labor (Northcott, 1996: 49–50).

This policy created masses of dispossessed, unemployed people, whose relation to the world through the work they did was subjected to the irresistible logic of the industrial revolution in the development of the assembly line. In an economic system that uses the term "surplus labor" to describe human beings, language separates the individual from the work s/he does as well as from the moral, political, social, and emotional consequences of doing it and from those who would be impacted by the activity. In this atomistic view, the world is made up of subjects and objects, "I" as the individual agent, and everything else as objects. Linear visual perspective, formalized in Renaissance art, confirms this monocular and remote individualism, drastically different from the communal sense of space in the premodern oral culture that it had superseded (Lowe, 1982; Romanyshyn, 1989). Out of these separations in all sense modalities of subject and object, dichotomies of matter/spirit, body/mind, sacred/profane, man/nature, public/private, male/female, arranged in hierarchies of power, replaced the interdependencies of the natural world.

While we may not consciously treat others instrumentally, as objects in our service, and would be horrified at the suggestion that we do, our perceptions and behavior are affected nonetheless. Such dualism is encoded in the competitive language of a self-preserving individualism, so that the entire social discourse is distorted and inevitably dysfunctional. This is "the consequence of a cultural fixation, an addiction," evidenced in "the profoundly degraded ecological situation, [of] a deadening or paralysis of some parts of human intelligence" (Berry, 1999: 182, 115). Rachel Naomi Remen (1999), clinical professor of family and community medicine at the UCSF School of Medicine, teaches a course called "The Care of the Soul" to medical students and a postgraduate course for physicians called "Relationship-Centered Care." Remen notes that standard medical education perpetuates this distortion, which she identifies as a cultural valorizing of individualism and devaluing of the quality of compassion: "Recovering compassion requires us to confront the shadow of our culture directly. We are a culture that values . . . self-sufficiency, competence, independence, mastery and control. In the shadow lies a profound sense of isolation from our human wholeness. We have developed a contempt for anything in ourselves and in others that has needs, and is capable of suffering. In our isolation, we tend to develop a suspicion of anything beyond ourselves, anything that falls outside of our control" (35). When "that which is capable of suffering" is repressed and relegated to the shadow of both the individual and of the society and culture, what happens?

Cultural symbols, collective images accepted by civilized societies that have been used to express "eternal truths," still carry their original numinosity, a psychic charge that evokes deep emotional responses. But rationalism has destroyed our capacity to understand these symbols, and with neglect or repression, their psychic energy disappears underground, beneath the level of consciousness. There they give life to whatever is uppermost in the unconscious. "The dark side, the toxic waste, was denied, ignored, hidden from sight, buried" (Berry, 1999: 112), within the logic of consumerism. Where our engagement is instrumental rather than reverential, what is lost is Earth as symbol of the sacred. With it goes anything else associated with nature, along the fault lines in the man-nature duality mentioned earlier, including actual females and the feminine generally, qualities, traits, habits of mind such as receptivity, and the intuitive. This denial of the external aspect of the dark side is matched by suppression of the inner life. So into the shadow go our vulnerabilities such as grief for all kinds of losses, including the devaluation of the work of our hands and minds, the destruction of our habitat, the fear that anyone will discover these things about us, and hope that things might change.

The psychological mechanism of denial, a survival function whose purpose is to keep the organism in homeostasis through interpretive means, operates at both the individual and the social level to maintain false consciousness about a given situation. Berry points to signs of "a deep inner rage of Western society against its earthly condition as a vital member of the life community . . . a disturbance sanctioned by the very structures of the culture itself in its present phase . . . [requiring] a corresponding deep cultural therapy" (1999: 165). This rage may be a cover-up for the grief of which there is so much and of which we are so ashamed. In psychoanalytic process, rage marks a state of heightened resistance to the approach of anything that might disrupt the carefully constructed shell of meaning covering the denied or suppressed reality. But when the strategy of denial becomes too costly, the individual or group is compelled, most often through the advent of some crisis, to undertake the next step. That is the point that the human species finds itself now.

It could be said of our dilemma that we are reenacting collectively the Heroic Journey, a pattern Joseph Campbell calls the Monomyth (1968: 36). Our genetic coding informs those psychic energy constellations that take shape as the primary archetypal forms deep in the unconscious realms of the human[6] and find expression in symbols such as the Heroic Journey, the Death-Rebirth cycle, the Sacred Center, and the Great Mother (Berry, 1999: 160). These can serve as a source of guidance in the creation of the new story we need now. The cathedral labyrinth, in both form and function, recapitulates all these symbols, especially the Journey, which "contains experiences that characterize all of our mythic journeys into the farthest regions of ourselves in search of the real thing: awakening; the call; the journey; the descent; the darkness; a healing crisis; an epiphany; the ascent; accepting unlikely companions on the journey; new visions of self and world; the long integration of the experience into daily life. These are developmental steps we all face" (Woodman and Mellick, 1998: 17).

Denial, in the Journey schema, is analogous to the Refusal of the Call (Campbell, 1968: 36). Such resistance can be taken to pathological extremes of stasis. The classic symbol is the ruler of a realm that has fallen into disorder or into a kind of trance. All life-forms in the realm suffer until the hero or heroine undertakes the journey in search of the remedy, which is often a trifle but an extremely inaccessible one. The subsequent adventure of Departure, Initiation, and Return, with all the tests and encounters involved, comprise the Journey or Individuation crisis. A culture in denial is analogous to the labyrinth in its maze-prison aspect: The unrecognized Shadow aspect of denial connects our contemporary crisis of transformation to the labyrinth myth, especially in regard to the strangely immortal Minotaur, target of all projected horror.

Campbell's (1968) analysis of what happened at Crete clearly embeds the mythic in the social and the psyche: King Minos was a judge and ruler of a great trading kingdom. His queen, Pasiphaë, had developed a passion for a magnificent white bull of mysterious origin. With the help of the inventor Daedalus, she contrived a model in the shape of a cow in which to conceal herself. Thus disguised, she deceived the bull, seducing him and consummating her passion. Out of this unnatural union was born a monstrous child, the Minotaur, with the head of a bull on a human body. The queen took the blame but the king was aware of his own guilt in all of this. The bull had been sent from the sea by Poseidon as a sign of Minos's right to be king. In return, Minos had promised to sacrifice the animal as an offering and symbol of service. But Minos reneged and, thinking Poseidon would not notice, had kept the bull and slaughtered another one from his herd instead. The monstrous child was hidden away in the labyrinth as prison-weapon.

The primary guilt was the king's: He had taken personal advantage of a public event. The return of the bull was meant to symbolize his selfless submission to the obligations of his role. By keeping it, he subverted this greater purpose, giving in to self-serving impulse. Thus the divinely anointed king became the tyrant "Holdfast," looking out for himself, an example of the principle "as above, so below" in the realm of the self. "As the rituals of initiation lead the individual through the death of the old self and rebirth into the new, so the great ceremonials of investiture divested him of his private character and clothed him in the mantle of his vocation" (Campbell, 1968: 15). To refuse the rite was a sacrilege that cut the individual off from the community, thereby breaking the whole social body into fragments, who then, as the many, made war on each other, each out for himself and governable only by force. This universal monster resonates in folk tales and legends as well as in the news of the world and its nightmares: the hoarder of the general good, greedy, claiming all rights to "mine"; inflated ego-aggrandizement, yet fearful, defensive of the aggressions coming at him from his own projections, isolated in self-made independence; making a wasteland wherever he is.

By contrast, the hero has achieved submission. In the story, Theseus undertakes the journey, receiving help to go underground and confront the frightful monster hidden in the dark. He delivers the Athenian people from the terror of the Minotaur, restoring the balance to society from the disorder Minos's greed had caused. Campbell (1968) links this action with collective redemption from the wasteland using the metaphor of a spring[7]: "Dream is the personalized myth, myth the depersonalized dream. Thus the hero is the man or woman who has been able to battle past his personal and local historical limitations to the generally valid, normally human forms. Such a one's visions, ideas, and inspirations come pristine from the primary springs of human life and thought. Hence they are eloquent, not of the present, disintegrating society and psyche but of the unquenched source through which society is reborn" (20).

When we overcome our fear and resistance to the initiation process, it becomes possible to face the disowned, unloved, and unlovable monstrous Minotaur parts of ourselves. The labyrinth is technology designed to do this, instantiating architecturally the enveloping safety of the compassionate receptive principle, which is activated when we, as heroes, penetrate the site. The ring of spiky lunations around the outside circuit, a sign of both warning and protection to the initiates, articulates visually: "Do not enter unless you are willing, and once you do, know you are safe in here." People show up to walk in the labyrinth when they are troubled, when they need to feel their way through a dilemma, when something in their lives is not working any longer—when the spring has been blocked and the realm is barren, when the old ruler's time is up. They come because they have heard a call, in spite of how embarrassing it might seem to be following a voice no one else can hear, or how absurd it appears to be trudging in circles around and around the parish hall. Crossing that threshold, the one that separates you from caring about what other people think, is the first obstacle. The crisis unfolds with the decision to make that commitment and take the first step: "What if I go in and find nothing there for me? What if God is not there for me? What if there is nothing there at all?"

The thread running through this is the desire humans have to be whole and no longer fragmented. Campbell's interpretation of the Minos myth evokes the origins of the word "religion" in the Latin *relígere,* to tie [back] together. By giving in to his impulse to self-serving greed, Minos concretized a symbolic aspect of relationship between the divine and the people that ought to flow, causing it to distort and become corrupt. By his grasping, he objectified it as the product rather than harmonizing himself with the process: He mistook the bull, a creature he could own, for the power of which the bull was the symbol, which can only be divinely bestowed. Arrogance around divinity results in Minotaurs in the "basement." What needs "tying back together" from the disarray is the same everywhere: all the lost parts, demonized, projected outward, and loaded with hurt and rage, the ignored and disowned, the perceptions and wisdom of the animal body, the reports from Earth, all the other names and frames for subjectivities and communions unheard, unseen, and unimagined in the wake of the juggernaut.

A PHENOMENOLOGY OF PILGRIMAGE

I was well prepared to be on a pilgrimage to the labyrinth, or so I thought. For months I had studied the history of the pattern and the writings of the pilgrimage leader, an Episcopalian priest and psychotherapist, the Reverend Dr. Lauren Artress. The long flight across the continent ended at the edge of the world. I had time before the first meeting of the group to visit the cathedral and walk both the indoor and outdoor labyrinths by myself, so the place was familiar. Everything was under control.

At dusk on the second day of the pilgrimage, we walked silently in procession from the rectory to the cathedral nave, carrying candles. The neo-Gothic cathedral was lit with hundreds of votive candles. A group of musicians played medieval chant music with Oriental overtones; some were singing. We stood in a circle around the perimeter of the labyrinth, and Dr. Artress invited us in one by one; I was in the middle of the group to enter. I expected to have a serene meditation in a beautiful setting. As I moved along the last few turns before reaching the center of the labyrinth, a strong feeling of sadness welled up in me and tears poured down my face. They did not stop; it was more than dampness, it was a torrent. I was thick with grief throughout my chest and throat and clenched my teeth to keep from making a noise. I was very embarrassed by being so snotty and unprepared with no tissue to cover my face and mop up the mess. The thought in my mind was of an old family sorrow. I was shocked that I had come so far to such a place and still that grief was with me. I was irritated with myself that I could not "get over it." Then I heard all around me people sniffling, and I saw many of them covering their faces. We all had something we could not get over. I saw that the flood of tears was a River of Sorrow and all of us humans were in it forever, that tranquil joy was a rare thing to be treasured: This was the meaning of life.

In the months that followed, I puzzled over my notes and memories, trying to understand what had happened. The experience was complex; it unfolded on many levels. At the time, I had felt constrained by a sense of having to remain "objective" (and in control), so that I was never quite sure if I was being "present in the moment." Did I even know what that meant? Now, years later, such a split is hard to imagine. I believe that the inherent nature of the labyrinth process is so powerful that it overrides such either/or constructions; that is one of the lessons. Teasing out the subtleties of the experience through the lenses of various epistemologies, looking now for signs of differentiation, subjectivity, and communion, the prevailing image of the event is of intersecting transparent spherical orbits, the meshing of world-lines. While the sensory realm was rich and elaborate—intricate and evocative music, flickering candles, gleaming fabric drapery and banners in brilliant colors and the thick soft carpet underfoot, soaring architecture full of shadows, cold marble and polished wood and metal, smells of beeswax and burning incense—the nonsensory world, the transpersonal, the presence and movement of the Spirit in us and among us, was mysterious. With what sensors do we pick up on this? It reminds me of Christo's "Running Fence," the long curtain across California to the shore of the Pacific, in which you "see" the movement of the wind in the rippling fabric. That is how you know it has been and gone.

I wondered about the influence of instinct on the ways people behave in the labyrinth. By oneself, alone in the labyrinth, the experience is very different from walking in a group of people, many of whom may be strangers to each other. Ethnologists define instinct as a disposition toward a pattern of behaviors. Closed instincts are genetically fixed, like the honey dance of bees. Open instincts can be described as "programming with a gap" to be filled by a range of experiences depending on the circumstances within which the creature finds itself. Aggression in humans can be seen as an open instinct, and is related to dominance over territorial space. The engagement can be symbolic, the submissive party making themselves "small." A symbolic response might be a friendly gesture. This exchange is often highly complex but the capacity is innate in a species (Midgely, 1980: 53).[8]

One of the things some people did that evening was to take advantage of the scarves provided, to use in various ways but most often to shroud their faces. I did not take a scarf, preferring to be unencumbered. But, as a sensory cue, the profile of veiled people was beautiful to see, mysterious to an onlooker, like a pupa with something going on invisibly concealed within. After I began to cry, I very much wanted to be veiled, to be physically inside such an enclosure with my snot and misery. I did not want to be seen so apparently out of control of my emotions. This was a high pitch of self-consciousness: In retrospect, objectively, who would care besides me? But in misery the body becomes an encapsulated world. I wanted to hide or run away. But the labyrinth obliged me to stay. Hearing other people sniffling and stifling sobs was a revelation. Here we were in public, a collection of individuals not managing! Then to see other people apparently having a wonderful time, dancing and swaying and smiling, was a shock. How could they be so happy? Then the image of the River of Sorrow arose, and the understanding that though we were all on it forever, it was possible to experience great joy and gladness, perhaps even because of the continuo of sorrow enhancing the contrast.

Confronting the socialization around aggression and competitiveness becomes necessary on a very narrow path with people whose chosen pace is different from your own. Each individual must deal with the feelings that arise when the issue of "appropriate" behavior comes up, particularly accommodation and politeness inwardly resented, when people with varying degrees of inhibition about body gesture are together in the contained labyrinth space. Some prefer, or feel constrained into, a slow dignified pace, and are shocked and offended by those who assert their right to more expressive and expansive movements. Facilitators give permission by emphasizing at the outset of each event that there is no one "right" way to walk. But often in the labyrinth I want to blow off steam by dancing, and I was brought up short when my training supervisor suggested that the more uninhibited also needed to be aware of and sensitive to the needs of others. Wasn't that "policing"? I had to examine the source of the steam, and the possibility that my resistance, both political and psychological, was blocking my compassion. In the labyrinth, everything is metaphor.

The reason for this resonance may lie in the analogy between the functioning of the brain and the layout of the labyrinth. The triune brain, vertically and laterally interconnected, finds "the appropriate stimulus and supportive environment" (Pearce, 1993: 49) in the labyrinth experience. Parallel-processing in/of/between reptile-brain, limbic system, and neocortex "entities" seem to be diagrammed by the layout of the labyrinth path itself: the concentric layers of the three transquadrant circuits identifiable among what appears at first to be a jumble of twists and turns. Clockwise and counterclockwise circumambulations characterize the labyrinth—a corkscrew or augur whereby one walks "inward" counterclockwise and then brings "up" or "out" and consolidates the psychic "sample" on the clockwise circuits.

There may be an analogy here also to the images and experiences of emotion/feeling that arise on the stages of the path. Jung talks about the formal elements in the process of development in the psyche, and many are represented in the labyrinth pattern and so retraced by the walker: the chaotic multiplicity and order; the opposition of right and left, in the alternation of the turns; the radial arrangement around the quadraform, rotation, and centering process—the climax in his view (cited in Chodorow, 1984: 46) and also in the labyrinth sequence of the Threefold Path, on reaching the middle. The welling of emotion that occurred during the event in San Francisco came as I moved toward the center. That part of the path I know to be located in the last and outermost, "deepest" of the transquadrant circuits. So I had walked through all the depths of the labyrinth. Perhaps I had simultaneously traversed the depths and lateral breadth of the three brain levels. Reptile-brain reported on the "safety of the tribe," the limbic system was full of compassion, the neocortex was vexed at my messy public behavior; I was "trapped" and suspended in public exposure yet, paradoxically, safe.

Rosenstiehl's (1987) graph theory analysis of the labyrinth is a left-brain schematic, linear and programmatic, while a sketch of his "infinite-field" diagram illustrates "the flux of global patterns of activation over the entire network" in Ashbrook's description of right-hemisphere activity (1989: 343). This evidence of coherence between left- and right-hemisphere patterning and product has a numinous glow about it. I am drawn to it; I think the answer might lie there. Body-knowing and concept-forming and rational/scientific description of the labyrinth experience overlay each other in alignment: "The order was there before we were." Maybe my tears were a response to, an expression of, the sense of homecoming. Maybe the public humiliation was a necessary unmasking. It is both private and social. Unmasking is the collective great fear, the dark side of the culture's brutal competitiveness, which appears to be a function of left-hemisphere lopsidedness, a cruel dualism, one bridged by the labyrinth experience.

Insights from yoga practitioners provide a radical interpretation of western mythological heritage as literal, though concealed, descriptions of body functions as sensors and moderators of human existence. The dynamics between mythological deities and entities are analogues to breath, senses, perceptions, cognition (Sansonese, 1994). Awareness of our inner workings, freed from our entanglements of desire with external objects/goals by means of the practices intended to attain trance state, will bring us in line with the harmonies apparent in all creation (as expressed by all ancient wisdom traditions and modern quantum physics). The archaic origins of trance state lie in repetitive motions, evolving later to control of breath and other body functions. In labyrinth practice, the repeated turns and runs have a pattern of concealed logic that subverts rational linear thinking—the mind "gives up" to being lost and compelled through space. Western concepts of relative value—for example, "ahead" and "behind—lose their meaning. One falls back on proprioceptive awareness of one's own pace, in time with one's own breathing. Strong emotions and realizations lie suppressed beneath normative social behaviors such as following a pace not your own. The process of adjustment can be a shock. The result, eventually, is a profound and energizing sense of relaxation.

In western civilization, mind and body were polarized and attended to by different professional disciplines, and ranked in value. The mind was associated with spirit, clarity, objectivity, and differentiation. The receptive feminine principle was located in the living body and that psychological repository for rejected impulses and qualities, with its personal, cultural, and collective aspects, which Jung (1971) called the Shadow. This dimension of human existence was demonized and devalued, suppressed and consequently distorted in eruption and projection. One result is that body gesture is under cultural surveillance. That is why the first step into the labyrinth, the public crossing of the threshold, is an emancipation. In this place the abandoned are found, brought back, restored.

In the Cathedral experience, the insight that joy was a rare treasure in the eternal flow of grief was offered outward in compassion. It had calmed me; maybe by offering it, the collective energy was transformed. This points to its potential as a ritual healing tool. The *communitas* context is significant in labyrinth practice. Clarissa Pinkola Estés (1992) describes weeping as a call for comfort and support. Weeping on the labyrinth does not necessarily mean you will receive gestures of comfort, though I have seen complete strangers fall into each other's arms when a family in crisis reached an emotional climax at the center and called soundlessly for help from the other participants. There is no "protocol" for this; it was spontaneous. Congregation—even a temporary one like the participants at an event—is "small-scale society" with frames for understanding and containing transformative process, if the individual chooses to engage with the group. Perhaps if we were more explicit about this, it would be more widely recognized as an alternative to cultural "pathologizing" of the symptoms of initiation to the rites of passage.

FINDING BALANCE

To give up on reason as the sole means of discernment allows the underlying truths of our cultural heritage to return from the wastebasket of modernity, opening up for me the possibility of simultaneity of consciousness. Math, science, mythology, dance—all the ways that humans have understood the world—telescoped into the awareness of each consecutive moment, once I learned to "listen" to the body-self, my own and others, and to "read" its messages. It has been fruitful to undertake this learning while maintaining a relationship to the pavement labyrinth as a constant amid shifting viewpoints and levels of awareness. I could at least keep track of the evolution in my responses to this unchanging spatial device.

The labyrinth is unchanging but the individuals coming to it are in ceaseless flux. Humans need some sensory cues in the immediate atmosphere to assist them in moving into a state receptive to such alternation in consciousness. Time of day, light levels, number of people present, the evocative effect of candles, music, incense, all make a difference to what ensues, even for the same individuals. The intentions of the group gathering to "create" an event can be influenced by what is said and done in the introduction, when the ringing of bells signals the onset of sacred time. Otherwise the group gathers on the surface of the labyrinth to drink coffee and chat after a service elsewhere in the building, with no intention of engaging with it whatsoever—only the children running on what fragments of the path are free of Sunday-school tables remind us it is there. Then I wonder how as a species we rely on a leader to say "Look! Listen! Touch!" so as to break through consensual oblivion, to draw the line between the sacred and the profane.

Here I am encouraged by Joan's report on the impact that walking the labyrinth had on her life. This proves to me what a powerful tool it is, overriding all our comforting (or despairing) assumptions about our evolutionary distance from what we once been able to "know" and the "bodiness" of knowing it. I had no doubt that the wisdom we could gain from it was accessible through the experience of the body rather than the machinations of reason. Joan's story suggested that the physical event of walking the labyrinth activated a template of tacit meaning: In mythological terms, she had gone into the darkness at the center and, by integrating and reviving her inner Theseus, had rescued her own "maiden" self. I saw then that we had evoked mysterious invisible powers beyond my understanding.

I am astonished when it all comes together and takes off, when all in the group feel some peripheral vascular "rush" simultaneously, spontaneously. One midsummer night's event, Taizé chanting in rounds had fragmented into some auditory architectonic structure, acentric and fluid. I had worked hard as facilitator to arrange an event, and I felt like celebrating its conclusion. I had walked alone into the center with everyone singing around the outside. The labyrinth was vibrating with accumulated energy and my feet spun vortexes on the turns. It reminded me of the Flamenco circle where everyone supports the one in the middle, clapping and exhorting until the performer is filled with the wild spirit the Rom call *duende.* I felt myself relax finally into trusting the sense of *communitas,* dancing and singing all the way to the centre, where I sat for a long time staring into the red roses and candlelight and incense smoke. Afterward I asked some tentative questions of the others. Everyone volunteered some version of the same physical and imaginal experience: an "electric" feeling, a surge of energy, a glowing vision of a net spread out between us, a glorious and majestic freedom in the improvisational harmonies of the music we were creating, a sense of contribution to and return from that creation.

NOTES

1. Kearney adds, "We no longer appear to know who exactly produces or controls the images which condition our consciousness" (1988: 3).
2. "We appear to have entered a post-modern civilisation where the image has become less and less the expression of an individual subject and more and more the commodity of an anonymous consumerist technology" (Kearney, 1988: 6).
3. "They presume a double perspective: maze-treaders, whose vision ahead and behind is severely constricted and fragmented, suffer confusion, whereas maze-viewers who see the pattern whole, from above or in a diagram, are dazzled by its complex artistry" (Doob, 1990: 1).
4. In, respectively, Brunés (1967), Critchlow, Carroll, and Lee (1973), and Lawlor (1982) regarding the medieval sacred geometry underlying the labyrinth's layout, and Rosenstiehl (1987) on the twentieth-century algebraic.
5. Compare in Capra: "The central insight of the systems theory of cognition, sometimes called the Santiago theory, is the identification of cognition, the process of knowing, with the process of life. . . . A brain is not necessary for mind to exist. A bacterium, or a plant, capable of perception and thus of cognition, has no brain but has a mind. Mind and matter no longer appear to belong to two separate categories but represent different aspects, or dimensions, of the same phenomenon of life. The relationship between mind and brain are clear. Descartes' characterization of the mind as 'the thinking thing,' (*res cogitans*) is finally abandoned. Mind is not a thing but a process—the process of cognition, which is identified with the process of life. The brain is a specific structure through which this process operates. The relationship between mind and brain therefore is one between process and structure" (1996: 172–73).
6. Jung (1971) defines archetypes this way: "In addition to our immediate consciousness, which is of a thoroughly personal nature, there exists a second psychic system of a collective, universal, and impersonal nature which is identical in all individuals. This collective unconscious does not develop individually but is inherited. Whereas the personal unconscious consists of contents, which have disappeared from consciousness through having been forgotten or repressed, the collective unconscious is made up essentially of pre-existent forms, which have never been in consciousness, the *archetypes.* The archetypes are the unconscious image of the instincts themselves, in other words, that they are *patterns of instinctual behaviour* . . . the forms which the instincts assume" (60–61).
7. The spring beneath Chartres Cathedral can still be seen in the crypt on the north side. It is very deep and lined with bricks. The tour guide tells with relish the story of the Christian martyrs slaughtered by the invading Vikings and thrown down the well shaft.
8. Compare with Jung's identification of archetypes as unconscious images of instincts, note 6.

REFERENCES

Ackrill, J. L. (Ed). (1987). A new Aristotle Reader. Princeton, N.J.: Princeton University Press.

Ashbrook, James B. (1989). The human brain and human destiny: A pattern for old brain empathy with the emergence of mind. *Zygon Journal of Religion and Science, 24* (3): 335–56.

Berry, Thomas, M. (1999). *The great work: Our way into the future.* New York: Random House.

Brunés, Tons. (1967). *The secrets of ancient geometry and its use.* Vols. 1 and 2. (Charles M. Napier, Trans.). Copenhagen: Rhodos.

Campbell, Joseph. (1968). *The hero with a thousand faces.* 2nd ed. Princeton, N.J.: Princeton University Press.

Capra, Fritjof. (1996). *The web of life.* New York: Doubleday.

Chodorow, Joan. (1984). To move and be moved. *Quadrant, 17* (2), 39–48.

Critchlow, Keith, Carroll, Jane, and Lee, Llewylyn Vaughn. (1973). Chartres maze, model of the universe? *Architectural Association Quarterly, 5* (2):11–20.

Compton, V. J. (2001). Experience and meaning in the Catherdral labyrinth experience. Unpublished master's thesis. University of Toronto.

Doob, Penelope Reed. (1990). *The idea of the labyrinth from classical antiquity through the middle ages.* Ithaca, N.Y.: Cornell University Press.

Estés, Clarissa Pinkola. (1992). *Women who run with the wolves: Myths and stories of the wild woman archetype.* New York: Ballantine.

Grant, George. (1998.) The computer does not impose on us the ways it should be used. In Sheila Grant and William Christian (Eds.), *The George Grant reader* (pp. 418–34). Toronto: University of Toronto Press.

Griffin, Susan. (1996). Can imagination save us? Thinking about the future with beginner's mind. *Utne Reader* (July/August): 43–46.

Jung, Carl Gustav. (1971).The concept of the collective unconscious. In. Joseph Campbell (Ed.), *The portable Jung* (pp. 59–69). New York: Viking Penguin.

Kearney, Richard. (1988). *The wake of the imagination: Ideas of creativity in western culture.* London: Hutchinson.

Lawlor, Robert. (1982). *Sacred geometry, philosophy and practice.* London: Thames and Hudson.

Lowe, Donald M. (1982). *History of bourgeois perception.* Chicago: University of Chicago Press.

Midgley, Mary. (1980). *Beast and man: The roots of human nature.* London: Routledge.

Northcott, Michael S. (1996). *The environment and Christian ethics.* Cambridge: Cambridge University Press.

O'Sullivan, Edmund V. (1999). *Transformative learning: Educational vision for the 21st century.* Toronto: University of Toronto Press.

Pearce, Joseph Chilton. (1993). *Evolution's end: Claiming the potential of our intelligence.* New York: HarperCollins.

Remen, Rachel Naomi. (1999). Educating for mission, meaning, and compassion. In Steven Glazer (Ed.), *The heart of learning: Spirituality in education* (pp. 33–50). New York: Jeremy Tarcher.

Romanyshyn, Robert D. (1989). *Technology as symptom and dream.* New York: Routledge.

Rosenstiehl, Pierre. (1987). How the "Path of Jerusalem" in Chartres separates birds from fishes. In H. S. M. Coxeter, M. Emmer, R. Penrose, and M. L. Teuber (Eds.). *M. C. Escher: Art and science* (pp. 221–30). Amsterdam: Elsevier.

Sansonese, J. Nigro. (1994). *The body of myth: Mythological shamanic trance and the sacred geography of the body.* Rochester, Vt: Inner Traditions International.

Swimme, Brian, and Berry, Thomas. (1992). *The universe story: An autobiography from planet earth.* San Francisco: Harper & Row.

Woodman, Marion, and Mellick, Jill. (1998). *Coming home to myself.* Berkeley, Calif.: Conari Press.

CHAPTER 10

Spiritual Knowing and Transformative Learning

GEORGE J. SEFA DEI

INTRODUCTION

This chapter discusses the place of spirituality and spiritual learning in the promotion of transformative education. In highlighting the importance of taking spirituality seriously in the politics and ontology of educational transformation, I locate my discursive framework in the discussion in the challenges of critical teaching to a diverse school audience in North American contexts. I bring an anticolonial reading to what it means to engage spirituality in the political project of transformative learning. My understanding of transformative learning is that education should be able to resist oppression and domination by strengthening the individual self and the collective souls to deal with the continued reproduction of colonial and recolonial relations in the academy. It also must assist the learner to deal with pervasive effects of imperial structures of the academy on the processes of knowledge production and validation; the understanding of indigenity; and the pursuit of agency, resistance, and politics for educational change. Dei, Hall and Goldin-Rosenberg (2000) have argued for working with "Indigenous knowledge" as a strategic knowledge base from which to rupture our academies (schools, colleges, and universities). In this discursive politics, the notion of Indigenous is understood as the absence of colonial imposition of the knowledge that is unique to a given culture or society. Such knowledge reflects the commonsense ideas and cultural resource knowledges of local peoples concerning everyday realities of living. It is knowledge referring to those whose authority resides in origin, place, history, and ancestry. (See also Dei, 2000.)

Arguably schooling in our societies today is at a crossroads. I say this because of what I see as pressing and pertinent demands being made by many critical educators for schooling to pave the way for a transformed view of our world in which all human subjects are able to assert their agency and work collectively to achieve goal of justice, peace, and harmony. But, of course, this noble objective rests on how as a community educators and learners can marshal their inner and outer collective strengths to work for change in a world today dictated by the unfettered needs of, and access to, global capital. We need a more critical, nuanced, and complex analysis of the state of schooling and how the impact of economic globalization (created by the corporate intrusion into education) has served to disenfranchise we as learners from our spiritual beings and made us objects of a social system that avoids meeting its social responsibilities to a collective. Amid all the decay and decadence (social, economic, material, and spiritual) we see around us today, I am heartened by the existence of one fact. The future is itself being contested through the promulgation of radical critique and countervisions of different social, economic, and ecological relationships.

I have chosen to devote my contribution in this collection to a discussion of the place of spirituality in transformative learning. I want to speak about the embodiment of the self, the spiritual sense of self in transformative learning, and how such body of knowledge not only assists us in developing a new vision of society but also strengthens individual and collective souls to survive the harsh realities of human existence. To know is not always determined by what one does know or possess. There is knowledge inherent in the body that may not easily manifest itself in how we make explicit the understanding of our social existence. There is a place for spirituality and spiritual knowledge in the construction of subjectivity and identity. The self (as identity and subjectivity) is itself linked to schooling and knowledge production. It is this awareness that also makes knowledge situational, positional, and contextual. The assertion of spirituality as a legitimate aspect of students' learning and knowledge is daring. But as Butler (2000) notes in the context of North American education, spirituality in modern schooling encounters two dangers: In a knowledge economy, spirituality can easily be commodified, rendered simply individualistic and solipsistic, rather than as emerging out of community/human struggle for justice and dignity. Also, approaches to spirituality need to be attentive to and critical and respectful of different religious faith traditions, including secular thought. A failure to do so could make the theorization of spirituality in schooling fundamentalist. There is more potential for students to develop a sense of spirituality and strength from local communities outside the school. Therefore, for schools to be able to foster spirituality, teachers who understand the value of the spiritual and emotional development of the learner are needed.

A critical understanding of spirituality must not evade power questions. That is, spirituality cannot be discussed outside the contexts of power. For example, we need to explore how certain spiritual values come to dominate over others. Simply, educators must eschew a liberal understanding of spirituality that separates the material from the nonmaterial existence. Because of the predominance of western science and dualistic thinking in schools, introducing spiritual discourse into various social imaginaries have been viewed with suspicion. (See Magnusson, 2000.) The spiritual discourse has been negated, devalued, or at best marginalized by western philosophical traditions or scientific thought. In fact, important intellectual, cultural, and political movements that have claimed to resist the violence of western science and colonizing knowledges have been weary of anchoring their analyses and debates in the spiritual foundation.

In bringing spirituality to the discussion of transformative education/learning, I pose the following question: How can educators organize talk, learning, and pedagogy in the context of spiritual education in our schools in a way that can motivate and enhance learning outcomes for all students? The strategies of empowering students/learners could make spaces in our schools for spirituality to be discussed and, particularly, for spiritual knowledge to be taken up as part of a school's curriculum.

Spirituality and spiritual discourses broach ideas and ontologies that emphasize connectedness, belongingness, identifications, well-being, love, compassion, peaceful co-existence with nature and among groups. Spirituality always has been part and parcel of the schooling experience for many learners, whether this is acknowledged by educators or not. (See Dei, James, James-Wilson, Karumanchery, and Zine, 2000.) Also, elsewhere (Dei, 2002) I have highlighted how, for example, students', educators', and parents' understanding of spirituality is evoked in the context of Ghanaian schooling.

To the skeptic, spirituality has no place in education. In North American public and academic discourses, the mention of spirituality and education is countered with a reminder of the separation of church and state. But as Groome (1999) points out in another context, " . . . an established religion should not mean excluding common spiritual values from our educational system. Proselytizing on behalf of a particular religion is very different than allowing spiritual values to permeate our approach to education" (1). Hence the evocation of "spirituality" is not in the context of subscribing to any high moral/religious order. The discussion of spirituality is not necessarily in association with any religious denomination or particular dogma. Religion may be a way to strengthen the spiritual sense of self. In this discussion, I allude to a meaning of "spiritual" that is not necessarily synonymous with religion. At interest is how spiritually influenced education is pursued in North American educational contexts as part of the processes of schooling—that is, how educators and students work with an understanding of the self and personhood as a basis to engage in schooling collectively. Thus, my interest is a theoretical examination of how schooling serves the spiritual development and/or unfolding of the learner and her or his community. The most important questions in such a discussion include, for example: In what ways does spirituality manifest itself in schooling? What are the challenges for promoting genuine educational options for our schools?

PERSONAL SUBJECT[IVE] LOCATION

As a critical learner and pedagogue, I am placed, located, and implicated in the struggle to affirm diverse knowledges in order to transform schooling and education. The identity form that I speak is crucial not only in terms of contextualizing what I have to say but also for the reader to develop an understanding of how and why I am producing a particular kind/form of knowledge in order to make sense of my world. I am Ghanaian by birth, now resident in North America. Speaking from a personal educational history and experience, I can say that formal schooling in Ghana for me was very western and disconnected from the life of the community of which I was a member. Yet it is this education that has enabled me to pursue graduate studies in a western academy and even teach in such an institution. One of the interesting things about the "colonial encounter" is that as western (formal) education strove to dislodge the learner from her or his community, the moral code of the cultural community sought to assert the self/learner within the group to which she/he belonged. Today it is this attraction of formal schooling that many learners speak of as desirable, and, they do not want to see ongoing locally initiated reforms placed at risk (Dei, 2002). For example, students want to learn the local knowledge and Indigenous experiences, but they also want to come out of their formal schooling with degrees and diplomas that are "transportable" in globalized contexts. For all students struggling in such educational contexts, there have been some losses along the way or at best the parallel moving along of two fragmented knowledge streams.

Consequently, as an adult pedagogue, I have located myself in the struggle to affirm a transformation of schooling and education that is inclusive of diverse knowledges and ways of knowing. I have further interacted with these knowledges in ways that inform, challenge, and affirm the self, culture, and community. I would assert that these pedagogical struggles and tensions occur on many educational fronts, one of which is spiritual.

What many see as the amputation of the self from the community is one of the consequences of the importation of colonial and imperial forms of education. (See also Carnoy, 1974; Mazrui, 1975.) This is why many voices call for a renewed spiritual grounding of the African identity and consciousness in opposition to Eurocentric knowledge. (See Asante, 1987, 1988; Ani, 1994; Dove, 1998; Asante and Abarry, 1996; Mazama, 1998.) While undoubtedly schooling in Africa today faces many challenges (e.g., curriculum [ir]relevance, [in]adequate physical and material support and infrastructure), it also should be pointed out that a closer look at local situations reveals how educators, students, parents, and community workers are responding to the educational challenge. These subjects are relying on local knowledges and conceptions of the self, community, culture, learning, and spirituality to fashion genuine educational alternatives.

THEORIZING "SPIRITUALITY"

Let me say from the outset that what I spell out as African views of spirituality may well be shared by other Indigenous, nonwestern peoples of the world. I make no claims of Africans having a monopoly of these ideas. In fact, any critical reader of Indigenous worldviews and knowledge systems will note that the concept of "ownership" of knowledge is an antithesis of what such counterknowledge forms imply. Furthermore, I acknowledge the diversity of views of spirituality even within a community. In Dei (2002) I point out that among cultures, there are varied meanings of spirituality. Some contested meanings of spirituality see the individual as essentially spiritual. Spirituality is also understood in connection to humility, healing, the value of wholeness, self and collective empowerment, liberation, and "reclaiming the vitality of life" (Palmer, 1999: 3). In this sense, teaching sacredness, respect, compassion, and connecting the self to the world and the self to others *is* spiritual education. In this context, I borrow from Rahnema (1995) whose view of spirituality encompasses a "sensitivity, the art of listening to the world at large and within one, from the hegemony of a conditioned 'me' constantly interfering in the process; the ability to relate to others and to act, without any pre-defined plan or ulterior motives; and the perennial qualities of love, compassion and goodness which are under constant assault in economized societies" (130).

Spiritual education embraces humility, respect, compassion, and gentleness that strengthen the self and the collective human spirit of the learner. The self is a complex, integrated being with multiple layers of meaning. The individual as a learner has psychological, emotional, spiritual, and cultural dimensions not often taken up in traditional/conventional processes of schooling. Holistic education that upholds the importance of spirituality recognizes this complexity by speaking to the idea of wholeness (Miller, 1989, 1997, 1999). Context and situation are important to understanding the complex wholeness of the individual self or being. The individual has responsibilities to the community, and it is through spiritual education that the connection between the person and the community is made. To promote education for change is to view the self as a resisting subject. It is also to destroy the self/other dichotomy, rendering the self as not autonomous but connecting to a larger collective. Spiritual education stresses the dominance of individual spirit and the power of self and self-mind for collective empowerment. The subject as the person upholds an inner understanding of oneself and of one's relationship to others.

Arguably, there are local and contextual variations in the understandings of knowledge, reality, subjectivity/objectivity. For example, African spiritual knowings are intimately bound up with the affirmation of self and Indigenous subjectivity. Spiritual and emotional involvement are inseparable in the production of knowledge. Many African ways of knowing (and, in fact, those of other Indigenous peoples) affirm that personal subjectivity and emotionality must be legitimized rather than devalued. Such knowing also asserts that the subjectivity/objectivity and rationality/irrationality dualities or splits are false. In fact, while spiritual knowledge challenges subject-object dualism, it simultaneously upholds "objectivity" to the subjective experience and, similarly, some "subjectivity" to the objective reality. The subjective is capable of comprehending the "objective universe." Subjectivity is a personal interpretation while objectivity can be simply the material and immaterial reality. It is important to reiterate that this understanding of spirituality (unlike liberal conceptions of spirituality) does not efface questions of power and power relations (self/other/group relation) nor does it falsely dichotomize or separate the material and the nonmaterial.

The individual can be both spiritual and nonspiritual; spiritual in the sense of acknowledging the power of the inner will to know and understand the self and to be able to interact with the outer world and the collective. To be nonspiritual is a failure to show individual humility and to work with the knowledge that comes with knowing the self and inner spirit. African spirituality stresses mind, body, and soul interactions. Such spirituality is about values, beliefs, and ideas of integrity and dignity that shape individual consciousness into a collective and unified existence. Spirituality need not be imposed. The individual develops spirituality through the engagement of society, culture, and nature interrelations.

Closely connected to spirituality is the view of emotions as an important source of knowledge. Emotion is that body of knowledge gained in and out of both the subjective and objective forms of existence. Emotional knowledge develops alongside intuition. It is knowledge that is embedded in the self and speaks to compassion of the human sense and mind. Read in a broad context, the notion of "emotional intelligence," as espoused by Cooper (1997), is very useful. Cooper sees emotional intelligence as "the ability to sense, understand and effectively apply the power and acumen of emotions as a source of human energy, information and influence. Human emotions are the domain of core feeling, gut-level instincts and emotional sensations. When trusted and respected, emotional intelligence provides a deeper, more fully formed understanding of oneself and those around us" (13).

Using these definitions of the spiritual and the emotional, an important ontological question is: What working assumptions about the nature of reality do educators and students hold? In the Ghanaian/African systems of thought, the *ontological* viewpoint stresses that to understand reality is to have a complete or holistic view of society. This view stresses the need for a harmonious coexistence among nature, culture, and society. There is the idea of mutual interdependence among all peoples and that the existence of the individual/subject is meaningful only in relation to the community that she or he is part of. This view also stresses the physical and metaphysical connection and the fact that the subject cannot be understood in its atomistic sense.

Also relevant to this discussion is an epistemological question that emerges as: What ways of knowing about such reality are applied by educators and learners? The *epistemological* position holds that there are different ways of knowing and conceptualizing reality. Knowledge is seen as cumulative and as emerging from experiencing the social world. Knowledge emerges from the interplay of body, mind, and soul. The existence of a metaphysical realm also means an uncertainty of knowledge and the possibility that intuition and emotions offer sites of knowing for the human senses. The metaphysical realm is composed of the natural and spiritual worlds and we do not know everything about these worlds. We, therefore, have to rely on our intuition and, sometimes, our emotions to provide us with answers to certain questions. Such knowledge is beyond the so-called objective understanding of knowledge. Relying on intuition and experiential knowledge allows the self to know and understand the outer world. If practice and experience are seen as the contextual basis of self-generated knowledge, then knowledge and survival go hand in hand. Furthermore, the knowledge that membership in the community accords to rights suggests that important responsibilities tie the individual self to the collective.

A further body of knowledge ties epistemology with ethics. The *axiological* position maintains that within societies there are "disputational contours of right and wrong or morality and values . . . [that is] . . . presumptions about the real, the true and the good" (Scheurich and Young, 1997: 6). In Ghanaian and African systems of thought, the cultural, spiritual, and ideational beliefs, values, and practices are evaluated in the history and contexts of communities, as societies strive to set their own moral tone. While these ideas may be shared by other Indigenous peoples, it is the privileging of certain core social values for "reward" (e.g., responsibilities over rights; community over individual; peaceful coexistence with nature over control or domination of nature) that sets different knowledge systems apart. The understanding of the spiritual self allows one to define her or his moral tone within the broad contexts of the society that the individual is part of. The individual is a subject; the individual is part of a community; the cultural norms of the groups guide and influence human behavior and action; and the spiritual existence is central to material existence.

As already noted, the Ghanaian/African concept of spirituality is part of a worldview that is often difficult for a western-educated person to understand. The western educational discourse tends to regard the subject of spirituality with suspicion, thus missing a dimension that, many would argue, is of key importance in African education. A form of local spiritual knowing is connected to the land, to the people, to ancestors, and to community. For example, individuals are traditionally brought up to appreciate the community and to know that they live in a land bestowed unto them by their forebears. They perceive their ancestors as still guarding over the community, keeping a watchful eye on everyday practice and social activity. There is a belief that the individual living subject could be punished for going against the wishes of the ancestors and/or for not looking out for the interests of the larger group or community. This is an important knowledge base that has, unfortunately, been corrupted over time and has had consequences. Some of the consequences have to do with the (individual) appropriation (selling) of communal lands by traditional elders and leaders who are supposed to be guardians and custodians of communal property. There is a sense that the dead ancestors no longer watch our individual actions and people can go scot-free for breaking traditional values and norms. Similarly, we see a breakdown in the respect of the elderly who are considered the closest representatives of our ancestors.

In connecting "knowledge," the "spiritual," and the "religions," I am not suggesting that all local cultural knowledges are spiritual, or that "religion" is synonymous with "spiritual." Rather, the concept of spirituality that will emerge from this study embraces all three realms. Many Africans who grew up in traditional communities view with concern the continued erosion of the kind of education (not schooling) they received in their homes/families. By this I mean the separation of education from local communities that has left many poor, rural communities disenfranchised from schooling, both in access and in content. The separation of education from local communities must be understood in the privileging of schooling over education, that is, schooling purported to take place within formal institutional structures. It is indeed the formal school tradition that the various external and local reforms have tried to change. In this context, children from poor rural communities have been disenfranchised from opportunities for formal learning if their parents cannot afford to pay to send them to boarding school because they needed the children to work on the farm. Those who did go to "school" were uprooted from their families, cultures, and communities. Their formal learning was disconnected from the "land" and community to which they belonged. It is important, therefore, for learning, teaching, and administration of education to reclaim these understandings in the search for genuine educational options.

Transformative Learning: Teaching and Learning to Become Critically and Spiritually Grounded

I want now to focus on the roles of educators to promote transformative education to students. I begin by asking: What should our teaching strategies be in order to promote transformative learning? I caution that at this stage my interest is not on the specifics of such methods. I am more interested in the philosophical grounding for our approaches to teaching methods and, particularly, the method of teaching that seeks the spiritual and emotional engagement of the learner in her/his education. In this context, therefore, teaching, learning, and in fact education become emotionally felt experiences that spiritually transform the learner. The learner's self and social identities are key to engaging and producing knowledge for change.

History, Place, and Culture

It is important for teaching to ground the learner in a sense of place, history, culture, and identity. Identification with the social and natural environments in which teaching and learning occurs must be seen as key to spiritual and emotional self-development. Given that educational practices in some Indigenous communities are quite different from those in North America, transformative teaching must examine how notions of self, personhood, place, history, culture, and belongingness to community are manifested in specific cultural contexts/values. Such knowledge is fundamental not only in terms of providing some sense of what it means to be spiritual but also in understanding how learning happens in diverse contexts. For example, how are creativity and self-reflection expressed in different cultural contexts? How do educators encourage creativity (beyond mere cognitive competencies), critical thinking, and emotional engagement of the learner in the schooling process? Providing answers to these questions calls for a deep appreciation and awareness of the particular metaphysical, epistemological, ethical, aesthetic, and social values of learners, and how these manifest themselves in specific social and educational practices. For many learners, knowledge of history, place, and culture helps to cultivate a sense of purpose and meaning in life.

Acknowledging Difference

Critical teaching must see learners beyond homogeneity (i.e., learners are a diverse, heterogenous group) and the generic construction of "students," and candidly explore all the emerging contestations, contradictions, and ambiguities in people's lives. Doing this serves to give a spiritual identity to the individual self as a basis from which to make a connection to the group or the community of learners. Learners do not learn as disembodied persons. Their race, class, gender, and sexual identities assist them in pursuing particular social and educational politics. There is also the power of difference. That is, seeing students/learners as powerfully demarcated by race, ethnicity, gender, class, language, culture, and religion can implicate knowledge production. Thus, identity is linked to schooling.

Beyond Particularities

The diversity and contextual variations and differences in ourselves and our students must always be visible in our pedagogic practice. But that, in itself, is not enough. We must challenge the essentializing of difference. I say this because there is only so much we can know from the entry point of a unitary fragmentary around difference. If we begin to see everything about our world from the starting gaze of difference, we also lose the connections and interconnectedness among ourselves. It is this sense of connections that make for the wholeness of existence. The world today is all about difference. Yet difference is beginning to mean nothing much. So we must connect the particularities and the historical specificities to their broader, macropolitical, economic, and spiritual contexts. For example, it is important to see the learner within the globalized context and to show how globalized contexts can and are being resisted just as the "global encounter" still shapes and influences the specificities and the particular. For example, given the encroachment of globalization, what happens in a small village in Africa is not immune to global events. Events in the village influence and are influenced by global events (e.g., rise in OPEC oil prices).

Creating Relevant Knowledge

I make pointed reference to the power and efficacy of teaching relevant knowledge, that is, knowledge anchored in local people's aspirations, concerns, and needs. It is knowledge local peoples can identify with. It is based on the philosophical positions that we must understand our students, learners, and even the subjects of our study on their own terms. For example, writing about Africa and African peoples, Richard Sklar (1993) has rightly noted that those who seek to interpret Africa must develop a sympathetic understanding of African thoughts and values as well as history and culture. This is crucial in order for transformative learning to grasp the total human condition of a people. Teaching as a method and a means to create relevant knowledge is crucial if we are to succeed in constructing identities outside of that identity that has, and continues to be, built in Euro-American ideology and hegemonic knowings. It calls for developing a particular prism, one that frames issues and questions within a particular lens: Is this in the best interests of the learner? Educators cannot take a comforting escape route that says "No one knows what is in the best interests of our students." At the very least, we can initiate our teaching practice by posing—at the start—the relevant questions. We can begin by collectively identifying our shared and competing and sometimes contradictory/oppositional interests. Creating relevant knowledge begins by identifying, generating, and articulating a pedagogic theory and practice that uses lived and actual experiences of local peoples as a starting basis of knowledge generation.

Collaborative Teaching

I am aware, at all times, of the desires and perils of collaboration. And yet I make a case for collaborative teaching on many fronts—for example, in engaging students, local communities, in the process of knowledge generation, in teaching across disciplines and subject matters. Also, in collaborative teaching, a critical educator and student will see experience (and practice) as the contextual basis of knowledge. Such collaboration should challenge the split between "the sources of raw data" and the "place of academic theorizing." It must present communities as active, spiritual subjects, resistors and creators, not just victims of their own histories and experiences. Such collaborative teaching will attest to the power of identity and its linkage with/to knowledge production. Thus, who is teaching about Africa is equally important. In our teaching academies, physical representation of different bodies is significant to rupturing conventional (academic) knowledge.

Telling Success Stories

Teaching should tell success stories as well as failures and disasters. We must challenge the academic attraction and fetish of focussing on "failures." We must ask: What can we learn from the success cases; the sites and sources of local people's resisting and empowering themselves through their own creativity and resourcefulness?

The Dangers, Perils and Seduction of Romanticism, Overmythicization, and the Claim to Authenticity

In developing counterdiscourses, it is easy to romanticize about a people after so much negativity and selective misrepresentation of their history. But critical teaching must eschew this practice. For example, it is important to be aware of the dangers of romanticization and overmythicization as we speak and write about the colonized "other/encounter" in order to counter the negativity and untruths of the past. It serves well to frequently ask ourselves: Why write back? the same way. On a related point, I see all knowledges as contested. There is always a selective representation of the past (and knowledge). (See Briggs, 1996; Clifford, 1983; Keesing, 1989; Linnekin, 1991, 1992; Makang, 1997.) Teaching is a contested educational practice. And, I view with deep suspicion any claim to "authenticity" as possessing authority or authoritative voice that is not open to challenge or critique (Handler, 1986). As many others have argued, there is the impurity of any claim to an untainted past. The past is itself subject to colonial and imperial contamination. But in taking this critical stance to "authenticity," I do not dismiss the power of imaginary mythologies as part of the decolonizing project. As an anticolonial pedagogue I share Lattas's (1993) view that the past must be re-created "as a way of formulating an uncolonized space to inhabit" (254).

The Sociopolitical Contexts of Knowledge Production

For those of us who teach, research, write, study, and, in fact, "produce knowledge" about human communities, it is useful to know and remember that the sources and uses of data are not apolitical. There are always profound social and political contexts and consequences for our constructions of knowledge. All knowledges are contingent in particular social and political contexts. Therefore, in our teaching practices we must always be conscious of the socioenvironmental and political contexts of data gathering (knowledge production). In many parts of our world, people's freedoms have been taken away as they teach critically and politically.

Implications for Global Transformative Learning

I admit that the issues I have raised are by no means exhaustive nor the last (un)spoken words on the topic. Much precedes this discussion and there is room for further debates and counterstances. I refuse to think that each of our agendas is so powerful that there is not room to hear another. So, allow me to conclude with this question: Where am I going with all this merging of the political and academic? I would like to stress that what I am calling for is teaching and learning for spiritual and political emancipation. Transformative education needs a discursive prism from which to anchor a critical educational practice. That is, it requires the creation and use of African epistemologies, perspectives in our teaching practice based on the idea that there are important culturally distinctive ways of knowing. Within the myriad epistemologies, there is a powerful conception that all elements of the universe derive from a similar substance, the spiritual. Hence emotions and intuitions are effective ways of gaining knowledge (Agyakwa, 1998). Western science cannot dismiss this body of knowledge. Local communities must be understood not simply within so-called western rational thought but instead within an Indigenous, culturally contextualized genesis. It is about putting all learners at the center of our discursive and teaching practices and of our academic engagements. Critically teaching as decolonization must not simply deconstruct/interrogate/challenge imperial, colonial, and oppressive knowledges but also subvert the hegemonizing of particular cultural, symbolic, and political practices and significations. (See Wilson-Tagoe, 1997; Smith, 1999.) If anything, I see my project today as one conceived in the political and academic practice of imagining and creating shifting representations of knowledge that counter static and fixed definitions and interpretations of the subjects of our study.

The main argument of this chapter has been that spirituality and spiritual knowing are valid bodies of knowledge that can be pursued in schooling to enhance schooling outcomes of a diverse body. (See Sharma, 2001.) Spirituality encourages the sharing of personal and collective experiences of understanding and dealing with the self. Olson (2000) notes that issues of spirituality in learning and pedagogical situations are critical for transformative teaching given that much of what is "universal" in spirituality exists in the particulars of knowing and asserting who we are, what our cultures are, and where we come from. Hence, a discussion of spirituality in education must be read as the antithesis to the concept that the learning of curriculum is ever solely "universal," where universal means neutral and common to all. Spirituality is a form in which we identify ourselves and the universal and is, therefore, an implicit way of asserting ourselves collectively and individually as creators and resistors and as agents and subjects of change. Therein lies the critical understanding of spirituality as a powerful tool in resisting miseducation, domination, or oppressive forces of schooling. When educators deny or refuse to engage spirituality and spiritual knowing, their educational practices can be destructive of the goal of education to transform society. Spiritual knowing must be utilized to involve and energize schools and local communities.

What spiritual knowing shows is that in the politics of knowledge production, we as educators must recognize the limitations and possibilities in our pedagogical, communicative, and discursive practices. One such limitation can be the intellectual arrogance of thinking that we know it all. It is important to work with the power of *not knowing* and allowing oneself to be challenged by other knowledges. Also, we must recognize that all discursive positions have the tendency to be essentialist positions. To take the position of difference does not exempt us from this critique, whether we are antiracist, anticolonial, postcolonial, or postmodernist theorists. As Maliha Chisti rightly notes, we can fall "into the traps of complete unilateral fragmentation around difference" (1999: 16). Furthermore, it is important for us to be aware that how we occupy spaces and the meanings we inscribe on our bodies and onto these spaces are as important as our political practices. Thus, we need constantly to explore this question: What political space do we choose to occupy at particular moments and why? Finally, we must continually acknowledge the dangers of rigidity in our positions and seek to learn across different knowledges.

Our pedagogies and communicative practices can offer us many possibilities in the politics of transformative education. There is the power of human/intellectual agency. That is, for me, as a minority scholar operating from an anticolonial and antiracist lens, the capacity to project oneself into one's own experience, culture, and history instead of continuing to live on borrowed terms of the dominant is crucial. (See also Mazama,1998.) We must endeavor to see resistance as important in the small acts as well as in major actions that lead to transformation. We also can be aware in our teaching practices that the more differentiated we are, the more interdependent we must be. We can define our academic politics not simply in terms of who we are (identities) but also in terms of what we want to achieve (see Yuval-Davis, 1994) so that we work together not strictly on the basis of our racialized, gendered, and classed identities but also on the basis of the politics we wish to pursue.

REFERENCES

Agyakwa, K. (1988). Intuition, knowledge and education. *Journal of Educational Thought, 22* (3): 161–77.

Ani, M. (1994). *Yurugu: An African-centered critique of European cultural thought and behaviour.* Trenton, N.J.: African World Press.

Asante, M. (1987). *The Afrocentric idea.* Philadelphia: Temple University Press.

Asante, M. (1988). *The Afrocentricity.* Trenton, N.J.: Africa World Press.

Asante, M., and Abarry, A. (1996). (Eds.). *African intellectual heritage.* Philadelphia: Temple University Press.

Briggs, C. L. (1996). The politics of discursive authority in research on the "Invention of Tradition." *Cultural Anthropology, 11* (4): 435–69.

Butler, P. (2000). Class Presentation: *1921Y: Principles of antiracism education.* Department of Sociology and Equity Studies, Ontario Institute for Studies in Education of the University of Toronto.

Carnoy, M. (1974). *Education as cultural imperialism.* New York: D. McKay

Clifford, James. (1983). On ethnographic authority. *Representations, 1:* 118–46.

Chisti, M. (1999). Muslim women, intellectual racism and the project towards solidarity and alliance building. Unpublished manuscript, Ontario Institute for Studies in Education of the University of Toronto.

Cooper, R. K. (1997). *Executive EQ: emotional intelligence in leadership and organizations.* New York: AIT and Essi Systems.

Dei, G. J. S. (2000). Rethinking the role of Indigenous knowledges in the academy. *International Journal of Inclusive Education, 4* (2): 111–32.

Dei, G. J. S. (2002). *Rethinking schooling and education in African contexts: The lessons of Ghana's educational reforms.* Trenton, N.J.: Africa World Press (forthcoming).

Dei, G. J. S., Hall, B., and Goldin-Rosenberg, D. (2000). *Indigenous knowledges in global contexts: Multiple readings of our world.* Toronto: University of Toronto Press.

Dei, G. J. S., James, I. M., James-Wilson, S., Karumanchery, L. and Zine, J. (2000). *Removing the margins: The challenges and possibilities of inclusive schooling.* Toronto: Canadian Scholar's Press.

Dove, N. 1998. African womanism: An Afrocentric theory. *Journal of Black Studies, 28* (5): 515–539.

Groome, T. H. (1999). *Infuse education with spiritual values.* Internet source: A Reprint from the *Christian Science Monitor,* February 10, 1998 www.lightparty.com/Spirituality/Education.html

Handler, R. (1986). Authenticity. *Anthropology Today, 2* (1): 2–4.

Kessing, M. (1989). Creating the past: Custom and identity in the contemporary Pacific. *Contemporary Pacific, 1* (1/2): 19–42.

Lattas, A. (1993). Essentialism, memory and resistance: Aboriginality and the politics of authenticity. *Oceania, 63* 2–67.

Linnekin, J. (1991). Texts bites and the r-word: The politics of representing scholarship. *Contemporary Pacific, 1,* (1 and 2): 171–76.

Linnekin, J. (1992). On the theory and politics of cultural construction in the Pacific. *Oceania, 62:* 249–63.

Magnusson, J. (2000). *Spiritual discourse and the politics ontology. A proposal for a roundtable discussion.* Ontario Institute for Studies in Education of the University of Toronto.

Makang, J. M. (1997). Of the good use of tradition: Keeping the critical perspective in African philosophy. In E. C. Eze (Ed.), *Postcolonial African philosophy: A critical reader* (pp. 324–338). Cambridge, Mass: Blackwell Publishers.

Mazama, A. (1998). The Eurocentric discourse on writing: An exercise in self-glorification. *Journal of Black Studies, 29* (1): 3–16.

Mazrui, A. (1975). The African university as a multinational corporation: Problems of penetration and dependency. *Harvard Educational Review, 45* (2): 191–210.

Miller, R. (1989). Two hundred years of holistic education. *Holistic Education Review, 1* (1), 5–12.
Miller, R. (1997). *What are schools for? Holistic education in American culture.* Brandon, Vt.: Holistic Education Press.
Miller, R. (1999). *Holistic education and the emerging culture.* Internet Source: Transcripts, Ron Miller: Spirituality in Education On Line. *http//cfs.colorado.edu/sine/transcripts/miller.html*
Olson, P. (2000). *Comments on M.A. thesis:* Renu, Sharma: *Rethinking spirituality and education in North American contexts.* Department of Sociology and Equity Studies, OISE, University of Toronto.
Palmer, P. (1999). *The grace of great things: Recovering the sacred in knowing, teaching and learning.* Internet source: Transcripts, Parker Palmer: Spirituality in Education On Line. *http//cfs.colorado.edu/sine/transcripts/palmer.html*
Rahnema, M. (1995). Participation. In M. Rahnema (Ed.), *Development dictionary: A guide to knowledge as power* (pp. 116–31). London: Zed Books.
Scheurich, J., and Young, M. (1997). Coloring epistemologies: Are our research epistemologies racially biased? *Educational Researcher, 26* (4): 4–16.
Sharma, R. (2001). *Rethinking spirituality and education in North American contexts.* Unpublished M.A Thesis, Department of Sociology and Equity Studies, OISE, University of Toronto.
Sklar, R. L. (1993). The African frontier for political science. In R. H. Bates, V. Y. Mudimbe, and J. O'Barr (Eds.), *Africa and the disciplines* (pp. 83–112). Chicago: University of Chicago Press.
Smith, L. 1999. *Decolonizing methodologies.* London: Zed Books.
Wilson-Tagoe, N. (1997). Reading towards a theorization of African women's writing: African women writers with feminist gynocriticism. In S. Newell (Ed.), *Writing African women: Gender, popular culture and literature in West Africa* (pp. 11–28). London: Zed Books.
Yuval-Davies, A. (1994). Women, ethnicity and empowerment. *Feminism and Psychology, 4* (1): 179–97.

CHAPTER 11

African Women and Spirituality

Connections between Thought and Education

Njoki Nathani Wane

When I was asked to write a paper on African women and spirituality, I readily accepted, despite knowing that the task would be fraught with difficulty. Spirituality was not a topic easily woven into an academic framework, or so I thought. This notion was dispelled, however, by the call for papers in relation to a book entitled *Transformative Learning: Essays on Praxis.* I knew that this was going to be a text with difference. I also felt that my chapter would contribute toward the scant body of knowledge currently available on the subject on African women and spirituality. I also felt that the issue of spirituality has deep implications for teaching, a relationship that I hope to pursue in future research.

Asking for guidance both from my Creator and ancestors was my first step in beginning this paper. By acknowledging my ancestors, I gain their strength. I also stand tall because, in the words of a Yoruba proverb, "I stand on the backs of those who came before me" (Vanzant, 1993). Although I grew up in a highly spiritual environment, I never thought a great deal about spirituality until I moved to Canada, where I began my quest for my spiritual roots, drawing upon reflections from both my personal experiences and accumulated knowledge. Beginning with my childhood, I reflected upon the spirituality that had imbued the lives of my grandmothers and mothers. I ventured to explore into and beyond the harsh landscape of my memory, a landscape that had been blackened by the brutality and injustice of slavery and colonialism that smothered African women. Reflecting as well upon the yokes imposed by postcolonialism and neo-colonialism, I came to realize my memories needed to be enriched and balanced with more positive experiences. At this point, I began to think of the Embu rural women who had participated in my research. Although I had not realized it at the time of my research during the mid-1990s, their spirituality was the spirituality of all African women. As I continued to reflect, the image of a gemstone began to emerge (Two Trees, 1993), and I began to appreciate that there is no distance between us, the Earth, the people around us, and the universe (Afua, 2000; Arewa, 1998; Two Trees, 1993).

Without having realized it, I had always been on a personal journey of spirituality, despite having neglected to include spirituality as a subject in my academic journey (Two Trees, 1993). Bridging the personal and academic worlds challenges a writer to maintain a fine balance between subjectivity and objectivity. While I knew that spirituality played a role in who I am, until this writing I had always treated the personal and the academic realm as separate.

While collections on spirituality and religion lay thickly on shelves and continue to invade the Internet, information on African women and spirituality is scant. Spirituality, as I understand it and as my grandmothers, mothers, and aunts had practiced it, was neither spoken nor written about. Indeed, for the Embu rural women, spirituality was not affiliated with a particular religion but was an everyday ritual. It was a way of being, of connecting with the land, the universe, and creation. It was a state that was rooted in who they were as women, as people. It was manifested in their everyday activities and relations. Thus, everyday manifestations of spirituality formed the theoretical foundation of their spiritual rootedness.

I am not an elder assigned the privilege of speaking on the sacredness and spiritual connection of African women in general nor of Embu rural women in particular. As with many other forms of Indigenous knowledge, spiritual knowledge has been passed down both through word of mouth and observation; hence the absence of written works on African spirituality. It is thus my intention that this chapter will make a contribution toward the subject of African women and spirituality.

INTRODUCTION

> To My Mother
> Black woman, woman of Africa, O my mother,
> I am thinking of you . . .
> O my mother. You carried me on your back.
> You fed me, you watched over my first uncertain steps.
> You were the first to open my eyes to the wonders of the earth.
> I am thinking of you . . .
> Woman of the fields, woman of the rivers, O my mother,
> I am thinking of you . . .
> O my mother, who dried my tears and filled my heart with laughter.
> How I should love to be a little child beside you again!
> Black woman, woman of Africa, O my mother, let me thank you.
> Thank you for all that you have done for me, your [child].
> Though so far away, [I am] still close to you.
> (Laye,1990:5).

Laye glorifies his mother not only for her nurture but also for her connection with the land, with the universe, and with creation. His words capture the centrality of African women—the creators of balance, the providers of life, and the nourishers of fields, rivers, and mountains. He is grateful to his mother for having deepened his appreciation for the physical and spiritual wonders of African women.

Their spiritual strength enabled African women to remain strong when their children, husbands, brothers were taken away as slaves; to remain strong during slavery, even when being raped in the presence of their children, husband, brothers, and sisters; even when forced to carry the resulting pregnancy to term. It is this strength that sustains African women as they endure the daily crises that continue to scar the African continent.

This chapter, based on work-in-progress *African Women and Spirituality*, is not an analysis of what was or was not done to African women. Instead, it explores the realm of possibility and gives voice to the unspoken strength of African women—their spiritual strength. The questions that give shape to this chapter include: What is the nature of African women's spirituality? What informs their spirituality? Finally, in the contemporary world, how do we sustain our spirituality? Beginning with a brief historical account of African women's spirituality, the chapter continues with some personal observations, along with some contributions from my research on the Embu rural women of Kenya. Further insights on women's spirituality have been gained through interviews given by North American women.[1] In the conclusion, I explore the potential of African women's spirituality to inform and enhance education, learning, and human relations.

The Origins of African Women's Spirituality

Owing to the absence of written records, explorations by Africans of their ancestral wisdom teachings have been limited (Afua, 2000; Amadiume, 1997; Arewa, 1998; James, 1993). However, thanks to our profound legacy of rich oral traditions, gleaned from African grandparents, grandmothers, parents, and African scholars such as Caroline Shola Arewa (1998), Amina Mama (1997), Ifi Amadiume (1997), and Muthoni Likimani (1985), we have been able to tap into the true roots of our spirituality.

Our foreparents introduced us to the spiritual legacy left to us by our Nubian ancestors (Afua, 2000; Arewa, 1998). The ancient teaching of the Nubian ancestors introduced the notion of spiritual guidance embodied in the Creator, the giver of life, harmony, balance, cosmic order, peace, and healing. Contrary to popular misconception, our ancestors did not worship numerous gods and goddesses. Rather, they viewed the Creator as the undifferentiated One/All Divine from which all life emanates. Our spiritual guidance was embodied both with the Creator and in the Great Mother African spirit and culture giver, who represents truth, balance, harmony, law, and cosmic order; and also in Het-Hru, the aspect of Divine love, beauty, and nurture; in Hru, the sacred warrior of light, the aspect of will and, the aspect of expansion; in Sekh-met, the lion-headed patroness of healers; and finally, in the sacred lotus, Nefer Atum, that aspect identified with the highest ascension and unlimited potential. In the ancient Afrakan[2] spiritual tradition, there is deep respect for the Mother Creator and the female priestess as well as for the Father Creator and the male priest. This respect was rooted in the understanding that the Divine attributes of the Creator possessed both feminine and masculine traits. This way of viewing the Divine, a way that filtered down into everyday life, was called "maat/riarchal" (*maat* meaning harmony and balance). Women enjoyed a measure of equality unduplicated in modern time, and tradition accorded equal respect to the mother-father aspects of Creator (Afua, 2000: 5).

The people of ancient Khamit[3] have much to share with their modern counterparts, despite their existence thousands of years ago. If the Khamit model were adopted and applied, it could contribute toward saving our planet from mass destruction. The ancient Khamitic people had respect for creation, for the universe, and kept in touch with their inner selves. If any of their actions resulted in disequilibrium, they had ways of purifying themselves, their environment, and their relations. The Khamitic masters arrived at this knowledge by paying attention to their relations with self, universe, and creation. Today we have different ways of purifying ourselves, and, as we do so, we grow closer to our divine self. Such a transformation can bring us to a state of constant bliss, peace, and wellness (Afua, 2000; Arewa, 1998; Redfield, 1997; Vanzant, 1993).

Following the path of body and mind, spirit purification can bring us to the place where we will be touched by and united with the great-great-great-grandparents of the Earth—our ancestors (Afua, 2000; Arewa, 1998). They can bring us the gift of restoring energy and to healing broken bodies through the "laying of hands." Laye captures this very clearly when he recalls a personal incident:

> I know when I describe [my mother's] powers, you will smile unbelievingly. . . . But I can tell you what I saw with my own eyes. Are there not things everywhere around us that we cannot explain? Late afternoon some men came and asked my mother to use her powers to get a horse on its feet. It was lying in the field and refused to move. . . . She went up to the horse, lifted her hand and said slowly and loudly: "From the day of my birth I had knowledge of no man until the day of my marriage. And from the day of my marriage I have had knowledge of no man other than my husband. If these things are true, then I command you, horse, rise up!" And we all saw the horse get up at once and follow its master quietly. . . . Where did these powers comes from? (1990: 30–31)

This gift has been explained as an endowment from Ngai (meaning "Creator" in Kiembu/Kikuyu[4]). Living the philosophy of harmony and respect for nature and people, our ancient ancestors held reverence and honor for the Creator and the accompanying divine manifestations in women, men, and all nature. These ancestors saw nature and their environment both as an expression of the Creator and as an inspiration for self-healing. They actively incorporated the use of the four natural elements (water, earth, air, and fire) into their daily and ritual practices. Using the element of water, they underwent purification rites through baptisms, fasting, and enemas (Afua, 2000; Arewa, 1998). Using the element of earth, they identified healing foods and also used herbology and aromatherapy to purify and rejuvenate the body. The element of air was applied through Ari Ankh Ka, now known as Hatha Yoga[5] (Arewa, 1998). The various sacred movements and poses of Hatha Yoga may be found on the walls of ancient temples and pyramids that were carved by our ancestors thousands of years ago. Last, the element of fire, based on the powerful rays of the sun, recharged and purified people through solar rituals. Fire foods, such as radishes, leeks, onions, and cayenne, were used to cleanse the inner system. Also, by rubbing frankincense and myrrh, cinnamon and other spices, herbs, and essential oils upon their bodies, people destroyed the negative spiritual entities that were attracted to the body. Through the blessing and consumption of the four elements, our ancient ancestors protected and strengthened their body, mind, and spirit from disease (Afua, 2000; Arewa, 1998).

Continuous purification, meditation, and wellness maintains the harmony and balance of the soul, body, and mind. Everything they did was sacred, Creator-directed, and emanated from their work to their clothing, from their relationships to their homes, and from their temples to their government. Like the Hunzas of the Himalayas today, my ancient ancestors were for thousands of years a nonpoliced, peaceful people who, through the maintenance of balance among all relationships, sustained a harmonious environment (Afua, 2000; Arewa, 1998).

There is the creative life force within us, around us, beneath us, and above us. This force is dynamic and resides everywhere and in everything (Arewa, 1998: xvii). To be spiritual is to maintain an awareness of this dynamism as it moves through and around our being. It is the practice of utilizing this creative force for the collective good of humanity (Arewa, 1998; Afua, 2000; Redfield, 1997). Arewa (1998) eloquently captures this when she quotes Siva Samhita: "In your body is Mount Meru encircled by the seven continents; the rivers are there too; the seas, the mountains, the plans, and the Gods of the fields. Prophets are to be seen in it, monks, places of pilgrimage and the deities presiding over them. The stars are there and the planets, and the sun together with the moon; there too are the two cosmic forces: that which destroys, that which creates; and all the elements; ether, air and fire, water and earth. Yes in your body are all things that exist in the three worlds, all performing their prescribed functions around Mount Meru. . . ." (2).

Mount Meru is a sacred place in Indian mythology. However, this mountain is not merely mythological, as is often assumed but actually exists in Tanzania, the birthplace of humanity. This sacred mountain has remained central to Indian spirituality until today. In Sanskrit *Meru* means "the center of the universe." To the ancient Egyptians, *Meru* meant "all things surrounding love." In my language (Embu/Kikuyu), *meru* means "ripe, something at its highest peak." To the ancient Indians, Mount Meru relates to the spine—Merudanda and the chakras. *Meru* is therefore the heart of all things. Also, the existence of Mount Meru and its associated mythology and sacred practices suggest that spiritual knowledge dates back to the early beginnings of the human race in East Africa (Arewa, 1998). Indeed, DNA tests indicate that all humans alive today are descended from a single female ancestor who lived in Africa 140,000 to 280,000 years ago. According to Dr. Albert Churchward, writing in 1921, people originated in the Great Lakes region of Tanzania. In 1959–60, Louis and Mary Leakey discovered tools that were 2 million years old in the Olduvai valley in northern Tanzania (Arewa, 1998). Their finding supports the theory that humans originated in Africa around the lakes and mountains of Tanzania, Uganda, and Kenya (cited in Arewa, 1998: 49). Hence, the origins of African women's spirituality enjoy a long lineage that continues up to the present day.

> Mother Africa blessed people with an everlasting love of the Eternal spirit. Mother Africa taught her children to respect the earth and all that dwell on her. She taught them the laws of nature and helped them understand the cosmic rhythms. Her people, who created rituals to communicate with these forces, which they then deified and praised, knew the celestial realms and elements. Her rich oral traditions form the foundations of world mythology and religion. Africa is a large continent with many different people and languages, yet in the ancient spiritual systems we can find unity. Belief in the External Spirit and respect for our Ancestors is probably the world's oldest spiritual practice. Ancient Africans believed in the continuation of Spirit. For this reason they returned the dead to the womb of the earth, ensuring a long lineage and unbroken cycle of life. . . . Ancient Africans lived closely with nature and they realized that internal energies are also governed by the elements. Knowledge of the elements forms

> the basis of religion-spiritual practice in many parts of the world. This inner science is still of relevance today. (Arewa, 1998: 38–49)

African Women and Spirituality: Revisiting My Birthplace and the Embu Rural Women

The culture of our ancestors was spiritual rather than religious. Religion is modern whereas spirituality is primordial (Afua, 2000; Arewa, 1998). Spiritual immersion was the mode of existence, a view that empowered our ancestors. They did not entertain any doctrine of salvation by proxy, and every member of society was responsible and accountable for his or her actions. The women of my birthplace, Kenya, and the Embu rural women,[6] who participated in my research work exemplify a cohesion of thought that was previously disregarded by some scholars. From my personal observations and textual documentation, the foundation for understanding African women and spirituality has been strengthened.

In my village in Embu, Kenya, a neighbor, Waitherero, whom I have known since I was a small child, had a reputation for talking to herself all the time. This exercise was most noticeable in the mornings, while she was working on the farm, and in the evenings. She talked to the plants, cows, sky, and all natural living objects. When I returned to my birthplace to carry out my research on the daily routine of Embu women in relation to their food processing activities, I asked Waitherero about her daily routine.

> I wake up before sunrise, give thanks to Ngai [Creator in Kiembu] for giving me another day to celebrate my life, my gifts from the land, and my relations with my family, friends, neighbors and strangers. I always remember those people who are travelling, the sick and those who have no food. Every morning when I stand at my threshold, I look at the rising sun, at Mount Kenya, at the sky, and then look down and touch the soil and any plants around me and say thank you out aloud. . . . I take in all that in one breath and thank Ngai for flowers, plants, air, singing birds, the morning dew. It is in that same moment I am able to tell whether the day will be good for harvesting if it harvesting time or weeding, if it is weeding time. . . . If there is anything that needs to be said, I say it then, and my words are carried by the wind to their destination. With that one breath, I am ready to begin my day. When I do this, I rekindle my energies within me.

Waitherero was not the only woman I spoke to who practiced morning or evening "rituals." Rukiri, a sixty-year-old Embu woman, also practiced her daily morning and evening rituals by conversing with nature. "I always 'talk' with the rising sun and to the setting sun," she told me.

Although uttered more than five years ago, the words of these women had not disclosed any significant meaning to me until recently. Waitherero's words underline her connection with Earth, the universe, the Ngai's creation. She maintains a strong connection to her land and expresses her gratitude to the Creator through her thoughts, speech, and deeds. Land is considered sacred for these women, all of whom live off the land. They understand that a reciprocal relationship of harmony and reverence maintains a balance between life-sustenance and loss.

For Waitherero and Rukiri, their expressions of gratitude also enrich their spirituality. Land, universe, and creation are treated as rare commodities. By honoring these commodities through simple words of gratitude, they feed their own spirit, which is the source of their strength. Although not succinctly expressed in terms such as "honoring the land" or "feeding my spirituality," their words of reverence for natural creations are expressions of their spirituality. Their exhortations are neither forced nor exploitative. Mugonda, another Embu woman, echoes this union of spirituality and nature when she says: "I am very close to the land. From this land, I can feed my family. This land has always been a part of me, a part of my family—land is very sacred to us . . . when I touch it, it feels good, when I touch the plant that comes underneath the soil, I know there will always be life coming out of the soil. The feel of the soil revitalizes these connections. . . ."

For Embu women, nourishing their spirituality through meditation continued throughout their activities. No specific time allotment was given and, as Wanjira once said: "I think about land, children, relations all the time. I do this while I am walking to the farm, to the river, to the woods to collect firewood. I think all these things even while eating . . . this is my way of sustaining my connections." In a similar manner, Mugendi, a Canadian woman of South Asian descent whom I interviewed, nourished her spirituality on a daily basis through her tasks and observations:

> I find nourishment of my spirit through my body and mind: breathing deeply, through writing, through silent retreats; through watching my child, through the rituals that I do throughout the day from brushing my teeth to lighting a candle to wishing someone well. Some people call it mindfulness, others Taqwah: I call it wonderment and delight at being alive, conscious, breathing, feeling, thinking, and loving. I sustain this sense of spirituality through my loving and nurturing relationships, speaking, creative writing, thinking, reading, taking walks alone and with others, listening to music, and being alone to contemplate. My roots, my sense of faith, my family's religion, my upbringing are key components of what informs my spirituality.

This brings healing and balance to our spiritual bodies. It is necessary to meditate in order to evoke the resurrection of our soul (Afua, 2000; Arewa, 1998). By its very nature, spirituality is a personal enterprise. An understanding of spirituality cannot be obtained from books. As one visiting Embu elder explained, upon reading the words of Waitherero and Rukiri, spirituality is part of the natural process of life. It is evident that these rural women have much to share that can to enrich our collective psyche. They share the same ancient Nubian traditions. These traditions can teach all people to connect with the land and to have meaningful rather than exploitative relationships with the universe and other creations (Afua, 2000).

> I believe these words indicate that spirituality is not something out there, something that you read in your books, something that you can learn in your university. It is in your heart, it is in you. Spirituality is when you wake up in the morning and call on Ngai to guide you and to be with you. Spirituality is when you stand on your threshold and acknowledge the land you are standing on, the air you breathe, the creation surrounding you. Spirituality is when you give thanks before you eat or drink to Ngai first and to your ancestors, and to the land. In those brief moments when you pour a libation and welcome people to share what you have. For me that is spirituality.

Owing to spirituality's intangible and nebulous qualities, defining it is extremely difficult. To facilitate an understanding, what it means, it is better to view it in the context within which it is sustained and cultivated. Land, for the Embu rural women, is recognized for its tangible importance. From land and nature the women connect and nourish their spirituality. To most Embu elders, land is sacred (Wane, 2000). One of the elders, Mukuru, once said to me: "Walk cautiously on the land. Walk like a chameleon. When you do that, you watch where you are going, you do not kill any of Ngai's creation and you do not 'abuse' the land. Never bite the hand that feeds your body and spirit. . . ." These words carry multiple meanings. Mukuru could have been cautioning me about my interactions with people or could have been telling me to honor that which feeds my psyche and surrounds me. The sky above and the earth beneath, the rivers, the moon, the changing seasons, the birds on the bush, sunrise and sunset—all are woven into a rich and sustaining "tapestry" that speaks continuously of the glory of the Creator. What surrounds us communicates a wonderful sense of time, almost a sense of timelessness (Riordain, 1996). The bond between nature and spirituality is strong for those who work intimately with the land. To this day, I remember the moment at which our family land was taken away and given to another family. I can still remember the pain in my mother's eyes, the unspoken questions, and the many times she touched the soil and the plants. I can still remember my mother shaking her head as if asking herself whether the new owners would have the same reverence for this land as she had or as her ancestors had had for generations; also, whether the new owners would ever know the land as well as her family did.

Many authors have written on this interconnection between nature and spirituality. The North American writer and meditator Elaine St. James recommends the following movement: "Start . . . with a deep invigorating breath of fresh air, and an appreciation of weather . . . make a point of delighting in the trees and birds and flowers and plant life. . . . Let the glories of nature energize your body, heal your psyche, and uplift your spirit. . . . Spend a few moments each day appreciating and drawing energy from nature. . . . Use that time to notice the patterns of clouds in the sky or the dew on the grass or take time at the end of the day to acknowledge the closing of the day with sunset. . . . Use the power of the cosmos to get in touch with who you really are. Linking with the sun increases our vitality and elevates our consciousness thereby contributing to our inner growth" (1995: 12–13).

Those words resonate with the statements and the relationship that the Embu women have developed with nature. To intertwine oneself both with nature and spirituality strengthens the relationship and encourages harmony and growth. This relationship was further reinforced when I accompanied the Embu women to their farms. Their actions always amazed and intrigued me. For example, if it was harvest time, the women carried seeds with them and, before any harvesting began, uttered some "seed words." Then they would cast forth some seeds from the previous harvest. At the time, their ceremony made little sense to me. It never occurred to me that these women were giving back to the land as they were taking from it. In this way they honored the land.

Even food stocks that were uprooted or pruned were treated with much care and respect. If the women were harvesting bananas, they gave the trunk to the cows and used the green leaves as a cover to keep food warm or for earthen pots, or to wrap roasted food, such as bananas, yams, or arrowroot. They stored dry banana leaves and used them later to rekindle fire, while they used the dry bark for making robes, mats, or trays. At the end of the day, some of the food would be given to the neighbors. Routinely, and before any foods or drinks are consumed, a libation is performed. When I asked Mumenyi what all of this meant, she replied: "We have always done this to honor our Creator, to acknowledge our ancestors and give thanks for the yields from the land. If you don't give thanks, or if you do not give back to the land, how do you expect to have any yields the following years? It is unfortunate because not many people follow this practice. As a result our land does not provide as much as it used to. We rarely give back, we rarely honor our land, or even our neighbors. When we harvest, we rarely give small portions of it to our neighbours. Most us are more concerned at how much we can make when we sell our crop."

Mumenyi emphasized the importance of honoring Creator. Her words reflect Redfield's perspective that "The belief seems to be growing that the act of giving engages a metaphysical process totally congruent with our knowledge that the universe is responsive" (1997: 175). These words emphasize the importance of reciprocity and the significance of creating and sustaining connections not only with the land but also with one's neighbors and ancestors. Many cultures throughout history have regarded nature as an integral and necessary part of their lives. However, many of us today have lost contact with the restorative, healing, and inspirational powers of nature.

In summary, the Embu women enjoy an intimate relationship with nature and its processes. The women enhance and sustain their spirituality both by expressing simple words of gratitude on a daily basis and by returning to the land a portion of what they have taken. This reciprocal relationship suggests the importance of continuous reverence for the balance of life and the harmony of spirituality.

Spirituality: North American Perspectives

Several of the North American women I interviewed provided a wide range of definitions for "spirituality." They defined spirituality according to their culture, experiences, and practices. The following accounts were related by seven women[7] of varying cultures and ages who articulated different understandings of spirituality.

Marie, a woman of European descent who has practiced spirituality for many years, emphasizes the personal domain within which her spirituality resides: "Spirituality is the means of tapping into the universe itself. It is a life path or practice, not religion. I find spirituality to be personal, unique, individual to individual—yet there seem to be similar archetypical 'spiritual experiences' within humanity. Are these experiences a source of universal truth? I do not know. All I understand is that my spirituality inspires me to live my life in accordance with the truths I sense within myself."

Monique, a woman of African descent, sees spirituality both as individualistic and linked to positive attributes: "Spirituality is bigger than consciousness. Spirituality is the power within an individual. It is available for us—it is up to the individual to open his/her heart or soul to receive the spiritual power. Spirituality can be turned on or off by the individual. Spirituality is integrity, humility, connectedness, sincerity, deeper faith, and trust in your beliefs."

Both Barbara Waterfall (a Metis woman) and Malaika (an African American woman) emphasize the womb as the origin of spirituality. Waterfall, reflecting on Native women's spirituality, emphasizes its personal nature:

> I am only an authority on myself. . . . These thoughts and understandings . . . come from within. They represent an epistemology of inner knowing. They come from a very old and deep place. They come from the womb and as such come from the place of creation. This place is the starting point and it is the place where all life begins. It is the place of Wombma's knowledge. . . . Wombma's spirituality is based upon the centrality and sacredness of life itself. This sense of spirituality values life and as such values the diversity in life. This sense of spirituality does not see us as separate from the Earth or as separate from other living beings. It understands us as belonging to and being one within a universal system of kinship ties.

Malaika, reflecting on her spirituality as an African American woman, said:

> Spirituality is a feeling. There are no answers as to how you get it. It is an experience. I cannot put it in a container and say, here it is. It is an awakening. It is unexplainable invisible thread that binds us together. Spirituality keeps us in the realm. It is the essence of how you overcome. It is the backbone of all nations. Spirituality is housed in our wombs. It is passed down to our children, male and female, through the umbilical cord. Many African American women do not know their spirituality because they get exposed to other forms of beliefs—religious beliefs, social, economic, and political dogmas—as a result many of us are kept away from our spirituality by these forces. We concentrate too much on them and give no time to listen to inner voices to our spiritual feelings.

Zainab, a woman of eastern-Filipino descent, describes spirituality as being inextricably connected to her relationships and experiences in daily living: "The word 'spirituality' can have different meanings. When I talk of my spirituality I am referring to my everyday reflection concerning my life, my surroundings, and especially my relations with those who are around me."

Other women could not agree with defining spirituality as a feeling that has guided them and their ancestors over an infinite period of time. Muiretu, a woman of African descent, views defining spirituality as a futile exercise: "Doing so would be an attempt to confine, limit, and possibly misrepresent such a complex, dynamic, and ancient philosophy that guides the lives of many Africans all over the world. I can only say that there is a deep respect for and guidance from African ancestors that is rooted in both the soil and souls of the continent."

Similarly, Amal echoes the difficulty of defining spirituality: "Spirituality is difficult to define in terms of words, writing, or through using a computer because spirituality is a feeling. It is the power that I inherited from my African ancestors. Our ancestors did not have to write about their spirituality because it was an everyday practice. Spirituality is the power that connects me to my Creator, to the universe, and that enables me to purify myself. It is the power that enables me to exist. This power comes from the Creator."

Rather than offering any concrete definitions, the women defined spirituality as something essentially intangible and indefinable. In my view, their insights manifest what spirituality is. Spirituality cannot be encapsulated in a neat definition because by its very nature it is personal, unique, and individual. Understanding spirituality means an understanding oneself.

As Malaika aptly observed, we cannot put spirituality into a container, for it is an invisible link that connects us all together. In short, spirituality is the wonderful sense of unity and rhythmic harmony with the world around us (Riordain, 1996).

Sustaining and Maintaining our Spirituality

Spirituality serves a myriad of purposes. According to Queen Afua (2000), when we embrace spirituality, we liberate ourselves from our old life and usher ourselves into a new way of being. Her sentiments are echoed by Marie: "The further one goes from the self, the less one knows. I heard this phrase when I was sixteen. I do not remember who said it, or from what context it came but these words inform and cultivate my spirituality daily. My spirituality was born first from self-knowledge. It grows and adheres to the truths life continually teaches me. Spirituality moves beyond statements like 'I believe' or 'I know,' living in realms of sensing and intuiting one's truth. I use words like moves, and lives, because spirituality is a living element within me. It breathes, it waxes and wanes, flows, rises and falls."

A spiritual journey requires a commitment to transform one's consciousness from ignorance, denial, or grandiosity into a harmonious alignment with one's divine nature. Malaika elaborates on the nature of African women's spirituality:

> The nature of Black women's spirituality is a lineage connection from ancestral wisdom passed down in words and deeds, mainly deeds. The connection takes place at the spiritual level, consciously before the physical birth of the person. Black women carry this ordained position of mightiness in manners different from other races of women. Being sought after for connecting, the rearing of children, matters of relations. . . . The nature of our spirituality is based also on the innate ability to nurture. The ability to nurture from the higher levels of humanity. In order to nurture the human flesh and spirit, your insight must be above the regular concepts of regular issues and the people with them. The nature of our spirituality is always present, always felt, we walk with it, we talk with it, our body language, eye movement, vocal opinion. We wear it like protective attire. Once you possess it, you can never disown it or have it not possess you, you become it, and it becomes you. It is an undaunting power, if used correctly for good, no one and nothing can disturb it, or you who possess it.

According to Malaika, spirituality has no beginning or end. It is within us:

> What informs our spirituality is usually personal, and each individual has their own justification. Most often people think that they're rightly guided by a Supreme power, and some think that they were born with this gift of enlightenment; still others think that the Supreme Being, Creator, God, or whatever other name given to it at the genetic level of their existence. . . . Spirituality comes like wisdom, slowly at times, and quite often quick and sudden, surprising, as it enlightens and teaches. Higher levels of spirituality, like wisdom, comes with age and exposure over a certain amount of time and involvement. . . . Ninety to 98 percent of our spirituality is based on us wanting to be correct in actions that we are allowed to govern over, for example, being good mothers, healers, wives, confidantes, socially being accepted in our families and our communities. Spiritually sensitive individuals begin to change their personal surroundings, and they avoid the people and things that cause a conflict within them. With heightened mind and keen sense, an individual can soar above the average person when dealing with perception and awareness. . . . The more advanced your mind becomes the more power, force, and insight you'll have. Spirituality becomes like infinity, it goes on and on once you become aware of what it is.

As for Zainab, where she grew up:

> God was more than just a fixed entity with a certain set of characteristics but rather a more encompassing, infinite Being who took a variety of forms and faces. [Spirituality] also taught me to be more open-minded about other people's beliefs and ideas. . . . I learned early in life that God was not something that was limited in terms of how He/She represented Himself/Herself to different people . . . probably the two biggest influences on my spiritual journey are my parents. My mother was raised in the Philippines for most of her life and grew up in a very devoted Roman Catholic family. . . . My father, on the other hand, was pretty much of a "lax" Hindu . . . but I saw him pray every morning in front of his altar where he would bring his breakfast and perform *puja.* This observation had a very important impact on me later, as I began to realize that spirituality was something that needed to be practiced regularly and maintained as part of one's daily life. . . . Who am I? What am I about? Where am I from? This was a major turning point in my own spiritual journey.

According to Monique: "Spirituality comes from God Almighty and has been practiced by our ancestors and passed on to the next generation for those who pay attention or believe in it. Personally, I found out about spirituality from my grandmother. I watched her make a difference in the community, using the power of healing with herbs, her dreams, and interpretations that led to miracles for the village community."

In many cultures, many people are aware of their spiritual self, aware that their life is being influenced by spiritual energies and a spiritual presence that either can help or hinder them in their quest for happiness (Evan, 1993). For most participants, spirituality is a transformative process by which we stop resisting the reality of the spirit. Spirituality enables us to uncover our spiritual unconsciousness, filled with repressed energies. It is a process of personally transforming one's underlying motivation and orientation but a crucial part of this process is a growing openness to the spirit (Evan, 1993). According to Evan, the spirit could be understood in reference to spiritual energy or even as spiritual entity that animates one's action from within. This interpretation resonates with that of Monique, who relates spirituality to "integrity, humility, and deeper faith in your beliefs." For Nyokavi and Mailaka, spirituality comes from the womb. It is passed down to them from their mothers.

Today contemporary seekers from various traditions are invoking the spiritual wisdom of their ancestors in order to revitalize those traditions through understanding and practising spirituality in modern life (Afua, 2000). Whether they are the Natives of North America reconnecting with the Creator, Chinese drinking freely from the wells of Taoism, or Jews debating their Talmudic roots, it is quite clear that we cannot become whole unless we understand the wisdom of those on whose shoulders we stand (Afua, 2000; Vanzant, 1993).

The creative life force is within us, around us, beneath us, and above us. This force is dynamic and resides everywhere and in everything (Arewa, 1998: xvii). To be spiritual is to maintain an awareness of this dynamism as it moves through and around our being. It is the utilization this creative force for the collective good of humanity (Arewa, 1998; Afua, 2000; Redfield, 1997). Such a creative force also can become instrumental in creating a meaningful education.

SPIRITUALITY AND EDUCATION

In attempting to define spirituality, several themes emerged. Both the Embu rural women and the interviewed participants acknowledged individualistic and personal nature. Spirituality often was associated with the natural processes of life: the land, the universe, and creation. However, spirituality cannot align itself to a society whose values have become increasingly profit-driven. In such a culture, human beings become spiritually disconnected with nature. O'Sullivan (1999: 262) points to the massive destruction wrought by humans who conceive themselves, as a result of this destruction, to be entities separate from the Earth. Developing the spiritual nature of students through our teaching will establish and enhance their awareness of their relationship with other species and with the land. Rather than viewing humans as occupying a hierarchical level above all other species, students would see the interconnectedness and interdependence that all living things have with one another (O'Sullivan, 1999: 70, 263). Such a vision would broaden their sense of community and help them maintain the respect for life.

Valuing the spirituality of students also means valuing the uniqueness of individuals, regardless of race, sex, creed, or ability. In education, what is nonmaterial often is disregarded, and individuals are judged only on the basis of their measurable and quantifiable skills and talents. Those who can follow the rules obediently are rewarded, whereas those who do not are both perceived and labeled accordingly (Miller, 1991: 7). Those who fall in the latter category are slotted into a system that disempowers them, perpetuates their inaccessibility to resources, and reinforces the relations of power and dominance held by certain groups.

To create an educational system that develops a student's spirituality, educators and teachers have to understand the holistic nature of education. First, they have to acknowledge that all students have spiritual potential, meaning that one must look beyond the scores and labels that serve to identify and segregate them. Scoring students only on their measurable skills and talents limits the educator's appreciation of a student's rich and diverse spiritual dimension. To acknowledge a student's spirituality means to recognize the interior depth that every student has. This starting point opens the doors to understanding the interconnectedness we have with each other. Linked by this commonality rather than separated by differences yields a more accepting and inclusive environment and way of thinking.

The next step is to construct a pedagogy that encourages spiritual growth. The challenge of linking spirituality with traditional systems of pedagogy has been an exciting yet frustrating journey for one educator at the postsecondary level—Two Trees (1993). While she is excited at the prospect of infusing spirituality with learning because of its enriching opportunities, she also is frustrated at the many scholars and students who have serious doubts about deviating from the norm. Academia has not yet developed a structure that would support instructors who seek to provide elements of other forms of learning or knowledge. Two Trees captures this dilemma when she writes about her own experience of teaching from a nontraditional approach: "We are in a place of seeking/learning, able to pass on what? Information, experience, knowledge, wisdom? Maybe we are locksmiths passing out the keys we have found to remember. Part of this task is to remind students and each other that this is a journey of mind and spirit. When the mind is stimulated by information, that intellectual excitement is the sign that a key has been found. . . . The student is looking at the key to remembering. . . . It is then the task of the student to take the key and look for the doorway or window, which it fits. This fit is a resonance experienced at the center of being-insight, intuition, gut feeling, and heart" (1993: 17–18).

The feelings described could easily be attributed to any educator or teacher whose goal is to nurture a child's spirituality with respect to his/her relationship with nature.

Whereas the Embu rural women could readily identify with the natural processes that surrounded them, students in urban areas, surrounded by concrete playgrounds and treeless lots, are discouraged from forming a connection between their spirituality and nature. Yet the pedagogical process could link up with the natural processes if even on only a small scale. Engaging in a seed-planting project (by using egg cartons) can cultivate responsibility and understanding of reciprocity. (The seed will grow into a plant that cleanses the air, and the child, in turn, will continue to nurture the plant for life.) Connecting this exercise with expressions of gratitude by way of reinforcing the spiritual connection can foster and strengthen a child's spirituality. On a larger scale, the class could embark on a tree-planting exercise on school property or in neighborhood park. The contribution that planting trees would make to the school community, to the neighborhood, and even to local businesses would enhance the spiritual development of the child.

The holistic nature of the spiritual journey often appears to be located outdoors. Nonetheless, through a series of inventive steps, a bridge between the outdoors and the classroom can be built. Two Trees (1993) brings natural artifacts that are imbued with meaning, such as branches and rocks, into the classroom. Next, she moves the physical space and location of the class into a more inviting and accommodating shape, such as a circle, without the interference of chairs or tables. Finally, the entire classroom space becomes transformed when the students themselves engage in activities that explore their spirituality.

The time is ripe for education to "resist being restricted by the shallowness of the market mentality" (O'Sullivan, 1999: 269). Because of the loss of the spiritual connection between humans and the land, humans continue to exploit and destroy the Earth, unbalancing and dismantling the life systems to extinction. Exploring spirituality within the context of education would create new pathways of understanding, both for educators and students. After all, everyone is endowed with spiritual dimensions that can enrich the learning experience. By weaving spirituality and nature into teaching practice, educators can instill spiritual growth while at the same time strengthening a reverence for and continued propagation of nature.

CONCLUSION

The oral tradition underlining the origin of African women is spirituality. From the beginning their spirituality has been closely intertwined with nature and its processes. This connection between spirituality and nature continues today among the Embu rural women, as they maintain their connection with the land, universe, and creation through thought, word, and deed. Because of their intimate relationship with the land, they understand and respect the reciprocity between both parties that is essential for maintaining the balance of life.

In contemporary North American education, a child's spirituality is not cultivated. This is unfortunate, given that the full experience of an education begins with an acknowledgment of spirituality. O'Sullivan (1999) remarks that the loss of spiritual connection between humans and nature has led to the wanton exploitation and destruction of resources. His words echo those of the ancient Khamits and Nubians who understood the importance of aligning one's spirituality with nature. Without nurturing the spirit, we continue to overlook the dialectical relationship between spirituality and nature. This relationship, like the land itself, needs continual cultivation and enrichment. In a world increasingly dominated by market-driven economies, the spiritual dimension is rapidly losing its foothold. The educational system must communicate the necessary connections all humans share with all species and with nature. To neglect the wisdom imparted both by ancient and modern African women will only hasten the demise of the living, beautiful but increasingly damaged world.

NOTES

1. The women who participated in this conversation come from different ethnical, cultural, as well as racial backgrounds. The following backgrounds were represented: African, South Asian, South Asian Philippines, Aboriginal Metis, European and African American.
2. *Afraka* means flesh (*Af*) and soul (*ka*) of the hidden sun (*ra*). Africa comes from the Arabic word that means separate, divide, or conquer—this name was imposed by the foreigners.
3. Khamit refers to the original Nubian Maatian culture of the Nile valley.
4. Kiembu and Kikuyu are dialects of Bantu, one of the languages of Kenya.
5. Hatha Yoga involves stretching as well as focusing on breath awareness. It requires concentration and results in relaxation. Among its numerous positive outcomes, Hatha Yoga is reputed to tone and strengthen the skeletal system and to improve the cleansing effect of the immune and circulatory systems. Spiritually, this yoga serves to "draw together the polarities of *Ha*—solar energy—with *Tha*—lunar energy. This unification leads to enlightenment" (Arewa, 1998: 120–21).
6. The Embu is a group of people belonging to the Bantu-speaking people. They occupy the slopes and surrounding areas of Mount Kenya. Although the Embu are comprised of different clans, they unite together for social and defense purposes (Berg-Schlosser, 1984: 46; Mwaniki, 1973: 1, 19). Bantu is one of the languages of Kenya and is used to denote the people who speak the language. Bantu is composed of approximately six hundred dialects (Ochieng, 1985: 7).
7. Pseudonyms for the names of women who gave interviews have been used to protect their privacy. Two of the women requested to have their real names used.

REFERENCES

Afua, Queen/Helen O. Robinson. (2000). *The sacred woman: A guide to healing the feminine body, mind, and spirit.* New York: Ballantine.

Amadiume, Ifi. (1997). Women's achievements in African political systems: Transforming culture for 500 years. In I. Amadiume (Ed.), *Reinventing Africa: Matriarchy, religion and culture* (pp. 89–108). London: Zed Books.

Arewa, Caroline Shola. (1998). *Opening to spirit: Contacting the healing power of the chakras and honoring African spirituality.* London: Thorsons.

Berg-Schlosser, Dick. (1984). *Tradition and change in Kenya.* Paderborn, India: Ferdinand Schoningh.

Churchward, Albert. (1921). *The origin and evolution of the human race.* London: Allen and Unwin.

Evan, Donald. (1993). *Spirituality and human nature.* Albany: State University of New York Press.

Hegy, Pierre. (1996). *Feminist voices in spirituality.* Lewiston, N.Y.: Edwin Mellen Press.

James, Joy. (1993). African philosophy, theory, and "living thinkers." In Joy James and Ruth Farmer (Eds.), *Spirit, space and survival: African American women in (white) academe* (pp. 31–46). New York: Routledge.

Laye, Camara. (1990). *The African child.* London: Thomas Nelson and Sons.
Likimani, Muthoni. (1985). *The Passbook number F.47927: Woman and Mau Mau in Kenya.* London: Macmillan.
Mama, Amina. (1997). Sheroes and villains: Conceptualizing colonial and contemporary violence against women in Africa. In M. J. Alexander and C. T. Mohanty (Eds.), *Feminist genealogies, colonial legacies, democratic futures* (pp. 46–63). New York: Routledge.
Miller, Ron. (1991). *New directions in education.* Brandon, Vermont: Holistic Education Press.
Mwaniki, H. S. Kabeca. (1973). *The living history of Embu and Mbeere to 1906.* Nairobi: East African Literature Bureau.
Ochieng, William R. (1985). *A history of Kenya.* London: Macmillan.
O'Sullivan, Edmund. (1999). *Transformative learning: Educational vision for the 21st century.* Toronto: University of Toronto Press.
Redfield, James (1997). *The celestine vision: Living the new spiritual awareness.* New York. Time Warner.
Riordain, John J. (1996). *Celtic spirituality: A view from the inside.* Dublin: Columbia Press.
St. James, Elaine. (1995). *Inner simplicity: 100 ways to regain peace and nourish your soul.* New York: Hyperion.
Two Trees, Kalylynn Sullivan. (1993). Mixed blood, new voices. In Joy James and Ruth Farmer (Eds.), *Spirit, space and survival: African American women in (white) academe* (pp. 13–22). New York: Routledge.
Vanzant, Iyanla. (1993). *Acts of faith: Daily meditation for people of color.* New York: Simon & Schuster.
Vanzant, Iyanla. (2000). *Until today: Daily devotion for spiritual growth and peace of mind.* New York: Simon and Schuster.
Wane, Njoki Nathani. (2000). Indigenous knowledge: Lesson from the Elders—A Kenyan case study. In G. Dei, B. Hall, and D. Goldin-Rosenberg, (Eds.), *Indigenous knowledges in global contexts: Multiple readings of our world* (pp. 95–108). Toronto: University of Toronto Press.

CHAPTER 12

Journey of our Spirits

Challenges for Adult Indigenous Learners

Renee Shilling

Introduction

as I offered the carrot with my flat palm he jerked away
so briskly that even his mane flickered in the wind
he kept his eye on me and continued to step clumsily backwards
still he had interest in what I had to offer

he nodded to me, almost in irritation
I do wonder what it was that he had to say
he stepped forward as to engage again
bursting with knowledge was this Arabian so connected to his history

the depth of his eyes, mysterious and painful
there is a story in this horse
his ancestors alive for centuries
memories flood forward to a time when the past is here

for a split second the worlds merge
every world there is
only to be separated again
and then the moment is over.

For over a century, Indigenous nations in Canada have been subjected to foreign systems of education. They have been controlled by oppressive and abusive policies that have completely undermined the social, cultural, political, and economic functions of their nations. The colonial experience is embedded in the spirit of the people and in the fabric of society.

Indigenous movements in education will need to consider the impact that colonization has had on their lives. Within this, we must acknowledge that the traditional forms of education and learning styles have been severely altered. I recall that an Elder once said, "In this time we will need to merge the traditional and modern ways of doing things." I think what she meant was that we need to maintain our Indigenous ways of thinking while we develop new ways of coping with our problems because our conditions are new. This process is part of decolonization, which is the deconstruction of the changes since colonization and the ability to look critically at how it has impacted on our families, communities, and nations. The effects of colonization must never be minimized.

In this chapter I will share some of my own learning experiences and acknowledge the perspective that I come from and the various factors that have influenced my life. I realize that adult learning in Indigenous communities requires new and creative ways to work through the existing pain and collective consciousness. It is not enough to develop programs that are based on foreign systems of education. Learning must be seen as an essential component of decolonization. Moreover, it is through the learning process that we may grow and heal our historical and contemporary wounds.

Memories

There are many moments of learning in our lives. With each connection we move forward to a new level of consciousness. Each time we gain another piece of the wisdom and knowledge that life has to offer. Our consciousness contains a vast collection of memories of our spirits, our ancestors, and ourselves. And with this, my learning journey is very much about reflection. As an Anishinaabe[1] Indigenous learner, I strive to create ways to integrate my experiences and learn from the processes. This is not always a comfortable place to be, but it is where the Creator put me.

In my search for meaning and understanding, I have encountered numerous experiences that have shaped my way of thinking and the way that I see the world. My connections with Anishinaabe knowledge and ceremonies have helped to create a perspective that acknowledges the complexities of our reality, our relationship to each other, and our relationship to the rest of the natural world. Within this, I realize that the layers of experience are often far beyond our existing memory and understanding.

In my own decolonization process, I have had to take a serious look at the formal and informal non-Indigenous education that has contributed to my knowledge base and ways of seeing the world. Through this, I embarked on a journey that sought out spiritual knowledge. I thought I was doing fine until a friend said, "You have to remember that we are not human beings having a spiritual experience, we are spiritual beings having a human experience." This statement instantly transformed my way of thinking, and I began to understand the depth of my being.

In a very significant way, my spirit awakened and began to listen to the world that I lived in. Most important, I began to listen to my spirit. I searched for meaning in my human existence. Through this, I recognized that my blood and cells inherited all the joy, the happiness, the pain, and the fear of my ancestors. I began to understand that my spirit and my body had histories of their own and that I, through thought, was to make sense of my current existence. There became a new meaning to harmonizing the body, mind, and spirit. And my life was no longer just about me.

I started to realize the impact that colonization had had on my family, community, nation, and myself. It was overwhelming to think that the entire colonial experience was inside of us. However, it was a relief finally to understand why we struggle so fiercely with our dysfunctions. Historical trauma response has been identified and is delineated as a constellation of features in reaction to multigenerational, collective, historical, and cumulative psychic wounding over time, both over the life span and across generations (Duran, Duran, Yellow Horse Brave Heart and Yellow Horse-Davis, 1998: 342). This concept must be brought into healing movements so that Indigenous people can begin to recover from colonization.

There is a tremendous amount of unresolved grief in Indigenous communities, and by "community" I refer to any collection of Indigenous people—be it urban, rural, or on Indian reserves. In discussing intergenerational grief, Duran and Duran (1995) wrote that "Some medicine people have equated the treatment process as one in which we not only treat the client but are also treating our ancestors, since it is only in this plane of existence that we get to accomplish resolution of life events. If we do not work out a resolution for our ancestors, we are then only ensuring that our children will be left to continue struggling with the problem" (154). We have been given this knowledge to help us move beyond the legacy of colonial destruction. The decolonization process calls upon us to look within ourselves and bring forward all the experience, knowledge, and wisdom of our ancestors. This includes traditional healing practices and ways of relating to each other.

In the fall of 2000, I was invited to participate in a three-day women's gathering for residential school survivors and their descendants. A powerful instinct and sense of urgency took over and I felt that I needed to attend the gathering. Instantly I felt the energy flow through my body, and I became very excited about the prospects of being part of this healing movement. As I sat and listened on the first day, I held back the tears and tried to block out the words that were being spoken. I wondered what was happening and why I was being so emotional. I even resorted to counting so that I could refocus my energy. Before I knew it, I had walked into the circle and was handed the eagle feather. I couldn't figure out what I was doing or what I thought I had to say. I listened as the words rolled off my tongue. I talked about my mom and acknowledged how she entered the spirit world before she had an opportunity to tell her story of residential school that began at the age of four. A car accident took her life fifteen years ago. She was driving, and the postmortem showed high levels of alcohol, not that it matters anymore. I had never even thought for a moment that she kept her pain in silence. It has only been in the last decade that people have begun to share their stories of residential schools and the changes in the communities. So I hoped that through the gathering I could heal for her too so that her spirit could rest easier.

Indigenous education can draw on many creative ways of learning that will help heal the wounds that are a result of colonization. Art forms such as writing, theater, and dance are the voices of the past as they speak through us and share their experiences. The memories inside of us need to be explored so that we can move on and begin to understand the impact that colonization has had on our families, communities, and nations.

THE SPIRIT OF CHAOS

In 1992 the initial "Dialogues between Western and Aboriginal Scientists" was held in Kalamazoo, Michigan. This group, which consisted of several Indigenous thinkers and western physicists, began meeting to discuss various concepts relating to chaos theory (Ross, 1996). Chaos, as I understand it from an Anishinaabe perspective, is constant motion—neutral until some form adapts it. It is said that out of chaos change will come. This change includes shifts in consciousness. The impact of colonization has created destructive chaos among Indigenous peoples, and the spirit is still deep within the communities today. This environment poses extreme challenges for adult learning because of the constant pressures that distract and dampen the desire to learn.

Indigenous peoples often experience a high level of collective stress in their daily lives. The constant energy of poverty, violence, sadness, family breakdown, abuse, death, assaults, accidents, suicide, chronic illnesses, unemployment, and intergenerational trauma paralyzes a community in a state where crisis is more likely than progress. This collective stress has been around for generations and continues to collect momentum. In fact, according to chaos theory, the only way for this movement to be reversed is by redirecting the energy. The healing initiatives and movements will definitely have an impact on the social chaos that resides in the communities. However, many causes contribute and perpetuate the continuance of such dysfunction.

It is particularly important to understand some of the external factors that have had devastating effects on Indigenous communities. One of the most significant is the change in traditional food source. This was compromised by the loss of land, traditional hunting grounds, traditional economies, and ways of life. The third International Conference on Diabetes and Indigenous Peoples: Theory, Reality, Hope (1995) reported high levels of diseases resulting from dietary and lifestyle changes. These diseases include diabetes, dental caries and tooth loss, alcoholism and fetal alcohol syndrome, cardiovascular disease, gall bladder disease, certain cancers, decreased fitness, obesity, anemia, and failure-to-thrive syndrome in infants. Inadequate healthcare, poverty, and lack of education and resources exacerbate the health issues and concerns in many Indigenous communities.

One of the leading social concerns is alcohol and substance abuse. Traditionalists believe that the ingestion of such substances leaves the person open and vulnerable to negative spirits, resulting in an unclean mind and body. It also has been said that the good spirits will not come around if alcohol is present. Some say that when you try to leave alcohol behind, it will come looking for you. It will even sneak into dreams of the tempted. Michael Raymond recalls, "My grandmother thought of alcohol as a bad spirit. She treated it like it was a spirit. She talked to it like it was a person standing close to her" (cited in Duran and Duran, 1995: 93). Indigenous people often call on the spirit of alcohol to replace the pain and discomfort of colonization.

Many Indigenous communities are overwhelmed by social chaos. Alcohol has the power to medicate those who cannot begin to face reality. It has become a tool to escape cumulative pain. In such situations, there is little concern for the intergenerational causes and virtually no desire to cleanse the body and spirit. The intensity of colonial damage exhausts the physical being, and the body becomes infested with sickness. A high percentage of deaths can be attributed to alcohol-related illnesses and accidents. This fact contributes to the collective grief and stress that Indigenous communities live with.

The spirit of alcohol has changed the natural functions of the body. Fetal alcohol syndrome continues to disable community life. In an article entitled "Bridging the Cultural Gap: Non-Aboriginal Educators and Aboriginal Adult Learners," Margaret Fisher Brillinger and Debra Cantrell (1993) point out that it is suspected that a high percentage of Aboriginal learners are affected to some degree with fetal alcohol syndrome. This can result in behavioral problems, physical abnormalities, poor health and judgment, and memory impairment. Recent dialogue has been initiated to raise awareness regarding second- and third-generational effects of fetal alcohol syndrome.

The contamination of Indigenous communities with various social dysfunctions has had severe effects on the general well-being of the people. Many types of disorders contribute to the instability of daily life. Such disorders can be spiritual, mental, emotional, and physical. Traditional perspectives on health and balance would conclude that if even one of these was out of balance, it would affect the other and the person's entire health would be jeopardized. The public medical care system and traditional Indigenous methods differ greatly in diagnosis and treatment. Where the public system would be more likely to medicate with a prescription, a traditional healer would conduct a ceremony using sacred medicines and calling on the help of spirits.

The use of traditional medicines and ceremonies has been of great benefit to the Indigenous community. In many senses, it has revived interest in cultural traditions and been very successful in the treatment of colonial damage. As Waldram, Herring and Young (1995) write: "Both the ability to heal and protect oneself from disease or illness were predicated upon the assistance of these other-than-human beings. However, those individuals with stronger than normal power could actually use that power to cause illness. Most Aboriginal healing traditions recognized that the power to heal also entailed the power to cause harm, illness, and misfortune to befall a particular victim" (101). This power to cause harm is often referred to as a "curse or bad medicine" and it can leave victims in a very ill state. Unless they seek out traditional methods, they will not be healed. It is important to acknowledge that traditional belief holds that all illnesses are an intrusion of the body.

The social chaos that exists within Indigenous communities continues to create barriers and generate much discomfort for the people. It extends beyond the social realities and into the cultural, political, and economic functions of the community. The decolonization process will give us the opportunity to explore the concept of crises that are a result of colonial intrusion. The challenge will be to create adult education programs that address the community's issues and encourage personal and collective transformation.

ACCEPTANCE, INTEGRATION, AND TRANSFORMATION

Indigenous movements in education will need to develop new and creative ways to work through existing pain and collective consciousness. Indigenous knowledge systems contain many valuable tools that can be brought forward and utilized in the decolonization period. It is important to look critically at the impact that colonization has had on the lives of Indigenous communities and begin to move forward to heal the wounds of the people. Through this process, Indigenous peoples will be able to transform their historical circumstances and become active participants in their own health and well-being as well as make significant contributions to global existence.

The initial step is to accept ourselves and our lives by honoring our physical and spiritual beings. It is essential to connect in a very profound way and begin to appreciate the opportunity that we have been given to heal on behalf of our ancestors. It is also essential to integrate our experiences into our whole being and allow for a union to come about. This union will be representative of many experiences, intergenerational relationships, and inherent connections with creation. Most important, this union provides an opportunity to connect and begin the healing process, both individually and collectively. As Peter Reason (1994) says: "To heal means to make whole: we can only understand our world as a whole if we are part of it: as soon as we attempt to stand outside: we divide and separate. In contrast, making whole necessarily implies participation: one characteristic of a participative worldview is that the individual person is restored to the circle of community and the human community to the context of the wider natural world. To make whole also means to make holy: another characteristic of participatory worldview is that the meaning and mystery are restored to human experience, so that the world is once again experienced as a sacred place" (10).

The legacy and pain of colonization is deeply embedded in Indigenous people, and many have passed on and will pass on, leaving the work to those who continue in this world. It is not for us to judge or criticize the struggle of any other person. Nor is it appropriate to try to convince anyone of a specific role or responsibility. The Creator has given us each a responsibility that will assist in the evolution of our spirits.

There are many ways that we can give voice to our ancestors. In doing so, we are able to retain the knowledge and share the wisdom with those around us. This type of integration allows us to utilize the experience and apply it in ways to our lives so that we can complete our life work. Lenore Stiffarm (1998) has developed a concept called "Spirit Writing," which is a healing method that utilizes automatic writing to cleanse and heal our spirits.

Transformation begins by accepting our circumstances and integrating the lessons that have come our way. Through the decolonization process we will need to take a critical look at non-Indigenous influences and begin to reconstruct our knowledge base with traditional thought processes and ways of thinking. With this, Indigenous people will reconnect with themselves, their families, their communities, and their nations. Fyre Jean Graveline (1998) has made a significant contribution to the notions of transforming European consciousness. She has successfully presented an Indigenous model that can be utilized in many forms and contexts of adult learning and education.

And finally, there is a need to acknowledge and appreciate the painful struggles of many people. Beck, Walters and Francisco (1996) looked at various journeys of spiritually gifted Indigenous peoples and found that "These individuals are born with or develop special sensitivity and interest in the elements that make up the sacred. Often, these persons are exposed to greater hardships than most people: personal injury, fright, anxiety and loneliness. If they succeed in their journey or quest for knowledge and in their work as sacred practitioners, they then have greater responsibility than most people and for this reason they are respected" (96). The vast struggles of Indigenous peoples must be acknowledged and incorporated into daily life. People are healing all the time. Healing can take place anywhere or anytime. It can be at a community gathering, powwow, healing circle, or ceremony; in a classroom, a treatment center, while reading, writing, or dreaming; and especially through visits with friends and family.

Indigenous nations have survived hundreds of years of the destructive forces of colonization. Colonial rule has affected virtually every aspect of their lives. To move forward, it is essential that we accept our circumstances; integrate our body, spirit, and intergenerational knowledge; and allow for personal and collective transformation. It is time to transform our collective consciousness by developing creative ways to heal, learn, and support each other. It is through this process that we will unite our families, communities, and nations with the rest of the natural world.

Conclusion

There are many challenges for adult Indigenous learners. It is time to bring forward the historical memories into contemporary reality, if only to come to peace with the devastating grief and hardship that Indigenous peoples have carried for centuries. It is essential that we heal the memories of our existence by giving the past a voice so the ancestors can speak, heal, and be assured that their wisdom and experience continues to guide their descendants.

Education must be a leading force in the movement to decolonize. Indigenous peoples have a vast collection of resources within their communities that will assist in this transformative process. It is essential to look critically at the impact that colonization has had on the lives of Indigenous people. Moreover, it is important to understand the influence that foreign systems have had on traditional ways of thinking and living. With this in mind, Indigenous peoples will be able to create adult education programs that will address community issues and assist in decolonizing their nations.

Finally, it is important not only to acknowledge the survival of Indigenous peoples but to send prayers to all those who keep with the struggle and all those who return back to the spirit world. Let the spirits sing the travel song so that all will have a safe journey.

Notes

1. This translates as the singular form of the word "Ojibway." The plural form is "Anishinaabeg," meaning Ojibway peoples.

References

Beck P. V., Walters, A. L., and Francisco, N. (1996). *The sacred: Ways of knowledge sources of life.* Tsaie, Ariz.: Navajo Community College Press.

Brillinger, Fisher Margaret, and Cantrell, Debra. (1993). Bridging the Cultural Gap: Non-Aboriginal Educators and Aboriginal Adult Learners. *Proceedings of the 12th Annual Conference of the Canadian Association for the Study of Adult Education* (pp. 143–48). Ottawa: University of Ottawa, Faculty of Education.

Duran, E., Duran, B., Yellow Horse Brave Heart, M. and Yellow Horse-Davis, S. (1998). Healing the American Indian soul wound. In Y. Danieli, (Ed.). *International handbook of multigenerationa legacies ofl trauma.* (pp. 341–54). New York: Plenum Press.

Duran, E., and Duran, B. (1995). *Native American postcolonial psychology.* Albany: State University of New York Press.

Graveline, F. J. (1998). *Circle works: Transforming Eurocentric consciousness.* Halifax, N.S.: Fernwood Publishing.

Proceedings of the 3rd International Conference on Diabetes and Indigenous Peoples: Theory, Reality, Hope. (1995). Winnipeg, Manitoba, May 26–30, 28–29.

Reason, P. (Ed.). (1994). *Participation in human inquiry.* London: Sage Publications.

Ross, R. (1996). *Returning to the teachings: Exploring Aboriginal justice.* Toronto: Penguin Books Canada.

Stiffarm, L. A. (1998). *As we see: Aboriginal pedagogy.* Saskatoon: University Extension Press, University of Saskatchewan.

Waldram, J. B., Herring, D. A., and Young, T. K. (1995). *Aboriginal health in Canada: Historical, cultural, and epidemiological perspectives.* Toronto: University of Toronto Press.

CHAPTER 13

Toward Transformative Learning

Ecological Perspectives for Adult Education

Darlene E. Clover

Introduction

> *The argument over what we may and may not teach in order to reconcile socio-economic development with environmental protection and rehabilitation (in short, quality of life), and to accommodate scientific exploration and uncertainty leads to the need to identify the broader context in which we teach.*
>
> —Peter Sutton, "Environmental Education"

Edmund O'Sullivan (1999) notes that newer currents of adult education take a "critical standpoint on the global market place, gender and class inequity, and postcolonial perspectives that question the dominance of western cultural hegemony" (63). However, he also rightly suggests that in general, "one of the most prominent omissions is the lack of attention to ecological issues . . . [and] the pre-eminent emphasis on inter-human problems frequently to the detriment of the relations of humans to the wider biotic community and the natural world" (63–64).

Emerging in response to escalating contemporary environmental problems, as O'Sullivan himself so effectively does, are new ideas, new learning frameworks, that attempt to broaden the vision, scope, and practice of adult education and learning. Environmental adult education is a reconceptualizing of the theory and practice of adult education within a holistic ecological framework. A primary focus of this new stream is human-Earth relations: What they are; what they mean; how they are developed, nurtured or destroyed; what knowledge counts (and is necessary) and its place in "moral" discourse; and our understandings of how they shape the cultural, social, economic, and political aspects of our lives.

In this chapter I offer a brief sketch of some of the main features of the complex ecological perspectives being woven into the fabric of adult education. These perspectives—which include ecological knowledge and cultural identity; consumption, production, and cultural homogenization; environmental racism and classism; ecofeminist standpoints; the intrinsic value of the rest of nature; and sustainability—place "concerns for the planet . . . at the forefront" and encourage transformative learning by broadening our theory and practice as adult educators (O'Sullivan, 1999: 70).

ECOLOGICAL KNOWLEDGE

A major epistemological focus of adult education is "people's knowledge." Knowing is understood to be "an ongoing creative and dynamic act which is fundamental to the process of living" (Guevara, 1995: 10). Cevero and Wilson (2001) suggest that "adult education's particular role in society can be seen as struggle for knowledge which is interwoven with the struggle for power" (12). Learning and knowledge creation begin from where people are at, and new constructions of knowledge are used in struggles to maintain or extend control over people's lives. Given this mandate, it would seem appropriate that people's "ecological knowledge," that knowledge which is relational to the natural world and people's struggles for a sustainable and equitable future, should be "one of the engines that drive[s] and define[s] adult education" (Cevero and Wilson, 2001: 12).

Ecological knowledge is nurtured by, with, and through the "land," the life-world. It comes from age-old traditions as well as daily lived experiences in a changing world; it is a web of old and new knowledge. This knowledge that was "built over decades through a cumulative process among generations [was what] enabled communities to be dynamic and functional" (Minor, 1994: 5). The ecological wisdom of Indigenous peoples around the laws and rhythms of the rest of nature, the "characteristics and uses of many species of vegetation," was what enabled them to survive (LEAP/INFORSE, 1997b: 5). In India, farmers have begun to reapply traditional knowledge such as terracing and intercropping to challenge the destructiveness of the so-called Green Revolution that resulted in heavy dependence on pesticides and herbicides. These practices have lead to farmers making demands for land reform and sustainable agricultural practices. In addition, fisherfolk in the Philippines use their knowledge of the sea to explore the issue of resource "management" (LEAP/INFORSE, 1997a).

Many women around the world have a repository of knowledge acquired through daily lived experience that gives them an expertise essential to planetary survival. This knowledge ranges from understanding the use of plants for traditional medicinal use, to seed collection, to the maintenance of biodiversity and water management, to name but a few (LEAP/INFORSE, 1997a).

Even when people have been severed from their homelands and moved to new ecosystems, they often demonstrate "immensely valuable knowledge and mechanisms for coping" (LEAP/INFORSE, 1997b: 5). Immigrant communities develop their own approaches to nurturing an ecological consciousness of place. By taking into account the daily lived experiences of racism, linguistic barriers, unemployment, and poverty, they develop a "philosophy that ecological healing comes only through the revitalization of the vision, strength, and moral values of a community" (Tan, 2002).

While the mainstream focus in public education is "awareness raising," environmental adult education begins from the premise that people know. They understand that environmental problems exist, and experience has taught them to value clean air and water. They recognize the spiritual and psychological value of green space, as can be seen from women's struggles to maintain or protect public lands from Nairobi, Kenya, to Salt Spring Island, Canada (Clover, 2001a).

Ecological knowledge is a lived process of knowing that has extraordinary ramifications for individuals and communities in terms of what is important, what struggles people engage in, and how the Earth is manifest in their values, sense of self-identity, and culture. As adult educators, we cannot extract this knowledge from the knower or the environment as content separate from the lived process of knowing. We cannot engage in a process of denying recognition of this ecological knowledge that makes a valid contribution to how we live and act in this world. However, when working within a framework of ecological knowledge, we also must recognize that in this urbanizing and changing world, it is rapidly being eroded. An aggressive application of outside knowledge by dominant cultures is destroying the conditions that permit ecological knowledge to flourish.

Alienation and Contemporary Change

> *The so-called age of Enlightenment . . . forced all other learning and knowledge into darkness . . . [rendering] invisible other ways of knowing such as native or traditional knowledge, peoples' spirituality, and especially women's knowledge.*
>
> —Vandana Shiva, cited in Darlene Clover, "Learning for Environmental Action"

Change, for better or worse, occurs constantly. The world is not a static place, and therefore, we never "know everything at any one point in time." To be most useful, a pedagogical practice must be formulated in the context of its historical location, which recognizes that the rest of nature and "societies' awareness and understanding of [it] too changes and evolves" (NIACE, 1993: 220).

Educator Teresia tells a story about a village in Indonesia. Although things have changed dramatically as a result of pollution, development, and commercial overfishing for profit, people continue to believe that "the sea has always provided from the days of our forefathers [sic] and will always provide" (cited in ASPBAE, 1995: 27). People tend to cling to "the myth that natural resources do not diminish" (Al Agib, 1992: 76). Traditional knowledge can be undermined by contemporary changes so that what was once true is rendered false.

Although humans have always affected their environments, today's destruction "is on a scale and complexity unprecedented in history." Any "contemporary pedagogy of responsibility must be commensurate" with these contemporary realities (Mische, 1992: 10). For example, ecological adult literacy practices in Africa engage learners in old and new methods of farming the land, a holistic activity that incorporates "a multitude of [social and ecological] issues and develops skills and knowledge of life and land" (LEAP, 1998: 1).

It is argued that the human separation from the rest of nature is at the root of many of our emotional, psychological, and socioecological problems. For many human beings, living close to the land and actively participating with it has become quite alien. The rapid rise of scientific discovery based on the ideas of Bacon, Descartes, Kant, Linnaeus, Newton, and Darwin and the emergence of capitalism have fundamentally changed how humans interact with the rest of nature. The mechanistic reductionist scientific worldview disconnected tuberculosis from slum housing and cancer from industrial pollution (Rowe, 1990). The rest of nature was classified as a nonliving thing or "resource" to be conquered, managed, captured, tamed, and owned (Bhasin, 1992). People were alienated from the process of production and the land. Numerous small farms engaged in multicropping have given way to large-scale agribusiness engaged in "monocropping because it is easier to manage big plots [and] more efficient for profit" (Bhasin, 1992: 29). Today the majority of people worldwide live in urban areas, either in "unsustainable suburbs" or in housing projects and shantytowns. Although the rest of nature is there, people do not interact, not because they wish to avoid it but because "the noxious pollution and potential to violence . . . makes such experiences difficult" (Kahn and Friedman, 1998: 25). While we need to recognize the importance of ecological knowledge, we cannot afford to privilege this experience at the expense of developing a critical dialogue that integrates the meanings produced in the nature of changed experience which can ultimately destroy livelihoods, spirituality, culture, and identity.

Ecological-Cultural Identity

> *the Pueblo man speaks of the world we bring with us*
> *he speaks of the relation with one another through ecology*
> *of having an ecologically specific identity*
>
> —Peter Cole, "An Academic Take on 'Indigenous Traditions and Ecology'"

Embedded in identity distinctions among peoples is cultural identity: the songs, art, music, values, customs, and so on of a people or nation. Interaction and interdependence between human culture and the rest of nature has evolved over thousands of years.

For many, the land is culture and culture is embedded in the land. This relationship is at the root of activities to preserve its value as a source of material existence and spiritual life. Cultural diversity in Uganda is intimately intertwined "with traditional ecological knowledge" (Minor, 1994: 4). In Sri Lanka "a clean and beautiful environment is believed to be both right and a cultural artifact" (LEAP/Ecologic, 1994: 27). In the Himalayas, the forests play a role not only in the physical well-being of people's lives but also in their emotional and spiritual lives, particularly in the lives of women who use them as basis for songs of struggle and change.

Much human creativity has been inspired by, with, through, and about the rest of nature. Unlike western societies, Aboriginal peoples did not draw a hard line between artists and the other members of the community. Impressionist painters often focused on the rest of nature, as did poets and myth-tellers. In Newfoundland, Canada, music links lives to the sea. In other countries, dance is grounded in the mysteries and ordinariness of the everyday and the complex symbiotic relationship between humans and culture. Vio Grossi (1995) observes that, in fact, nature's culture is so much a part of human culture and identity that to disconnect them is to disconnect "from a sensitivity to things of a different quality and imagination" (42). Ecological cultural identity is based on the feelings and relationships people develop with the place they live, the patterns and changes in the landscape, and interactions with other forms of life. It is spirituality, the self, collectivity, self-esteem, values, customs, self-respect, and self-direction. It is both a cognitive and an embodied way of knowing.

Environmental adult education explores these cultural identities as the essence of the sacred, "the last possible relationship with the mystery which has always expanded the boundaries of humanity through the pathways of the uncommon, the imaginary, and desire" (Viezzer, 1992: 3)

The Problematic of Cultural Homogenization

Interaction and borrowing always have taken place within and between cultures. However, a global application of dominant culture is proving to be destructive. The Philippines shows signs of "western culture everywhere . . . stomping on culture. Families who had farmed and fished sustainably, simply and collectively for generations were forgotten . . . rendered second class on their own land" (Keough, Carmona and Grandinetti, 1995: 9). The process of cultural homogenization identifies "Third World, tribal or women's culture, industry, agriculture, medicine, and science [as] not efficient, not scientific, not good. The culture of the powerful is superior" (30).

More than any other, "U.S. culture has been taking over the world. Coca Cola, hamburgers, Barbie dolls, and Superman have become international. For profits U.S. companies have pushed junk food, junk drinks, junk culture everywhere. . . . It is not a cultural exchange, but a one-sided cultural dominance" (Bhasin, 1992: 28).

The media plays an exemplary role in creating "the minds to buy this junk. From a village in Dhaka to a school in Washington the children are watching the same thing. Do the children in the U.S. know the dolls Indian children play with or the stories they hear?" (Bhasin, 1992: 30). Around the world, this destructive force is working to erode values, lifestyles, and even imagination and memory and cannot go unnamed (Follen and Clover, 1997: 505).

CHALLENGING PATTERNS OF CONSUMPTION AND PRODUCTION

Contemporary consumerism "has become a powerful and evocative symbol of contemporary capitalism and globalization. Indeed, in light of the environmental crisis and social transformations worldwide, it is increasingly visible. Economic growth, lifestyles and modern identities, are in one way or another, tied inextricably to consumerism" (Clover, 2001b:73).

Consumerism taunts society with scarcity, appeals to competitiveness, and mimics the tensions of seasonal rarity (Griffiths, 1997). It "consumes our music, our colors . . . our resources and our relationships" (16). It is a discourse through which power is both exercised and contested; it is a gendered, classed, and racial practice. Consumerism is a cornerstone of the capitalist system; it is about production, production, and more production to seduction. It is a deeply ingrained ideological and structural problem given strength through advertising and marketing, a pervasive informal learning process that creates need and orchestrates ignorance.

People "are influenced to establish their identities through the ways in which they consume and multiple meanings attached to this activity" (Clover, 2000: 1). In

India, consumerism has become a "competition . . . amongst different segments of middle classes to acquire the latest electronic gadgets. We are constantly told by advertisements: 'Don't just envy it, buy it,' or 'Neighbour's envy, owner's pride'" (Chaudhry, 1989: 74). Around the world, one hears the seductive call of the consumer dream: "Step right up! the more you buy, the more you save! Buy now, pay later! It's the real thing!" (Keough, Carmona, and Grandinetti, 1995: 11). Perhaps "one of the deepest and most pervasive educative processes at work in the twentieth century has been learning to consume" (Welton, 1987: 55).

Mikkelson (1992: 72) argues that people in the West in particular must recognize their responsibility in the "obsessive growth" of material production and consumption. As Bhasin (1992: 30) argues, this "excessive and wasteful consumption by some deprives the majority of essentials and destroys ecology." New frameworks of adult education must raise the issue of consumption, but it must be understood as a political and collective problem and not simply an individual or behaviorial one (Clover, 2000). Seeing consumerism as an individual problem ignores ideological and structural practices at its root. It also ignores the gender politics of consumption (Seager, 1993). Advertisers and producers always have been aware that women are the primary shoppers and caretakers of home and family. This makes them extremely "vulnerable to consumer manipulation and subject to blatant coercive advertising techniques" (Seager, 1993: 257). It is critical that we do not fall into a trap of "blaming" women for their buying habits but that we discuss consumption within the proper contexts of poverty, sexism in advertising, media manipulation, and lack of choice.

ENVIRONMENTAL RACISM AND CLASSISM

she speaks of environmental racism
the omni-presence of toxic dumpsites on Nature, land

—Peter Cole, "An Academic Take on 'Indigenous Traditions and Ecology'"

Ecological degradation and human inequities often are rooted in the same underlying structural inequalities within societies. Environmental racism is "the exclusion of people of color from policy-making, the deliberate targeting of these communities for toxic waste disposal and the official sanctioning of life-threatening poisons and pollutants" (LEAP, 1996: 1). Its politics and practice are worldwide and very real.

"When minerals like gold, diamonds, platinum and coal are extracted but the surrounding communities are among the poorest in the country; when black townships [of South Africa] experience the leakage of water pipes and sewers without any hope of the local government attending to repairs; when black township streets lack sheltering trees for shade and ornamental flowers for beautification; when blacks-only areas are targeted as industrial sites . . . and when a town council decides to dump poisonous waste products from manufacturing in a blacks-only township, racism is clear" (Lengwati, 1995: 103). Often immigrants and people of color are marginalized from "mainstream" environmental activities. Tan (2002) tells the story of when "a team of African-Canadian youths engaging in a tree planting project in the Don River Valley on a hot summer day in Toronto was mistaken as labouring convicts by the largely white recreational visitors to the park."

Networks and organizations by people of color work to "deepen everyone's understanding of environmental racism and how it manifests itself so that communities can continue to direct action" (Lengwati, 1995: 103). Their efforts to organize against "the dumping of toxic waste is a crucial additive to the strengthening" social theory and adult learning practices (Lengwati, 1995: 103).

In addition to environmental racism, there also exists "environmental classism," and often the two are interwoven. Guevara (1995) has discovered that the urban poor of Manila who live off trash feel that they themselves are viewed as part of the "environmental" problem. The right to healthy food, shelter, environment, and lifestyle is most often believed to be the preserve or entitlement of the middle and/or wealthier classes (NIACE, 1993). For the poorer classes, a discussion on the resolution of environmental problems often is the equivalent of the "eradication of their sector" and an exercise in blame (NIACE, 1993: 30).

Meanwhile, the middle class, and in particular women, is heavily targeted through advertising. They are constantly encouraged "to acquire all the latest electronic gadgets" (Chaudbury, 1989: 74). It is, in fact, the educated class that has been "schooled and conditioned in patterns of unquestioning consumption, waste, and ignorance of their integral relationship and dependency on the Earth and each other" (Mische, 1992: 9).

A feminist analysis can be extremely useful in exploring the meaning of persistent and pervasive marginalization in contemporary society: "the subtle class biases of much environmental advocacy: the ways in which it reflects the perspective of an (over)-privileged minority while ignoring or devaluing the situation of the economically marginalized and the economically insecure" (Mische, 1992: 33). Race and class are fundamental inequities that must form part of the dialogue around human-Earth relations. In addition to these, there is the issue of gender.

ECOFEMINISM

> *New patterns of thought and belief are emerging that will transform our experience, our thinking and our action. . . . This emergent worldview is multifaceted: it has been particularly described as systemic, holistic, more feminine.*
>
> —Peter Reason, "A Participatory World"

Feminist environmental adult educator Moema Viezzer (1992) argues that "One cannot say that it is more natural for women to be more interested in the environment than men, but it can perhaps be understood through history that if in a social relationship where women have always been treated as subordinate to men, and have been totally dominated by them, a parallel setting would have been created for men to dominate and destroy nature as well" (2).

She believes that it will never be possible to solve contemporary environmental problems if oppressive historically ascribed gender roles are not challenged and changed. Transformative learning through environmental adult education means taking "a new approach to gender relations and relations with nature [by beginning] with a critical evaluation of the deterioration of the well being of both through patriarchal science and technology" (Viezzer, 1992: 2).

Ecofeminism, a term coined in 1974 by French feminist Francoise D'Eaubonne, is "a new word for an ancient wisdom" (Mies and Shiva, 1993: 13). It is an umbrella concept for the forthright attempts to weave feminism and environmentalism to articulate a feminist ecological ethic that goes beyond the concept of social justice alone, "re-weaving new stories that acknowledge and value the biological and cultural diversity that sustains all life" (Diamond and Orenstien, 1990: xi).

Ecofeminism offers something in both the political and the spiritual realm. In terms of the political, it provides a basis to understand historical, social, and philosophical connections between the domination of women and nature; it also "critiques . . . the ways in which social and ecological worlds are gendered," articulating alternative perspectives of the life-world (Cuomo, 1998: 19). Ecofeminism offers a broader analysis of power that makes visible the exploitation of the rest of nature and its impact on the planet which supports all life (Mies and Shiva, 1993). It challenges the ideology of "resource" control that promotes human separation from its life-support system, allowing its misuse and abuse as well as creating a climate of competition among people and animals for space and among people for resources. It also challenges the discourse of "resource conservation," asking questions such as "conservation for whom?" since acts of conservation seldom benefit the well-being of local people but rather enhance tourism (Gaard, 1993). Vandava Shiva (1989) has also coined the term "maldevelopment," arguing that it is a: "violation of the integrity of organic, interconnected, interdependent systems [which is] blind to the fact that a recognition of nature's harmony and action to maintain it are preconditions for distributive justice" (84).

Ecofeminism is also a "movement involved in developing nonpatriarchial, earth- and women-based spiritualities" (Warren, 1988: 64). There is also a strong sense that the "crisis that threatens the destruction of the Earth is not only social, political economic and technological but is at root spiritual" (Christ, 1990: 58). The framework of spirituality is "the rediscovery of the sacredness of life." This is in "everyday life, in our work, the things that surround us, in our immanence. And from time to time there should be celebrations of this sacredness in rituals, in dance and in song" (Mies and Shiva, 1993: 17–18). For many women in the South who are fighting to conserve the natural resources that they understand to be the very basis of survival, "the spiritual element is the icing-on-the cake." For them, spirituality is "rooted in the everyday subsistence production of most of the world's women" (19).

Ecofeminism urges us to develop a democratic politics and emancipatory practice of lifelong learning around life-centeredness. It urges us to work to develop "new paradigms of social relationships that give us a sustainable economy for human beings and for nature [based on] a new relationship between men and women" (Viezzer, 1992: 3).

Nature's Intrinsic Value

We cannot, as humans, precisely experience the living sensations of another form. We do not know, with full clarity, their desires or motivations; we cannot know, or can never be sure that we know, what they know.

—David Abram, *The Spell of the Sensuous*

At present, the most prevalent value orientation of the world is anthropocentric. Moreover, the implicit value system in much adult education "does not accord value to nature or to living systems in their own right" (NIACE, 1993: 19). But the rest of nature "has value in itself. It is not simply a means to our ends, but an end-in-itself. And since it has intrinsic value, it deserves our respect and that we treat it well" (Marshall, 2001: 25).

These ideas of intrinsic value, respect and fair treatment are a call, as Sutton (1989) argues, for "an ecological ethic of survival [that] demands a fundamental change in human attitudes towards ourselves and nature—perhaps a Second Copernican Revolution—which applies to ecology the insight that neither Earth nor man [*sic*] is the centre of the cosmos" (10).

It is imperative that contemporary education work toward new relationships based on values of caring and respect but also "beauty, diversity and [an] interest in life forms and natural systems" (NIACE, 1993: 19). Developing an effective partnership with the rest of nature "ultimately depends upon the widespread adoption of an environmental ethic or code of conduct reflecting" biological diversity and life-centered development (Strathy, 1995: 72). It is based on learning about "values in nature [and] how to link these with social justice" (LEAP/Ecologic, 1994: 11).

Environmental problems today demonstrate a "clear role for moral education, the teaching of good environmental development practices and of natural sciences imbued with humility" (Parker and Towner, 1993: 208). Raising value questions in educational discussions can be "a good way to build participation and relevance into our courses and usually produces the kind of lively debate that is characteristic of the best kinds of interactive adult education" (216). Although environmental adult education must be a space for "legitimate debate and discussion amongst groups and people that openly espouse and promote particular views on environmental values" (NIACE, 1993: 20), it cannot be neutral. Rather, it must foster values like "justice, ethics, morality, beauty and love [as well as] reverence for all life, simple living, living in harmony with nature, and respect for diversity" (Benavides, 1992: 35).

SUSTAINABILITY

Peter Sutton (1989) argues that the complex issue of development leads us to ponder "If we educators wish to teach [people] to be aware of environmental concerns, what precisely can we tell them? We cannot divide the world into parts—one part developed and one part underdeveloped. We cannot say that the high living standard in the industrialised West cannot be attained by the rest of the world" (7).

If, as Sutton notes, we "limit ourselves to encouraging a change in attitudes and a questioning of the merits of global development," then we run the risk of "being dismissed as unrealistic dreamers who would change human nature, while the rest of the world gets on with earning a living" (1989: 7). Perhaps the question is not, as Sutton asks, "how can we *tell* people [emphasis mine]" but rather how can we create opportunities for people to imagine and work toward life-centered forms of development?

The Problematic of Sustainable Development

> *While it is possible to conceive conceptually a sustainable development, it is not the form most generally advocated by world governments.*
>
> —Roger T. Cross, "Teachers' Views on Sustainable Development"

"The term sustainable development has its origins in the socio-political rationalisation of the increasing alarm over the state of the environment" (Cross, 1998: 43). Daly (1997) argues that following the publication of *The Brundtland Report* (World Commission on Environment and Development, 1987) the term "sustainable development" "rose to the prominence of a mantra—or a shibboleth" (1). The report defined sustainable development as the ability to meet the needs of the present generation without sacrificing the ability of the future to meet its needs. Daly suggests that while "not vacuous by any means, [the] definition [and purpose] were sufficiently vague to allow for a broad consensus. Probably that was a good political strategy at the time—a consensus on a vague concept was better than a disagreement over a sharply defined one" (2).

However, by the 1990s the initial vagueness of the concept was no longer a basis for consensus. Instead, it became the breeding ground for disagreement as it was soon perceived that sustainable development simply reaffirmed the ideal of the "good life" that could be derived from the methods of production of the technological society. It also implied a recognition of the power of modern technology to arrive at solutions and technological fixes for environmental problems.

Sustainable development is neither a radical nor a green concept. In its present manifestation, it does not critique capitalism but rather functions within the "assumption that the potential for environmental disaster is solvable within the present global systems. In popular understanding it appears to provide a way in which the economic development of the countries of the North may continue to occur, provided adjustments are made to take 'green' issues into account. . . . Actions such as recycling and energy conservation are typical trivial example" (Daly, 1997: 42–43).

One could very easily argue that Canada, in fact, practices sustainable development. Do we not replace the 1,000-year-old old-growth forests we cut with tree plantations? Yes, we do, and this practice provides us with new carbon sinks and new paper- and lumber-producing systems that are sustainable and do not jeopardize future generations. Do we not create fish farms, once we have devastated our natural stocks, so that we will have fish for export and home consumption in perpetuity? Yes, we do, so we practice sustainable development.

What people argued for at Rio, and continue to argue, is that development had a purely economic growth focus that needed to be deconstructed. Development had to be based on creating internal or local markets that are socially, culturally, and environmentally responsible and protected "from external and stronger economies" (African Regional NGOs Think Tank, 1992: 88). It had to be about a new partnership with the rest of nature and include gender, class, and racial equity; intrinsic value; "creativity, potential and satisfaction" (Bhasin, 1992: 32).

Shiva (1992) suggests that more emphasis must be placed on an approach to sustainability rather than the development of an overarching term applicable to all people in all societies. She also notes that the greatest problem with Bruntland's definition of sustainability is its continued view of the environment as a "resource" or commodifiable "good" to enhance the growth-oriented capitalist system "rather than something that supports lives and livelihoods and is the primary source of sustenance" (193).

A sustainability framework includes people and ecosystem health, human and species' rights to live, equity, and democracy. It is based on developing "creative alternatives of a reconciliation of socio-economic development with environmental protection and rehabilitation of quality of life" (Duhaylungsog, 1994: 4). It is life-centered and moves away from the notion of the rest of nature "as subordinate to man [*sic*] . . . [or] merely a provider of resources and absorber of wastes" (Bhasin, 1992: 31). Being life-centered pushes back the boundaries of the purely economic and conceptualizes a way forward that is "participatory and bio-regionally focussed" (Duhaylungsog, 1994: 4).

Implications for Praxis and Conclusions

> *[he spoke] of myth of reading nature*
> *of singing to the corn his brothers and sisters*
> *we converse with plants and stars*
> *we paint our faces, we dance*
>
> —Peter Cole, "An Academic Take on 'Indigenous Traditions and Ecology'"

The ideas outlined in this chapter have numerous implications for practice in both personal and political terms, but I outline only a few. First, the ideas provide the opportunity to encourage adult learners to talk about their relationship to the rest of nature in ways that are not purely about meeting physical needs. They can tell stories, write poems, and create drawings about lives that include the larger life-world in more cultural and spiritual ways. In my own work, I now tap into their rich store of ecological knowledge, this sensual relationship, and use them to create new knowledge. Second, it has meant for me the development of an education practice that is lodged in "place" and that uses the rest of nature and the community as teachers and sites of learning. Learning in place is more than studying interrelationships in our environment. Equally important is experiencing these relationships. It is an interchange among the body, the mind, and the senses. By learning in place we begin to see the community and its surrounding environment as a landscape of resistance; a source of regeneration, inspiration, beauty, and sensuality; and a site of critique, organizing, and envisioning with multiple networks of meaning. For as David Orr (1992) observes, "landscape . . . shapes mindscape" (128). Third, it helps me to understand that the expansiveness of environmental issues, it is not just about "trees" and their place in action to bring about change. There are race, gender, and class implications behind and within destructive environmental acts. By including these issues I am able to broaden the discourse of power and control. Those who control the world's resources control the world's power, which is the primary reason why the United States, with the collaboration of many other countries, placed the environment firmly in the "trade" camp at the Earth Summit in Rio, where it could not be tainted by laws of "human rights" or notions of "intrinsic value." Finally, these concepts make adult education more relevant to younger adults, which seems to be a major preoccupation raised at adult education gatherings. For example, young adults' actions around globalization fundamentally include the environment. They "have grown up on a diet of rainforest crunch cereal, eco-cartoons, blue box discipline and fleece jackets made from recycled pop bottles" (Fry and Lousley, 2001: 25). They are predominantly vegetarian—by choice and politics—and predominately female. If we pay attention to what they are saying, we realize that much of their focus is issues such as "consumerism," corporate/media propaganda power to sell the "consumer dream," genetic modification, excessive waste, and the pollution of water, food, and air.

Pat Mische (1992) argues that "a central challenge before us is to develop an ethical pedagogy of ecological responsibility" (9). The task is immense, and no single chapter can do justice to the richness of the ecological perspectives being woven into the fabric of adult education around the world.

Environmental adult education has at its core a fundamental transformation in human-Earth relations by weaving environmental issues into cultural, political, social, and economic discourses. It is about living to learn and learning to live in harmony and cooperation with each other and the rest of nature. Human-Earth relations include people's spiritual and cultural links to the environment, perceiving culture not merely as an instance for mediating the use of nature. Environmental adult education includes understanding the importance and place of our ecological identities and knowledge and the need to "build on people's knowledge and traditions . . . [and] avoid knowledge going into the hands of a few people" (Bhasin, 1992: 34). It challenges cultural homogenization, structures of consumerism, environmental classism and racism and sustainable development while promoting life-centered ecofeminist values, something which "the world needs heavy doses of" (32).

References

Abram, David. (1996). *The spell of the sensuous.* New York: Random House.

African Regional NGOs Think-Tank Consultation Committee. (1992). A common position: UNCED and beyond. *Convergence, 25* (2): 88–92.

Al Agib, Ibrahim. (1992). Environmental education in the Arab countries. *Convergence, 25* (2): 75–76.

ASPBAE. (1995). *Evolving an Asian-South pacific framework for adult and community environmental education.* Darwin, Australia: Environmental Education Programme of ASPBAE.)

Benavides, Marta. (1992). Lessons from 500 Years of a "new world order"—Towards the 21st century: Education for quality of life. *Convergence, 25* (2): 37–45.

Bhasin, Kamla. (1992). Alternative and sustainable development. *Convergence, 25* (2): 26–36.

Cevero, Ronald and Wilson, Arthur and Associates (Eds.). (2001). *Power in practice.* San Francisco: Jossey-Bass.

Chaudhary, Anil. (1989). Environmentally sound alternatives: Setting the context. *Convergence, 25* (4): 73–77.

Christ, Carol. (1990). Rethinking theology and nature. In Irene Diamond and Lee Quinby (Eds.). *Feminism and Foucault: Reflections on resistance* (pp. 58–69). Boston: Northeastern University Press.

Clover, Darlene. (2001a). Aesthetic activism: Women, community arts and sustainability. *Artwork, 49:* 1–7.

Clover, Darlene. (2001b). Youth action and learning for sustainable consumption in Canada. In *Youth sustainable consumption patterns and life styles,* (pp. 73–103). Paris: UNESCO/UNEP.

Clover, Darlene. (2000). *Youth action and learning for sustainable consumption in Canada.* Paper prepared from UNESCO. November.

Clover, Darlene. (1995) Learning for environmental action: Building international consensus. In Beverly Cassara, (Ed.), *Adult education through world collaboration,* (pp. 219–30). Malabar, Florida): Krieger Publishing Company.

Cole, Peter. (1998). An academic take on "Indigenous traditions and ecology." *Canadian Journal of Environmental Education, 3*:100–115.

Cross, Roger T. (1998). Teachers' views on sustainable development. *Environmental Education Research*, 4 (1): 41–52.

Cuomo, Chris J. (1998). *Feminism and ecological communities.* London: Routledge.

Daly, Herman. (1997). *Beyond the limits of growth.* London: Routledge.

Diamond, Irene, and Orenstein, Gloria Fenman. (1990). *Reweaving the world.* San Francisco: Sierra Club Books.

Duhaylungsod, Noel. (1994). Liberatory environmental adult education. *Report of the Meeting of Experts.* Toronto: International Council for Adult Education.

Follen, Shirley, and Clover, Darlene. (1997). Community revitalisation through critical environmental adult education. In *Crossing borders, breaking boundaries: Proceedings of the 27th annual SCUTREA conference* (pp. 505–510).

Fry, Kimberley, and Lousley, Cheryl. (2001). Girls just want to have fun—with politics. *Alternatives, 27* (2): 24–28.

Gaard, Greta. (Ed.). (1993). *Ecofeminism: Women, animals, nature.* Philadelphia: Temple University Press.

Griffiths, Jay. (1997). Consumerism consumes. *Resurgence, 179,* 16–17.

Guevara, José Roberto. (1995). *Renewing renew: A restoration ecology workshop manual,* Manila: Centre for Environmental Concerns.

Kahn, Peter H., and Friedman, Batya. (1998). On nature and environmental education: Black parents speak from the inner city. *Environmental Education Research 4* (1), 25–39.

Keough, Noel, Carmona, Emman, and Grandinetti, Linda. (1995). Tales from the Sari-Sari: In search of Bigfoot. *Convergence, 28* (4): 5–11.

Learning For Environmental Action Programme (LEAP). (1998). Ecological literacy in Africa. *Pachamama 1and 2* Toronto: ICAE.

Learning For Environmental Action Programme (LEAP). (1996). Living downstream. *Pachamama* Toronto: International Council for Adult Education.

Learning For Environmental Action Programme/Ecologic (LEAP/Ecologic). (1994). *A workbook to move from words to action.* Toronto: ICAE.

Learning For Environmental Action Programme (and the International Network for Sustainable Energy (LEAP/INFORSE). (1997). *Working conceptual paper on environmental adult and popular education.* Toronto: Transformative Learning Centre.

Learning For Environmental Action Programme (and the International Network for Sustainable Energy (LEAP/INFORSE). (1997a). *Case studies in environmental adult and popular education.* Toronto: International Council for Adult Education.

Learning For Environmental Action Programme (and the International Network for Sustainable Energy (LEAP/INFORSE). (1997b). *Environmental adult education: Awareness to action.* Paper presented at UNESCO's Fifth International Conference on Adult Education (CONFINTEA V), Hamburg, Germany, July.

Lengwati, Makkies David. (1995). The politics of environmental destruction and the use of nature as teacher and site of learning, *Convergence, 28* (4): 99–105.

Marshall, Peter. (2001). Liberation ecology. *Resurgence:* 24–27.

Mies, Maria and Shiva, Vandana, (1993). *Ecofeminism,* Halifax, N.S.: Fernwood Publications.

Mikkelson, Kent. (1992). Environmental and adult education: Towards a Danish dimension. *Convergence, 25* (2): 77–74.

Minor, Owor Peter. (1994). Uganda case study. *A workbook to move from words to action* (pp. 4–6). Toronto: LEAP/Ecologic.

Mische, Patricia. (1992). Towards a pedagogy of ecological responsibility: Learning to reinhabit the world. *Convergence, 25* (2): 9–23.

National Institute for Adult Continuing Education (NIACE). (1993). Learning for the future: A special issue on adult learning and the environment. *Adults Learning .4* (8).

Orr, David. (1992). *Ecological literacy: Education and the transition to a postmodern world.* Albany: State University of New York Press.

O'Sullivan, Edmund. (1999). *Transformative learning: Educational vision for the 21st century.* London: Zed Books.

Parker, Jenneth, and Towner, Eric. (1993). *Learning for the future: Adult learning and the environment.* A NIACE Policy Discussion Paper. London: NIACE.

Reason, Peter. (1998). A participatory world. *Resurgence, 186:* 42–44.

Rowe, Stan. (1990). *Home place.* Edmonton, Canada: New West Publishers Ltd.

Seager, Joni. (1993). *Earth follies: Coming to feminist terms with the global environmental crisis.* New York: Routledge.

Shiva, Vandana. (1992). Recovering the real meaning of sustainability. In David E. Cooper and Joy A. Palmer (Eds.), *The environment in question: Ethics and global issues* (pp.187–193). London: Zed Books.

Shiva, Vandana. (1989.) *Staying alive.* London: Zed Books.

Strathy, Kerrie. (1995). Saving the plants that pave lives: SPACHEE/Fiji Department of Forestry, Women and Forest Programme. *Convergence, 28* (4), 71–80.

Sutton, Peter. (1989). Environmental education: What can we teach? *Convergence 22,* (4), 5–12.

Tan, Sandra. (2002). Antiracist environmental adult education in a trans-global community: Two case studies from Toronto. *Global perspectives on environmental adult education: Justice, sustainability and transformation.* (forthcoming, Peter Lang Publishers).

Viezzer, Moema.(1992). *A feminist approach to environmental education.* Paper presented at ECO-ED, Toronto, Canada, October.

Vio Grossi, Francisco. (1995). Cambios paradigmaticos sociedad ecologica y educacion, *Convergence, 28* (4): 31- 43.

Warren, Karen J. (1988). Toward an ecofeminist ethic. *Studies in the Humanities.* Pennsylvania: Indiana University.

Welton, Michael. (1987). *Knowledge for the people.* Toronto: OISE Press.

World Commission on Environment and Development. (1987). *Our common future.* Oxford: Oxford University Press.

CHAPTER 14

Transforming the Ecology of Violence

Ecology, War, Patriarchy, and the Institutionalization of Violence

Eimear O'Neill and Edmund O'Sullivan

Riane Eisler, in the preface to her book *The Chalice and the Blade* (1988), puts the following question to herself and to her readers: "Why do we hurt and persecute each other? Why is our world so full of man's infamous inhumanity to man and to woman? How can human beings be so brutal to their kind? What is it that so chronically tilts us toward cruelty rather than kindness, toward war rather than peace, toward destruction rather than actualization?" (xiii).

One of the more facile responses to this question is that violence and cruelty are part of the makeup of the human. We will not pursue that line of thinking here. It is nevertheless necessary to pay close attention to the issue of violence if we are going to pursue its transformation with any degree of depth. At the other extreme of "essential violence" is the view of human nature as "infinitely perfectible" and in the process of evolutionary change. This is also a facile view of human nature, and the incredible violence of the twentieth century contradicts or at least questions our ideas about human progress and perfectibility.

Our own treatment on violence and destruction will follow a different path as we develop and weave various facets together. Violence will take on different shades and hues as we consider the subtleties of the issues and problems in which it is embedded.

Violence within the Cultural Context of the Human

Because of the incredible saturation of violence and hatred within our own period, it is frequently assumed that violence and force dominate the human endeavor. The violence of humans toward one another, both personally and culturally, belies any facile belief that the human race is any better off now than at earlier times. The increasing use of arms, the holocaust of Jews in Germany, the killing fields of Cambodia, the atrocities of Bosnia, the proliferation of massive amounts of nuclear weapons along with conventional weapons, to name just a few examples, can lead one to conclude that the human race trades in violence. Certainly, the twentieth century indicates that there is very little evidence that humans are a peaceful species; peace and nonviolence are rare exceptions. At this level of analysis, we are left with very depressing conclusions about the long-term possibilities of the human. Military establishments all over this planet have justified their existence by saying that war and violence are endemic to the human project. Within this view of the world, the question surrounding war and violence is not whether they will occur, but rather who will survive. From the military mentality, at the very best, it is said that the presence of war machines may act as a deterrence to warfare. The catchphrase here is "the best defense is a good offense."

Within the more recent developments of historical scholarship, this morbid interpretation of human history is being called into question (Eisler, 1988; Gimbutas, 1974; Stone, 1976). This scholarship is bringing about a new interpretation of western history that broadens our understanding of human experience and especially our understanding of some of the factors that attract or deter cultural violence. The most recent scholarship in the historical understanding of cultures is seen in the development of interpretations of patriarchy (Berry, 1988; Eisler, 1988 Gimbutas, 1974; Stone, 1976). In a historical time dimension, prepatriarchy refers to the matricentric period of Old Europe that apparently flourished around 6500 B.C. and extended up to the Aryan invasions around 3500 B.C. Patriarchy is coterminous with the advent and extension of western civilization, and its time location is the last five thousand years. Postpatriarchy is considered an emergent form of history taking place in the present and moving into a future that is said to be beyond patriarchal structures of extreme hierarchy and identified with total participatory governance and emergent forms of global participatory culture. The reader must be cautioned that the stuff of human history does not in reality divide into such neatly arranged sequences. Nevertheless, the pattern just suggested offers a provisional intelligibility that will allow us to further understand our topic at hand.

DOMINATOR CULTURES

The construct of "dominator culture" has recently been introduced in the writings of Riane Eisler (1988) whose work involves a historical interpretation of cultural violence. (See also Eisler and Loye, 1990.) Eisler draws on recent developments in historical scholarship especially from the work of Marija Gimbutas (1974) in her historical treatment of Goddess cultures. What comes out of her historical work is a very different understanding of the presence of human violence in the longer trajectory of human history. In drawing on the work of Gimbutas, Eisler ventures that when you look at early paleolithic and neolithic societies, where Goddess worship seems to have predominated, there appear to be no signs of sexual inequality or dominance. This finding stands in marked contrast to modern historical cultures coming to us in the Judeo-Christian synthesis. Our own historical prehistory stems from worship of a father-god. The Judeo-Christian heritage, with its emphasis on father-god worship, also carries with it a gender dominance of male privilege. This hierarchy of male dominance is accompanied by the presence of violent social structures. Eisler maintains that where one finds male dominance, one also finds the institutions of private property, slavery, and agrarian agriculture.

By contrast, archaeological data suggest that in the paleolithic and neolithic times, extremely egalitarian societies existed. These cultures also appear to have a lack of violence (Eisler, 1988; Eisler and Loye, 1990). One can draw from this work a more hopeful view of history where it is reasonable to hazard that war and violence are not inevitable to our human story. With this hopeful interpretation in mind, we nevertheless are led to a very critical perspective on our own historical legacy, which is deeply embedded in a hierarchical conception of power that comes to us in the structures of patriarchy. In its simplest interpretation, patriarchy is a system of power where men dominate. Our culture is the recipient of patriarchal power that is a legacy of modern history. Eisler (1988) names our own patriarchal structures of power a "dominator model." A dominator model expresses a hierarchy of power based on the threat or the use of force. She makes an important distinction between a "dominator hierarchy" and other hierarchies seen in nature, which she calls "actualization hierarchies."

> The term *domination hierarchies* describes systems based on force or the express or implied threat of force, which are characteristic of the human rank orderings in male-dominant societies. Such hierarchies are very different from the type of hierarchies found in progressions from lower to higher orderings of functioning—such as the progression from cells to organs in living organisms, for example. These types of hierarchies may be characterized by the term *actualization hierarchies* because their function is to maximize the organism's potentials. By contrast, as evidenced by both sociological and psychological studies, human hierarchies based on force or the threat of force not only inhibit personal creativity but also result in social systems in which the lowest (basest) human qualities are reinforced and humanity's higher aspirations (traits such as compassion and empathy as well as striving for truth and justice) are systematically suppressed. (204)

The "dominator society" is laced through all our social, cultural, and economic institutions. In looking at our western historical heritage, we can see the dominator form in the four patriarchal establishments that have been in control of western history over the centuries: the classical empires, the ecclesiastical establishment, the nation-state, and the modern corporation (Berry, 1988). All of these institutions have been, in their historical makeup, exclusively male dominated and created primarily for the fulfillment of the human as envisaged by men. Historically, women had minimal if any consistent role in the direction of these establishments (Berry, 1988). Currently there is a rising awareness of our predicament and of patriarchy's role in creating it. Concurrent with this rising awareness are attempts to dismantle or avert the dominator structures. It is to be noted that the strong feminist movement of the last two decades of the twentieth century has accomplished a great deal in raising our culture's awareness of the destructive effects of patriarchy throughout contemporary society.

PATRIARCHY AND THE ECOLOGY OF MEN'S VIOLENCE

The etymology of the word "ecology" is "eco," meaning home, and "logos," meaning the study of home. We live our lives in nested hierarchies of community. The Earth is our home, our community is our home, and the intimacy of our family is also our home. When we think of home, the idea of safe boundaries comes to mind. Ideally, our home is a place where nurturance, succorance, and affirmation are expected. It is not a spatial location but rather a region of nearness with boundaries that give us security and trust. We can say, then, that our house is our home. We can also say that our body is our home. Violence is, within this context, a process that violates those boundaries of trust and security. Rape is violent not only because it is a physical violation but, more so, because it is a violation of those boundaries of self that give us a trust and security in life. It is a deep invasion of the spirit, and violations of the spirit do great damage. Particularly damaging are those violations of boundaries in relationships of trust, intimacy, and expectations of care.

There is an ecology of violence. It is an inversion of our natural affinity for conditions that enable us to trust in the basic goodness of life. Patriarchy is a system of domination not only because it is an institution of power. It also violates boundaries. When this occurs, one can say that we are living in an institution or institutions of violence where males are the prominent boundary breakers. Therefore, male violence under conditions of patriarchy is an ecology of violence. An ecology of violence is a home in violation.

We have just examined the ecology of violence under patriarchy that has created systematic violation of the natural world at the planetary level. Let us now turn to the microcosm.

Intimate Violence

Data in the United States compiled by the Department of Justice in 1991 indicate that while women are less likely to be victims of violent crime than men, they are six times more likely to be harmed by an intimate (French, 1992). In Canada, we find that the vast majority of family violence is perpetrated by men on women and children in the household (Lynn and O'Neill, 1995). When violence takes the form of sexual abuse, it is found, in Canada, that 98.8 percent of the perpetrators are male and 1.2 percent are female. Incest is a form of violence that is only just becoming known in terms of its scope and magnitude. It is now becoming clear that it is widespread and is not class linked (French, 1992). Men of every class and educational level rape little boys and girls, girls being the primary target. The overwhelming majority of serial murderers are men, and most mass murders are committed by men, frequently focused on women (Mies and Shiva, 1993).

Lynn and O'Neill (1995), who minutely catalogue the Canadian data on intimacy and violence, feel that male violence cannot be understood unless you place family relations within a particular economic and political system. They venture the idea that with the development of industrial capitalist states came the separation of community life into the public world of work, law, and politics and the private world of family and intimate relationships. Following from this public and private fissure, there was a concomitant separation of men and women from each other in terms of their work, their relative access to the public world, and their involvement in the ongoing vicissitudes of family life. Lynn and O'Neill suggest that in an advanced capitalist country like Canada, men have lost the power over their own work. In the conditions of work under capitalism, men appeared to have been conditioned not to rebel against oppressive, exploitive work situations; instead they feel safer to express their anger in the home. The use of violence in the home appears to maintain the previously unquestioned sex-role hierarchy and is condoned therein.

These findings do not apply only in North America. The problem of men's violence is of global proportions. Angela Miles (1996) reflects on the global context: "All over the world women are beaten, raped, burned, sexually abused and harassed, mutilated, confined, forced into marriage and into pregnancy, sold into prostitution and pornography, aborted after aminocenteses, killed as infants and as adults, denied food and medical treatment and education, and forced to work without pay *because they are women*" (117). Miles notes that while the forms of abuse vary across cultures, classes, and nations, violence of men against women per se is universal.

The Integral Weaving of Gender, Race and Class

It is important to recognize that issues of violence, in their most complete interpretation and analysis, must include the complex issues of race, class, gender, and sexual orientations and that treatment must convey the integral nature of these factors. We cannot give full expression to these factors because of space constraints. All the same, we remain aware that in reality these social forces are, more often than not, interrelated. These forces, taken singly or combined, are integral parts of the dominator structures. Consider Marilyn Frye's (1983) description of a birdcage:

> If you look very closely at just one wire in the cage, you cannot see the other wires. If your conception of what is before you is determined by this myopic focus, you could look at that one wire, up and down the length of it, and be unable to see why a bird would not just fly around the wire any time it wanted to go somewhere. Furthermore, even if, one day at a time, you myopically inspected each wire, you still could not see why a bird would have trouble going past the wires to get anywhere. There is no physical property of any one wire, *nothing* that the closest scrutiny could discover, that would reveal how a bird could be inhibited or harmed by it except in the most accidental way. It is only when you step back, stop looking at the wires one by one, microscopically, and take a macroscopic view of the whole cage, that you can see why the birds do not go anywhere, and then you will see it in a moment. It will require no great subtlety of mental powers. It is perfectly *obvious* that the bird is surrounded by a network of systematically related barriers, no one of which would be the least hindrance to its flight, but which, by their relations to each other, are as confining as the solid walls of a dungeon. (4–5)

The birdcage provides us with a picture on how gender, race, and class interact with one another and also enables us to see that in a whole picture of any social reality of oppression and domination, these dimensions are woven together in an intricate pattern. Thus the dominator pattern of patriarchy does not exist separately from race and class domination. At the level of social and political reality, these patterns of domination are integral to one another and work together in intricate patterns of oppression.

Education, Equity, and Difference

The presence of difference and diversity within the Earth community and the human community is a matter of fact. It becomes a problematic issue for educators when issues of social power are recognized and addressed along lines of difference.

In a closely argued piece that connects cultural diversity and the ecological crisis, C. A. Bowers (1993) maintains that the two major changes that will dominate the educational/political scene over the coming decades are the growing awareness and pride in distinct cultural identities and the ecological crisis. Addressing the field of teacher education, he stresses the point that these two dimensions are not to be seen as pulling in separate directions. He states:

> While on one level the two phenomena appear to be separate and distinct, and thus requiring different responses, they are, in fact, related. Both, for quite different reasons, signal that the assumptions underlying modern culture are no longer sustainable. The viability of the earth's ecosystems is being seriously threatened by the technological practices and ever expanding demands on nonrenewable "resources" deemed necessary by the modern individual. And the myth of personal success that prompted generations of people to turn their backs on the web of relationships and patterns that constituted their ethnic heritage by entering the competitive and highly individualistic mainstream culture is now becoming increasingly illusory. Unemployment, drug use, alienation, structural poverty, stress. and toxin-induced illnesses represent the reality that now overshadows this myth. The flaws in this modern form of consciousness are now being challenged by both environmentalists and groups attempting to recover their ethnic identities—African Americans, Hispanics, and Native Americans (Sioux, Nez Perce, Cheyenne, etc.). To put it another way, the ecological crisis and the emergence of ethnic consciousness have their roots in the same dominant cultural values and practices that are now seen as increasingly problematic. (163)

We now turn to the complex issue of power and its relation to equity.

POWER EQUITY

Cultural focus on the quality and effect of the power relationships between human beings, individually or in groups, is both a theoretical and a practical preoccupation after the twentieth century's appalling human violence and devastation of habitat. Power differences among humans are the point of relationship that determines whether one person's being will harm, constrain, or enhance the other. Positions of domination and subordinance depend on individual and group power. The need to move toward more equitable power distribution within relationships of domination and to become aware of the mutuality of effect in those relationships are two emergent and intertwined tenets of feminist and critical thinking at this time (Spretnak, 1997; O'Sullivan, 1999).

Why do some human beings disregard, exploit, abuse, and seek to eradicate others? How do human beings subordinated in various ways survive and challenge those power relationships? What about those power relationships internalized as part of one's own sense of self? What transforms those internalized relationships of domination, given that these are as constraining as those external social structures we are all involved in? Why and how do those in relative power deny the reality of others' distress and protest? How can the interpersonal domain of intimate relationships be democratized, become a domain of relationship in which each participant has equal say, in a manner compatible with democratization in the public sphere?

Those in power clearly mark those subordinated as different from themselves. Questions about difference on many dimensions, of how and by whom "the other" is thought about, are crucial to understanding the increasing scale of interhuman violence and Earth devastation. In the last forty years there has been general movement toward more "deeply relational" understandings of difference and of power relationships. By "deeply relational" we mean considering each of us as always in evolving dynamic and ecological relationship to the full emergent network/web of human and nonhuman others. Each one of us is uniquely differentiated and related within the web of life. In such a view, diversity is necessary, valuable, and to be acknowledged in every relationship.

In the larger social context, the material and practical effects of subordination and dominance, of privilege and entitlement, of power and its abuses, are well documented. The effects of social power inequities on women and children, on this polluted, plundered, fragile planet are updated annually.[1] The power differences and dynamics between women and men, and among women and men, affect both the individual and the cultural sense of self. In this context, it is necessary to begin by defining what is meant by "power."

Power can be defined in multiple ways, depending on one's discipline. For example, in terms of new theories of complexity such as quantum physics, power is defined as movement in relation to others' movement, generally movement below an atomic or molecular level that produces or releases energy. When applied to relationships between human beings, power is described in a variety of ways, often with differentiation being made between "power-with" and "power-over" definitions. For example, "power over" is described by the politically minded from a variety of disciplines as "A exercising power in ways that affect B in a manner contrary to B's interests, and restrictive of B's options for action." As Lukes (1974) points out, A exercises "power over" B not only by getting her or him to do what s/he does not want to do but effectively continues to maintain "power over" by influencing, shaping, or determining B's very wants and hopes, B's internal landscape. Power-over relationships are those where one group or individual benefits at the cost of others.

"Power with" could be defined as the shared agency that empowers all of us inclusively, that which transforms both A and B toward more life-affirming movement both in terms of their internal landscapes and in terms of the realities of their daily living in relationship with others. In relational examples, "power with" is defined as "movement holding awareness of the immanent interconnected nature of human living" or "movement in relationships" (Miller, 1976, 1984; Starhawk, 1982). "The ability to take one's place in whatever discourse is essential to action and the right to have one's part matter" is the way feminist and social constructionist Carolyn Heilbrun (1988: 44) defines it. The current definition fits with relational, feminist, and postmodern or social constructionist definitions, with both "power-over" and "power-with" situations. The definition of power used here includes: (1) capacity for self-definition and self-regulation; (2) access to resources; and (3) degree of conscious participation in decisions affecting day-to-day living (Lynn and O'Neill, 1995).

Defining one's self (1) is very much about whose voices are included or excluded in the definition of who you are and what you do. Is it from your own voice and/or those of your parents or intimates, the culture's, or even the predominating voices of the times you live in? Self-regulation—that is, autopoetic emergence from one's context—refers to the ability to move, act, be agentic on our own behalf, as well as having clear self-definition or differentiation within our human community. The resources referred to in (2) are the broad range of resources that meet basic human needs, including support, nurturance, expression, information, education, mobility, community, healthy living environment, and personal space in addition to the money, technology, consumer goods, social services, income support, protection of the more vulnerable, community health, and child care options of modernity. And (3), the degree of one's conscious participation in decisions affecting day-to-day living includes political representation within and democratic governance of the groups to which one belongs in addition to personal daily living choices and the extent to which one participates in social and community activism to affect all of these.

The degree of participation in decisions affecting day-to-day living also is shaped by human consciousness, which can be disrupted by trauma and oppression. Power around participation also includes the degree to which you have a voice in decisions affecting your local environment and the health of the planet Earth itself given that subtends human life. "Self-regulation"—that is, doing things *for* yourself—needs to be differentiated from "autonomy," doing things *by* yourself. One can be self-regulating and connected, just as one can be autonomous without self-direction. As human beings, the degree to which we have that capacity for self-regulation shapes our access to any of the resources available to us, and both of these are affected by our conscious participation in decisions that affect our day-to-day living. And in terms of "access to resources," mutuality, the degree of two-way flow between individuals and groups, transforms access, sharing that which may have been previously inaccessible. We are mutually constituting and can be so in transformative ways.

Defining power as (1) capacity for self-definition and self-regulation; (2) access to resources, including mutuality; and (3), degree of conscious participation in decisions affecting day-to-day living, covers various dimensions of power, applying not only to "power over" and "power with" but also to power as experienced at the personal level, as ascribed by the society to individuals and groups, and to power as realized in daily living. The relationship of social power to consideration of these three dimensions is not simple or cumulative. For example, while access to certain resources including knowledge is available primarily to those in privileged positions in a culture, other resources are generated from and more readily available within the margins.[2]

The deeply relational aspect of such a definition of power is captured by the understanding that full "power with" would add capacity "capacity to see others as self-regulating," opening up access to resources to others" and "increasing conscious participation of others in decisions affecting there day-to-day living" to the working definition of power above. This expansion is useful when we come to consider how we might learn more collaborative and democratic participation in human living within the current ecology of violence.

When gender relations function within the structures of patriarchy, then all sorts of violence and selection are carried out within that structure of dominance. Within the norms of patriarchy, there are high correlations between position and gender. This positioning or favoring of males is legitimized by the culture through its political, economic, religious, and educational institutions. The institutions of patriarchy, like any other deeply violent social formations, combine direct, structural, and cultural violence in a vicious triangle. Patriarchy is not only an institutional structure that protects male privilege, it also serves as a legitimizing institution for the perpetration of male violence. In order to address the pervasiveness of patriarchal violence, there must be a deep and profound exploration of how dominance and subordinance operate along gender lines. There must be a deep questioning and study of the background cultural factors that feature in male violence and in their domination of women. The institutions of patriarchy have been challenged by the women's movement. That is an opening toward change that is necessary and, at the same time, long in coming. But change in this area, if it is to be effective, must be initiated by men in our culture who have come to see how male misogyny and privilege work against both men and women. What we need is an education that will challenge male privilege but also give us a vision of a partnership between the sexes that has power equity as its basis (O'Sullivan, 1999). The recent work of Riane Eisler and David Loye (1990) in *The Partnership Way* and the *New Dynamics* workshops inspired by the work of Carole Pierce and associates are creative approaches that move on a path to power equity between the genders (Pierce and Wagner, 1994). In the final analysis, nonsexist education must examine gender inequity at all the levels of the social life in which it is embedded. A break in the hegemony of patriarchal power will help us forge a vision of education where equity is the norm rather than the objective.

In summary, what is alternative and transformative within the ecology of violence is the need to consider more fully and relationally the effects of power difference on whose perspective, whose version of reality, gets internalized as constituting personal and cultural truth. Such truth, such visions, drive our actions. Integral approaches are based on the understanding that transformation of human violence requires seeing human violence as stoppable, as nonessential, as arising out of and spewing back into cultural contexts of dominance and patriarchy. In seeing men's dominance in intimate and deep cultural contexts, in seeing stances of dominance generally, as *risking* violence, certain principles of change become clearer. Changing the ecology of violence necessitates men's and women's involvement in integrated movement toward equity, toward antioppression, toward raising consciousness, increasing awareness of the effects of power differences on mutuality in relationships.

Race, class, sexual orientation, bodily ability, age, understandings of the sacred: All these are bases for dominance in North American cultures. We need understanding not only of overt violence but also of colonization, including colonizations of the human soul. We need understandings of colonization woven with those of feminism and socialism, of antiglobalization and indigenous knowledges. However, through all of this, the current reality is that most of us were raised in ecologies of male-female dominance. What we suggest as part of transforming those interwoven ecologies of dominance and subordiance among humans is operationalized understandings, description, and recognition of the dynamics of dominance and subordinance that are played out in relationships between men and women in intimate relationship. While same-sex intimate relationships and relationships among women and men of differing cultural groupings of power are clearly as central to transformation, dominance and subordinance in male-female relationships is the most familiar dominance-subordinance dynamic internalized in our culture. This is where most of us learn patterns of closeness, in relationships with fathers and mothers, husbands and wives, our selves and our siblings, lovers and friends, sons and daughters, brothers and sisters, relatives and cultures.

We need the concept of a journey, not always direct, often cycling back and forth between what is progressive and what is cancerous, what is merely survival and what is more creative evolution. Sometimes we are in defensive and unconscious survival mode, and at other times we spiral ahead, glimpsing more collegial, synergistic ways of moving ahead together. And on this journey we need to deal with the dichotomization of and judgment implicit in compulsory heterosexuality rather than the opening to creative possibilities inherent in a more continuum model of changing and diverse developments in sexual orientation over the lifetime. We also are aware of having to deal with the constrictions of the personal self inherent not only in cultural dichotomization but also in trauma. By trauma we mean the molded structuring of the self in oppression, the shaping by unavoidable and conflictual competing forces, in situations such as childhood sexual abuse or racism or dominant definitions of the sacred and profane.

In transformative approaches to the ecology of violence, we approach these aspects in ways beyond the solely cognitive, ways that weave the realms of the imagined and symbolic, the sensory and embodied.

NOTES

The ideas in this article draw heavily from O'Neill (2002) and O'Sullivan (1999). To contact the author's e-mail: eoneill@oise.utoronto.ca or *eosullivan@oise.utoronto.ca.*

1. The ideas expressed herein draw heavily from O'Neill (2002).
2. The ideas expressed herein draw heavily from O'Neill (2002).

REFERENCES

Berry, Thomas. (1988). *The dream of the earth.* San Francisco: Sierra Club.

Bowers, C. A. (1993). *Critical essays on education, modernity, and the recovery of the ecological imperative.* New York: Teachers College/Columbia University Press.

Eisler, Riane. (1988). *The chalice and the blade: Our history, our future.* San Francisco: Harper & Row.

Eisler, Riane, and Loye, David. (1990). *The partnership way.* San Francisco: Harper and Row.

French, Marilyn. (1992). *The war against women.* Toronto: Summit.

Frye, Marilyn. (1983). *The politics of reality.* Freedom, Calif.: The Crossing Press.

Gimbutas, Marija. (1974). *The gods and goddesses of Old Europe, 7000 to 3500 B.C.: Myths, legends, and cult images.* London: Thames & Hudson and University of California Press.

Heilbrun, Carolyn. (1988). *Writing a woman's life.* New York: Ballantine.

Lukes, Stephen. (1974). *Power: A radical view.* London: Macmillan.

Lynn, Marion, and O'Neill, Eimear. (1995). Families, power, and violence. In Nancy Mandell and Ann Duffy (Eds.), *Canadian families: Diversity, conflict and change* (pp. 271- 305). Toronto: Harcourt Brace.

Mies, Maria, and Shiva, Vandana. (1993). *Ecofeminism.* Halifax, N.S.: Fernwood Publications.

Milbrath, Lester, W. (1989). *Envisioning a sustainable society: Learning our way out.* Albany: State University of New York Press.

Miles, Angela. (1996). *Integrative feminisms: Building global visions.* New York: Routledge.

Miller, Jean Baker. (1976). *Toward a new psychology of women.* Boston: Beacon Press.

Miller, Jean Baker. (1984). The development of women's sense of self. Work in Progress No 12, *Stone Center Working Paper Series.* Wellesley, Mass.

O'Neill, Eimear. (2002) *Holding flames: Illuminating women's learnings about transformation.* Manuscript in preparation. Ontario Institute for Studies in Education of the University of Toronto.

O'Sullivan, Edmund. (1999). *Transformative learning: Educational vision for the 21st century.* New York: St. Martin's Press.

Pierce, Carol, and Wagner, David. (1994). *The male/female continuum: Paths to colleagueship.* Lacona,. N.H.: New Dynamics Publications.
Spretnak, Charlene. (1997). *The resurgence of the real.* New York: Addison-Wesley.
Starhawk. (1982). *Dreaming in the dark: Magic, sex and politics.* Boston: Beacon Press.
Stone, Merlin. (1976). *When God was a woman.* New York: Harvest Book Paperback.

CHAPTER 15

Transformative Learning and Cultures of Peace

ANNE GOODMAN

With the advent of the new millennium, a new phrase entered the international lexicon. The year 2000 was declared the International Year for the Culture of Peace and the ensuing decade the Decade for the Culture of Peace and Nonviolence for the Children of the World. Since that time the notion of a "culture of peace" has been embraced by many and ignored by many more, much maligned, and much celebrated. Above all, it has been interpreted in many different ways.

Developed by the United Nations Educational, Scientific and Cultural Organization (UNESCO) in the context of the post–Cold War world, the culture of peace seeks to fulfill the organization's original mandate to create a peace consciousness. I suggest the newly minted terminology not only describes concepts that have been around for some time and that are gaining prominence in many places and disciplines, but that there is great resonance between the culture of peace and transformative learning. In this chapter I describe my involvement in the culture of peace, how I have come to understand it, and how it relates to notions of transformative learning and teaching. Describing both what I call the "official" culture of peace and the "unofficial" aspects that lie outside its purview, I look at the contradictions, limitations, and possibilities of each and the relationship of each to transformative learning. I assert the need to adopt both a critical and a visionary perspective, something I see as a key feature of both the culture of peace and of transformative learning.

ENTRY POINTS AND INSIGHTS

My earliest entry point into the issues that are the topic of this chapter was my activism, first in the antiapartheid struggle in South Africa and later in the social movements and party politics in Canada. I came to understand the complex interlocking set of problems that constitute the global *problématique* and the inextricable links among sustainable human development, protection of the environment, human rights, disarmament, women's equality, and respect for diversity. Later, as a graduate student in education, I explored the theoretical underpinnings of these themes and realized it was not surprising that the issues were interdependent since they were all rooted in the taken-for-granted assumptions underlying our modern industrial western civilization, the scope of which now encompasses the whole world. My exposure to the concept of transformative learning as a doctoral student and then as an instructor allowed me to see its relevance for my peace work.

Another point of convergence for me was the coming together of the different levels or contexts I now understand as constitutive of both the culture of peace and transformative learning: the personal, the local/regional, and the global/planetary (O'Sullivan, 1999: 168). Initially distrustful of inner development, seeing it as a diversion from the real work of social action, I gradually came to pay attention to spiritual growth and to appreciate it as an essential complement to social change. As a former South African, I finally achieved reconciliation with my past, a process that has helped me immeasurably in understanding what a difficult and ultimately liberating process this can be. I see the same dynamics in my work with people from wartorn communities and in the insights I am developing that political work is inextricably linked with the inner work of healing and developing different understandings of situations. The reclamation of certain aspects of my own identity, especially those of being Jewish and African, has been important not only on a personal level but as offering insights into the paradigmatic changes required in our cosmology.

When I first read about the culture of peace in a monograph produced by UNESCO (Adams, 1995), I recognized it as work I had been doing for years, even if I had not named it that way. In the culture of peace I recognized a framework that encompassed all the diverse yet integrally connected projects and actions I was involved in as well as all the principles and priorities that drove my work. Working against nuclear power and for alternative sustainable energy was part of building a culture of peace, as was opposing the global market economy, working for disarmament, trying to increase the role of women in decision making, doing conflict resolution work in warring communities, working for a more democratic political system, and trying to get peace education into the schools and universities. I also had many opportunities to work on specific culture of peace projects, ranging in scope from the local to the global and covering a wide range of activities, including writing, speaking, conducting workshops, running conferences, teaching, and organizing

While I did not have to change the focus of my activities to work for a culture of peace, I noticed a subtle but significant shift when I began to explicitly name and situate my work that way. For a start, I noticed a new excitement and positive energy, especially when sharing the ideas with others. The culture of peace also gave a context for the work that I and others were doing, making it part of a global "movement of movements," as former secretary general of UNESCO, Federico Mayor, put it. Transformative learning works in much the same way. O'Sullivan (1999), following Thomas Berry, describes the essential qualities of the universe—and our role in it as human subjects—as "differentiation, subjectivity, and communion" (191–93). In an interconnected world, the uniqueness of each of us is situated in deep and respectful relationship to all living beings and Earth.

The Culture of Peace and Transformative Learning

While the culture of peace and transformative learning are both emerging concepts and we do not yet know how each will fully unfold, it is evident that there are many similarities. An exchange of views from each perspective would likely be beneficial. The culture of peace and transformative learning are deeply connected. In many instances, the phrases can be interchanged without changing the essential meaning, as in the working definition used by the Transformative Learning Centre.[1] Both concepts involve and appreciate complexity and intricacy, relationship, and interconnection. In both, context and meaning are seen as fundamental, and multilayer notions of the context: global, local, and personal—and the interconnections between them are recognized. Each articulates the need for fundamental changes in values, attitudes, and behaviors, to critique oppressive structures and to develop alternatives. Diversity is valued in both, as is an interdisciplinary approach.

Peace studies differentiates between *direct violence* (the most visible kind, where people are seen to be directly harmed), *structural violence* (the indirect violence inherent in the way social structures are ordered), and *cultural violence*. The last is the least visible but arguably the most influential, since it exists at the level of taken-for-granted assumptions about the world. There are parallels to each of these in peace. I suggest that both transformative learning and cultures of peace have to do with changes primarily in the deep culture. This is not to say that changes in the other levels are not sought or are seen not to matter, but rather that it is the deep culture that is primary. In transformative learning, a distinction is made between *reform criticism,* which deals with changes in the direct and structural levels, and *transformative criticism,* which "calls into question the fundamental mythos of the dominant cultural form and indicates that the culture can no longer viably maintain its continuity and vision" (O'Sullivan, 1999: 5). This basic, radical level of deep culture is also the primary, although by no means exclusive, focus of the culture of peace.

Another way of saying this is to recognize that key to both concepts is the notion of a shift of consciousness. UNESCO's constitutional mandate on which the culture of peace concept builds is described as "building the defences of peace in the minds of men and women" (Adams, 1995: 6); Berry's introduction to O'Sullivan's book acknowledges his insight that "while we will need a news way of living, we need even more urgently a new way of thinking." (O'Sullivan, 1999: xii).

THE OFFICIAL CULTURE OF PEACE

Through UNESCO in particular, and the United Nations (UN) system as a whole, have come the official declarations and program of action on the culture of peace as well as the normative instruments that represent mileposts on the road toward it. Behind the bureaucratic language are countless inspiring stories of the steady progression of ideas and initiatives. Take human rights, for example. While human rights violations in many parts of the world remain one of the most serious problems facing humanity, we should not forget that it was only fifty years ago that we officially recognized, through the adoption of the Universal Declaration of Human Rights, that everyone in the world has rights that deserve protection. And we have progressed in our understanding since then, adding another two generations of rights and beginning to see what it would take for the human right to peace to be realized. Implicit in the official declarations are other struggles too—for example, for women's equality, for sustainable human development, and for participatory communication and the free flow of information. Many of these issues were taken up in a series of world conferences in the 1970s and 1980s, and another series in the 1990s. Through these processes, civil society has become better informed and more active, in many cases holding "people's conferences" in parallel to the official conferences.

The official documents outline a plan of action to be undertaken by governments, civil society, and the UN. The plan, covering many areas, is detailed and specific. Its purpose is to transform the current culture of war and violence into a culture of peace, and our pictures of the dimensions and characteristics of these two cultures have developed in opposition and contradistinction to each other. Through seeing what needs to be changed, we can see what is. The culture of war and violence is characterized by such factors as power defined as violence, authoritarian decision making, male dominance, secrecy and manipulation of information, exploitation, and the image of the other as an "enemy." The program of action clearly defines what needs to be done to create a culture of peace. Implementing the actions will be far from easy, given that there is much opposition to be overcome and patterns and ways of doing things to reverse. But the agenda is clear, comprehensive, and precise, a cogent rebuttal to those who criticize the culture of peace as being a nebulous concept of all things to all people.

The official program is also important in that it reasserts the vision behind the charter of the UN—to save succeeding generations from the scourge of war—and of UNESCO's constitution—to construct peace in the minds of people. The official story has come to us not only through the UN. The Hague Appeal for Peace, a gathering of thousands of peace activists in 1999, a hundred years after the original Hague peace conference, delineated a global plan of action toward a culture of peace. Chapters of the story also have been written in the peace research and education literature and in the new concepts that are emerging in international and national government policy.

While UNESCO could be described as an agency committed to transformative learning, and one that produces documents and materials and supports projects leading in this direction, there are limits to what a giant bureaucracy can do, especially one bound by the constraints of the nation-states that give it its existence. Much of the work of the UN system as a whole is work of reform rather than transformation.

The Unofficial Story

While the official culture of peace is essential, it is not the whole story. For that we must go beyond the rational and the objective, calling on the artists and storytellers, the children and the mystics, on the cultures that traditionally had no word for peace since peace was seen to infuse everything. This aspect of the culture of peace cannot be quantified or easily recorded, and since its outcomes are not necessarily known in advance, it is more difficult to describe than the official program. Much of it falls outside the parameters of the modern conception of knowledge.

It is perhaps synchronous, rather than coincidental, that the International Year for the Culture of Peace was followed by the International Year for the Dialogue of Civilizations. In a way, this anticipated some of the missing dimensions of the culture of peace as it has been formulated. The traditional Chinese yin/yang symbol comes to mind, with its polarities in tension rather than in opposition, complementing and counterbalancing each other.

What do I see as the missing dimensions? The official culture of peace program emphasizes the first of each of these polarities, neglecting the second: west/east; reason/emotion; north/south; male/female; critique/vision; product/process. While the documents do call for equality between women and men, for the reduction of inequality between nations, and for the recognition of other ways of knowing (e.g., Indigenous wisdom), they are drafted in a particular way from a western, male, northern, and rational perspective that leaves out something vital. For a full appreciation of a culture of peace, we need to find a way to include these aspects.

In *Cultures of Peace: The Hidden Side of History*, Elise Boulding (2000) suggests ways to do this. She takes a long view over time, incorporating both the past and the future along with the present and a broad view over the globe. She examines not just the dominant cultural forms but the alternatives that exist: the Two-Thirds World along with the One-Third; the cultures of everyday life as well as the public sphere that enters the news and the history books; the hoped-for worlds that have yet to materialize alongside the values and social structures of the world we actually have. And it is not just human social structures Boulding looks at, but also the global contexts of the biosphere—the living world—and the technosphere—the world created by our technological extensions. Boulding sees cultures of peace all around us. The histories of the history books, the ones that tell of conquests, whether of other people or of nature, do not tell the whole story. In an earlier book, Boulding (1976) describes the largely untold history of women as the "underside of history." In much the same way, she sees cultures of peace as having always existed alongside, beneath, in opposition to, and in contradistinction to the cultures of violence that have mainly predominated. Much of everyday living—the nurturing of families, the celebration of events, human creativity—represents examples of peace culture hidden from official view. Alongside the holy war doctrines existing in almost all religions are "peaceable garden" cultures. Boulding (2000) is appreciative of the role of utopian thinking as offering both creative ideas of what the world could be like as well as a range of practical experiments, from small communities to massive national projects. Cultures of peace even exist in the dissatisfaction of those who are supposed to have it all, and Boulding points to the burgeoning men's antiviolence movement and the voluntary simplicity movement.

The Conference on the Search for the True Meaning of Peace held at the University for Peace in Costa Rica in June 1989 (Brenes-Castro, 1991) in many ways anticipated the culture of peace program. The Declaration for Human Responsibilities for Peace and Sustainable Development produced at the conference has much resonance with Manifesto 2000, a central organizing tool of the year and decade. The conference report, written before the collapse of the Eastern bloc, highlights the polarization between east (socialism) and west (capitalism) and between north and south. Much of the official culture of peace program can be seen as a way of reconciling the polarities by working, for example, to eradicate poverty and economic disparity and toward disarmament. But the report also examines these dichotomies on a deeper level, looking at the evolution of consciousness through the integration of the polarities. This thrust of the conference has not been taken up by UNESCO and the UN system in the same way and is, in large part, the agenda of the unofficial culture of peace.

A fascinating juxtaposition of two facets of the culture of peace occurred at an international meeting I attended in the Netherlands in July 1999. One of the participants eschewed hierarchical models, using metaphors instead: a tree representing a way of organizing, the sun representing truth and the rain love, and declared that his main purpose was "to be happy." Struggling to understand this, a woman working in a more bureaucratic setting wanted an emphasis on strategic planning and quantifiable goals. The communication between them was difficult, but the basis of respect and desire to listen and understand helped all of us to see the complementarity of the two approaches.

At the same meeting, I tried to build in space for the spontaneous. During the opening session, when participants brainstormed their hopes and expectations for the meeting, I noticed that the facilitator wrote down all the suggestions—except mine, which was to have no particular outcomes! In fact, the intangible qualities of communion, trust, and shared vision gave the meeting a particular energy that even reflected into a report I wrote about it. People who had not even been at the meeting commented on the report and sent copies to others.

I have noticed a similar energy at other meetings. I think of it as working "in the light," not ignoring or negating the problems of the world but somehow shifting focus so we align ourselves with the hopes of all of us who know that peace is possible. How else to explain that a meeting we had in Toronto started at noon and at 6:00 P.M. I finally had to tell people that the meeting was over and that they must go home! There was simply no sense of it being such a long meeting. The time flew by, dare I say with the quality of an altered state of consciousness, and no one wanted to leave. O'Sullivan's description of the work of transformative learning as "kindling the fires of the soul" (1999: 259–281) resonates deeply with my experience.

My organization, Voice of Women (VOW), has developed a workshop kit called "Creating a Culture of Peace." It incorporates the unofficial along with the official aspects of the culture of peace: parables, metaphors, and pictures as well as factual information. I like to tell stories when I use it, about how we chose the color we use in the kit, yellow, and how one of the exercises is connected to a friend doing peace education work in Sierra Leone. A quote by Thich Nhat Hanh, the Vietnamese Buddhist monk, inspired the sunflower image we use, an image that has simultaneously been taken up as a symbol for nuclear disarmament since sunflower seeds were planted in the Ukraine when the last nuclear weapons were removed. The metaphor that most captures our imagination and that of people we reach with our workshop is that of the Earthworm, and we explore why it so perfectly represents a peace promoter. Metaphors and stories have an elicative, evocative character that encourage new insights. A teacher who used our kit, for example, told me how some students who were studying the American Civil War made a connection between Earthworm work and the Underground Railroad that brought slaves to freedom.

There is a great creative energy to be realized by making use of serendipitous events. At a workshop I facilitated on the culture of peace, the participants were all feeling positive and empowered—until one of them looked out of the window during the lunch break and saw the following graffito reproduced in figure 15.1.

The energy level dropped palpably and we all felt dismayed, but reasoned discussion was not the way to deal this. I copied the graffito onto a flip chart, leaving it for the group to deal with when the time was right, and carried on with the workshop. At one point, a participant leapt up, grabbed a marker, and to the pessimistic slogan "World peace can't happen" vigorously added these triumphant words: " . . . without you!" Another participant joined in, adding tips to the arrows so they pointed in-

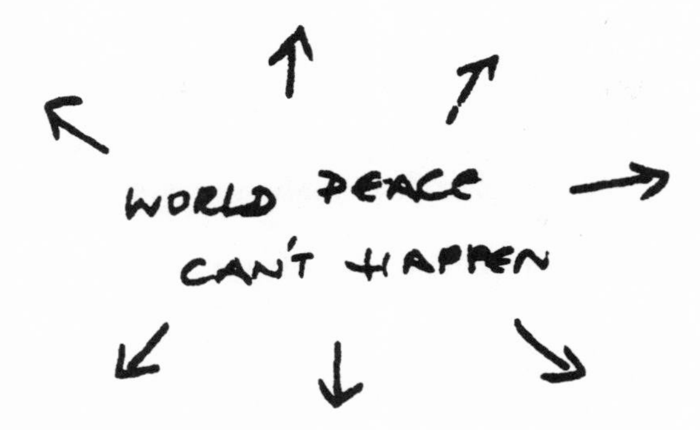

Figure 15.1 The Original Graffiti

ward as a unifying force, instead of just outward. She also added two-sided arrows connecting the different forces.

The altered graffito is presented in figure 15.2.

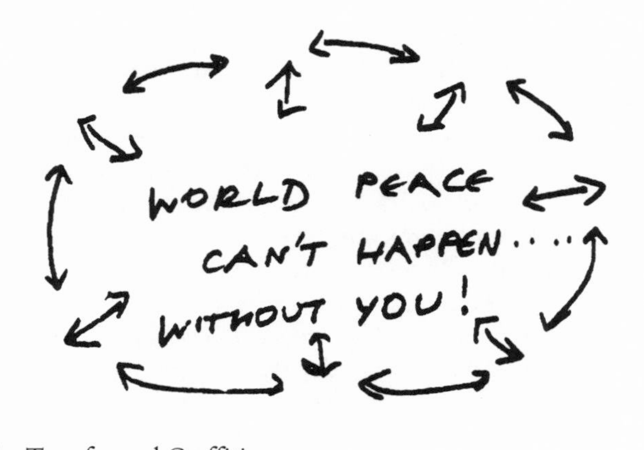

Figure 15.2 The Transformed Graffiti

Unpacking both the message of the altered graffito and the process of changing it is a rich learning experience. While spontaneous learning like this obviously cannot be planned for, it points to the need to leave space for the unexpected and to gain insights through different kinds of processes that tell us more than we know rationally.

Popular education, widely used in the global South, makes extensive use of nonrational dimensions. I was fortunate to attend a peace education workshop given by two grassroots educators from Sierra Leone at the pan-African women's conference for a Culture of Peace held in Zanzibar in June 1999. We did an exercise from their manual (UNESCO/FAWE, 1999) in which we made use of actual building materials in an experiential way and then went on to examine the concept of "peacebuilding" in light of our experiences.

MY PRAXIS

While my praxis is implicit throughout the chapter, I will look here at some of the underlying principles and also give some examples to make my particular approach more clear.

Coming as I do from two directions—an activist/peace worker and an educator—I have a deep sense of the complementarity and convergence of education and action. I cannot think of a situation where I do not use praxis: education, reflection, and action as a constantly developing spiral of learning, knowing, and being. I see consciousness as a causal reality, so that basic to my praxis is the sense of changing the world by changing how we understand it, but I also see structures and behaviors at all levels as mutually informing and resonating.

While I sometimes yearn for more stability in my work and less necessity constantly to invent and create what I do, I feel truly blessed by the freedom I have and the opportunity to work at what I understand as my life purpose. Working on the boundaries, for example, between academia and the community, or between different cultures in conflict, is intensely satisfying.

Depending on where any particular part of my work is situated, I do different things to expand the transformative learning possibilities. Within formal settings, like regular university classrooms, for example, I have found it necessary to change structures and roles as well as physical settings. I taught an "Introduction to Peace Studies" class with close to 150 students last year, and most of my work was to set up opportunities for small group learning, student responsibility for their own learning, and opportunities for action, for example practicums in the community. All this made for a great deal of work, but I did not believe that the alternative—a three-hour lecture-style class in an auditorium—was conducive to transformative learning, no matter how compelling the subject matter.

Another principle I have developed in my praxis is that of the interconnectedness of the different levels of context. The course I have been teaching through the Transformative Learning Centre, "Identity Based Conflict and Conflict Transformation," affords the participants (and me) the opportunity to examine the links between our learning at all levels—the personal, the regional/local, and the global/cosmological, and I am constantly inspired by the richness of the exchanges and insights. This multiplicity of levels is evident in another situation I am working on where learning is seemingly less prominent. A project called the Peacebuilding Communities Working Group that brings together people from different wartorn societies evidences this same interchange between the levels and has made me understand that deep-rooted and sustainable political changes will not be possible without changes of a personal nature. In the Voice of Women kit, questions we use to elicit small-group discussion point directly to the need to explore the different levels. Question 1, for example, reads like this: "How are you working toward a culture of peace? What is the focus of your work? Individual? Community? National? International? All of the above? How do these levels interact?"

Another basic principle is the need to incorporate both critique and vision, since I see neither on its own as sufficient. I also feel the need to use different ways and styles of learning. With the work I have been doing with different communities, and with my own self-development, I find myself using a variety of approaches, not just the rational and academic. While this is easier to do in nonformal settings, it is possible and very welcome in institutional settings. My peace studies students, for example, were responsible for arranging an "opening" to each class, and the approaches were amazingly creative and varied.

Probably the most important factor in my teaching is a sense of integrity, of bringing myself as a whole into any teaching/learning situation and at the same time encouraging that in my students. One of the most helpful things I have found is that when I share some of the issues that have caused me ambivalence—for example, what it was like to live in a country where I was cast in the role of oppressor—this opens up the space for an honest and open exchange among the students.

THE MEANING OF "CULTURE" IN THE CULTURE OF PEACE

"Culture" as art, music, and the like is clearly an aspect of the culture of peace. Culture is one of UNESCO's three modalities, and good use has been made of artwork and symbols in the official culture of peace program. In peace work more generally, there is a growing awareness of the importance of the arts. Drama, art, dance, music, and storytelling are used all over the world for peace-building, reconciliation, and trauma healing. For me, a potent indicator of structural violence is the reduction of arts programs in schools, with budgetary constraints being cited as the reason. One of my colleagues from VOW, a woman with a long history of teaching dance and art, calls this trend "criminal" and foresees violent consequences.

I was delighted when, from a VOW training workshop, a collective was born with an inspiring project in mind: to create a book of songs for peace from many different nations, cultures, ages, and philosophies (Novick and Berry, 2000). The person who suggested that every meeting should be a party, and every party a meeting, had a good idea. We must find ways to celebrate our humanness, enjoy ourselves, sing, dance, make art and music, since a culture of peace cannot exist without this dimension.

Beyond the arts, culture in the culture of peace is used in a much broader sense, too, as the way we think about things and do things. There is an early reference to culture in this sense in the Seeking the True Meaning of Peace conference report: a culture of violence is said to be present in the "many manifestations of our consciousness and way of life" (Brenes-Castro, 1991: 19). This deep level of culture is also the focus of transformative learning.

The all-pervasive yet invisible quality of culture is what makes the culture of peace such a difficult concept to explain or even to see, since it is inherent in the worldview and way of being of a civilization. In Canada, violence tends to be less overt and its effects further removed than in many other countries, and people often deny or fail to recognize the violence underlying our institutions, our spending priorities, our economy, the forms of energy we use, the way our news is reported, our very food and water.

The description of the word 'culture' in the VOW kit indicates there is more implicit in the culture of peace idea than can be defined and quantified in the official documents: "There's a beautiful fit between the word "culture" and the idea that the culture of peace denotes. Culture, according to its dictionary definition, is both a process and a condition produced; a noun and a verb. It relates both to the natural world and the development of a society, and its scale ranges from microorganisms to the sum total of attainments of a civilization. And most apt of all is its derivations. From the Latin roots *cultura* and *colere,* it means to care for" (Adelson, 1998: 15).

CRITICAL AND VISIONARY PERSPECTIVES

Much of the work of the social movements, including the peace movement, has been in opposition to an unjust status quo, and its focus is active and outward-directed. In contrast, a spiritually based futuristic movement with an inner focus, especially some "New Age" manifestations, posits a peaceful, inevitable transition to a new kind of future with a different type of consciousness. For me, a great strength of both the culture of peace and transformative learning is that they combine these elements.

The culture of peace is a vision of what could be. It is also, as Martin Buber (1949) puts it, a vision of "what should be . . . inseparable from a critical and fundamental relationship to the existing condition of humanity" (7). In other words, the vision is rooted in the actuality of suffering and the need for change.

Encouraging people to envision a culture of peace frees their imagination and empowers them as they see that they and others are already working toward its realization. Concentrating on the visionary aspects allows us to share best practices, for example, by researching and validating traditional peace-building practices from around the world. It also points out to people that the values and skills they use in some areas of their life are transferable to others. This is particularly important in the case of women who all over the world are socialized to be peacemakers and conciliators in the domestic sphere but are barred from using these skills in the public sphere because they are denied access to decision-making and official peace processes.

A critical perspective is at least as important as a visionary one—and extremely difficult to develop and maintain, since what we are looking to change are the foundational assumptions we take for granted, or what Schumacher (1974) calls "the ideas with which we think" (67). This poses several challenges to those of us wanting to effect change. We need to learn to address root causes instead of only reacting to the effects. We need to develop more systemic and holistic ways of looking at issues and more effective ways of working together globally. I believe those of us working for transformation also must develop a greater capacity to be self-critical and to examine whether what we are doing is contributing to genuine change or to perpetuating the system by reinforcing its values, however unwittingly.

Anthropologist George Spindler (1982) makes an insightful distinction between "changes in principle" and "substitute changes." The latter, while giving the appearance of change, actually serve to enable the continuity of the status quo, and I believe we need to be vigilant and discerning, especially in light of the great capacity of the culture of violence to absorb, deflect, and co-opt the forces for change. O'Sullivan's (1999) distinction between "reform criticism" and "transformative criticism" captures the same important nuances (4–5).

Apparently successful initiatives may work as substitute changes. In Canada, we were able to get a cabinet minister and several other high-profile politicians to sign Manifesto 2000. But without any concomitant changes in government policies and priorities, do the signatures signify anything more than an empty gesture at best, a denial of responsibility of Canada's contribution to the culture of war and violence at worst? UNESCO produced a forty-five-second media clip promoting the international year and has been quite successful in having it broadcast, particularly in Europe. But to achieve this, the advertisement had to be tailored to the requirements of the mass media, so the message is infused with the values of the dominant culture that women are largely irrelevant and that only high-profile "stars" make a difference.

On the other hand, what appear as failures may be opportunities for understanding more fully the constraints against change and what needs to be done. In Canada, various government ministries disavowed ownership of the culture of peace. (The relatively obscure Ministry of Heritage finally took it on.) The difficulty in finding an institutional niche enables us to recognize that the culture of peace is not a discrete program that fits into any mandate currently formulated, but rather it needs to become the lens through which all government policies are viewed.

In Canada, criticizing the government is not as dangerous as it is in some parts of the world, but it has its own challenges. There is often a gap between rhetoric and reality, the government tends to make high-profile gestures of peace that belie the larger thrust of its policies, and the effects of harmful policies are often far removed.

Canada has long enjoyed a reputation as a peacekeeper. Peacekeeping, the innovation that won Lester Pearson a Nobel Peace Prize, is a valuable contribution to the culture of peace, although the contemporary practice of sending peacekeeping troops to areas where there is no peace arguably exacerbates rather than alleviates tensions. Canada is also active in peacebuilding, the logical successor to peacekeeping. Referred to in a well-known paper by former UN secretary general Boutros Boutros-Gali (1992), peacebuilding is action in postconflict situations that will help prevent a relapse into conflict. The Canadian government has an innovative peacebuilding program jointly administered by its Ministries of Foreign Affairs and International Development. Current Canadian government policy emphasizes human security and soft power, concepts than have much resonance with the culture of peace. A recent high-profile initiative, the "Ottawa process," which brought about a global ban on antipersonnel land mines, was an unusual partnership of government and nongovernmental organizations. And Canada was the first country with nuclear capability not to develop its own nuclear weapons.

While it is important to recognize and support these encouraging trends, it is necessary and sometimes challenging to be critical and to point out the contradictions and discrepancies behind Canada's peacemaker image. Canada is a major contributor to the global arms trade and has a high military budget, of which peace building is a very small fraction. Our membership in the North Atlantic Treaty Organization (NATO) and close relationship with the United States means that Canada seldom takes independent positions on issues. While Canada officially has renounced nuclear weapons, we contribute to their development in several ways—through making components, allowing them to be tested over Canadian airspace, and, mostly, through the sale of uranium. At home, the historic and current treatment of Aboriginal people remains a serious human rights issue. We also see steady erosion of citizenship rights and a lack of transparency in government. A recent example: plutonium from dismantled nuclear weapons was brought in from the United States by helicopter, a move involving secrecy, deception, a disregard for the opinions of the local communities who had all passed resolutions against it, as well as an-all party committee that renounced the plan as "totally unfeasible" (Standing Committee on Foreign Affairs and Inernational Trade, 1998: 93). Another pressing issue is the deterioration of social programs and a growing disparity between rich and poor, issues that various levels of government blame on each other or on the forces of globalization.

Becoming Human

Is it impossibly naïve to think that a culture of peace could ever exist? I do not believe so. Despite the powerful interests maintaining the culture of war and violence and the sheer magnitude of the transformational task, I have come to see the culture of peace as our natural way of being and living. It has yet to be realized, to be sure, but it is there in potential, dreamed of by every religion and culture, built toward slowly and painstakingly, and hinted at in inspiring instances of peacebuilding and cooperation, even in the most difficult of situations. I recognize it, too, as the "Ecozoic age" described by O'Sullivan (1999) following Berry, an age where there is a new intimacy between all living beings and the cosmos and a far more profound understanding of our responsibilities and mutuality.

I am not a pessimist. Pessimism, exemplified by the original graffito statement "World peace can't happen," precludes action, since there is a sense that nothing can be done. I am not optimistic, either, since optimism, with its blind faith that everything will somehow work out, is a denial of the bloodshed and suffering of our time, the ever-present threat of annihilation posed by nuclear weapons, and the critical condition of the planet. The antithesis of the graffito would portray optimism. My attitude, instead, is that of hope, which I see not as a simplistic emotion but as a complex way of understanding and being, creating and resisting, an active process that both sustains and is sustained by a vision of a desirable future. Hope is the transformed graffito message "World peace can't happen . . . without you!"

This message of hope is the message of Manifesto 2000. Not a petition, it is a powerful and eloquent pledge for commitment to change. While I have some difficulty with the emphasis on collecting large numbers of signatures, seeing it as indicative of the kind of instrumental rationality the manifesto is trying to challenge, I have no doubt that if enough of us took the pledge seriously, we could and would change the world. The underlying rationale is that key to a culture of peace is the notion of a peace promoter. Simple, yes, but difficult; the UNESCO charter in another guise, and also the focus of the VOW workshop and kit.

In Yiddish, to be called a *mensch,* which simply means a person, is a great compliment. Why? The idea captures an important nuance. I see becoming more fully human as a key aspect of our evolution toward a culture of peace, and I believe that despite all the suffering, the violence, and the problems in the world, we are moving in this direction. This progress cannot be understood as linear and incremental; rather it should be seen as a staged developmental process, analogous to individual human development toward maturity.

In this conception, the modern age would be adolescence, with its concentration on individuation, self-assertion, and agency. Modernism was a necessary corrective to the overemphasis on community of the preceding medieval period. The age also ushered in the idea of human rights, the promotion and implementation of which remains an unfinished project of both the official and the unofficial culture of peace. Human rights are a vital aspect of becoming human, for not only are they the rights accorded us by virtue of our being human, they are the principles, the practice of which will make us more fully human.

The modern western conception of the individual, with its emphasis on autonomy and separation, is only one model of being human. In African culture, for example, relationships are central, and the humanity of each person is seen as integrally bound to the humanity of others. The concept is captured by the word *ubuntu*, which translates roughly to the phrase "person is a person through other people." In Berry and O'Sullivan's cosmology, the human can be understood only in the context of the universe, the primary subject, which is "an interacting and genetically related community of beings bound together in an inseparable relationship in space and time" (O'Sullivan, 1999: 185).

With the very survival of the planet and of our species at stake, this concept of interdependence is becoming increasingly accepted, and people are coming to understand that humans cannot exist except in connection with the planet and its life processes. The "logic" of militarism is also unraveling, having reached its zenith with Mutual Assured Destruction, better known by its all too appropriate acronym, MAD. New ideas are arising to challenge the concept that the security of one group can be obtained by threatening or undermining the security of others—ideas like common security, peacebuilding, and soft power.

The culture of war and violence is the antithesis of being human. A basic human concern, indeed a prime focus of all living beings, has been marginalized and ignored: the survival and well-being of the next generation. In "Earthworms, Sunflowers and a Culture of Peace" (Adelson, 1998:15), I put forth a vision of politicians the world over using as their yardstick the effect their decisions will have on children everywhere, rather than the abstract balance-sheet measures like the gross national product. In formulating their proposal and plan of action for a decade of peace and nonviolence for the children of the world, the Nobel peace laureates asked the same basic question: How can we create a world that reflects our humanity?

We still have one foot in the world where to ask such a question, to be concerned with the well-being of the world's children, is seen as hopelessly idealistic and even subversive. We do not yet know how to answer the question, although the program of action on a culture of peace has at least given us a blueprint. But the fact that the question has been asked at all, and with the unanimous support of all nations through the UN General Assembly, is a clear indication that the culture of peace is an idea whose time has come.

NOTES

1. Transformative learning involves experiencing a deep structural shift in basic premises of thought, feelings, and actions. It is a shift of consciousness that dramatically and permanently alters our way of being in the world. Such a shift involves our understanding of ourselves and our self-locations; our relationships with other humans and with the natural world; our understanding of relations of power in interlocking structures of class, race, and gender; our body-awarenesses; our visions of alternative approaches to living; and our sense of possibilities for social justice and peace and personal joy.

REFERENCES

Adams, David. (Ed.). (1995). *UNESCO and a culture of peace: Promoting a global movement.* Paris: UNESCO.

Adelson, Anne. (1998). Earthworms, sunflowers and a culture of peace. In Canadian Voice of Women for Peace. *Creating a culture of peace* (pp. 14–15). Toronto: Canadian Voice of Women for Peace.

Boulding, Elise. (1976). *The underside of history: A view of women through time.* Boulder, Colo.: Westview Press.

Boulding, Elise. (2000). *Cultures of peace: The hidden side of history.* Syracuse, N.Y.: Syracuse University Press.

Boutros-Gali, Boutros. (1992). *An agenda for peace.* New York: United Nations.

Brenes-Castro, Aberlardo. (Ed.). (1991). *Seeking the true meaning of peace.* Conference Proceedings and PostConference Contributions. San José, Costa Rica: University for Peace Press.

Buber, Martin. (1949). *Paths in Utopia.* London: Routledge and Kegan Paul.

Novick, Honey, and Berry, Susan. (2000). *Let's sing together of peace.* Toronto: Creative Visualization Studio.

O'Sullivan, Edmund. (1999). *Transformative learning: Educational vision for the 21st century.* London: Zed Books.

Schumacher, E. F. (1974). *Small is beautiful.* London: Abacus.

Spindler. George. (Ed.). (1982). *Doing the ethnography of schooling.* New York: CBS College Publishing.

Standing Committee on Foreign Affairs and International Trade. (1998). *Canada and the nuclear challenge.* Ottawa: House of Commons.

United Nations Education Science and Cultural Organization (UNESCO). (2000). *Manifesto 2000. www.unesco.org/manifesto2000.*

United Nations Education Science and Cultural Organization (UNESCO)/Federation of African Women Educationalists (FAWE). (1999). *Training module for education for a culture of peace.* Paris: UNESCO/FAWE.

United Nations. (1989). *A/44/626 Declaration for human responsibilities for peace and sustainable development.* New York: United Nations.

CHAPTER 16

Transforming Research

Possibilities for Arts-Informed Scholarship?

J. Gary Knowles and Ardra L. Cole

trans·form,' v.t. transformed; trans, across, beyond, implying change and forma form.
trans·form,' v.i. to be or become changed in form; to be metamorphosed.
trans·form'a·ble, a. capable of being transformed.
trans·for·mā'tion, n. the act or operation of changing the form or external appearance; the state of being transformed, a change in form, appearance . . . etc.
trans·form'a·tive, a. transforming or a tendency to transform.[1]

As researchers, whose university "training" was grounded in the traditions of the social sciences, we have transformed. We are now in a much different intellectual space than we were a decade and a half ago. John Dewey would say that we have been, and are being, educated in the ways of inquiry—given his belief that education is analogous to change. This in itself is hardly noteworthy, given our grounding in the principles of experiential learning and our perspectives on adult and lifelong learning. As professors within the academy we are enveloped in educative possibilities, especially by stimulating graduate students who challenge and mold our intellectual authority and development by their challenging questions, fresh perspectives, and sustained vigor. Transformational possibilities. Transformation.

research, s.b. the act of searching closely or carefully for . . . ; a course of critical inquiry; inquiry into things. Also (without article), as a quality of persons, habitude of carrying out such investigations.
research, v. to search into (a matter or subject) . . . ; to look again.

Our early, traditional experiences of the academy and of the social science disciplines remain a part of our intellectual heritage, but some of these experiences and perspectives are now pushed into the recesses of our thinking. With the support of colleagues and a good amount of courage, we have gradually transformed elements of our professional, scholarly work so that much of it rests in a domain that we have come to describe as "arts-informed research." Indeed, our essential view is that arts-informed research has the transformative potential to reach out from the academy, beyond its sacred halls to communities beyond.

As less mature faculty members who felt relatively secure within the academy, we quickly became disillusioned by the moat of "science and mysticism" built to keep researchers in and communities out of the ivory tower. Bolstered and challenged by our personal histories to build a bridge across the moat, we began questioning the pragmatic value of our traditional-looking scholarship and imagining some new possibilities. Such transformed scholarship was a long time coming to us and, certainly, into academe. For us it came partly out of a search for alternative modes of representation, expression.

Our transformed and transforming scholarship has roots in the traditions of qualitative inquiry, broadly conceived, and notes a reliance on and vital connection to the fine arts, also broadly conceived. These are, in conception and implementation, expansive views of the procedural and representational possibilities of inquiry. In this regard we have extended the characteristics of sound, qualitative research by imbuing inquiries with qualities of process and form grounded in the arts. How did this come about in our own work? And how is this movement connected to the wider academic community, especially to researchers within colleges, schools, and faculties of education? There are personal explanations to these questions as well as those associated with our location in institutional (or university) contexts and the academic community beyond.

scholar (with qualifying adj.), one who is quick (or the reverse) at learning; one who has acquired learning in the "Schools."
scholarly, a., pertaining to or characterizing a scholar; benefiting or natural to a scholar.
scholarly, adv., as befits a scholar.
scholarship, the attainments of a scholar (also, the collective attainments of a scholar); the status of a scholar at a school, college, or university.

Parts of our histories as individuals in the world are connected to our learned resistance to the status quo. We have been saturated with the subtle meanings associated with the characters, beliefs, and actions of family members—perspectives grounded in the struggles of working and lower-middle-class families striving to be heard in their communities. Our family histories are eternally rooted in a pragmatic ideology. We entered the academy at different times, each of us having parents who wondered about the purpose and value of our work and place therein; especially, our researching. As for it being bona fide work: "What bloody use is it?" "Does it make a difference? What good's it going to do?" "Who reads that stuff you write, anyway?" These are questions our parents have asked. In our families, in order to have value, time spent on activities needed to result in something tangible or useful. Work has a fundamental function, utility, or at least something analogous to what some researching colleagues would call a pragmatic validity. Within such perspectives there is no place for academic hedonism. (One of our mothers still awaits the day when we'll get "real jobs.") So it was that we came to the academy expecting that our research would count for something, make a difference.

More than once we have become more than just a little impatient, even indignant, at our family's skepticism about our work. But that has passed. Time tells. Even so, they were justified. Now our one remaining parent is resigned to the fact that she will never understand what it is we do. On reading some of our earlier work she is likely to exclaim something like "It's ruddy unintelligible!" or "Get to the point!" or "What's it got to do with me?" or "How will it make my life better?" "What's the point of that foolishness?" No politeness. No mincing words. Her long years, short attention span, and prickly nature translate into "I do not suffer fools gladly," believing that our work was, in itself, "foolish." She might grill us about the purpose of our work in the academy and wonder about the value of research that is never heard or understood by the public in other than in thirty-second mass media sound bites. She would no doubt challenge our standing; she would challenge our understanding. For her, and many others, research is "way out there," having little to do with "ordinary folk." It has little relevance, is difficult to understand, and has little impact on communities and neighbourhoods, the places where our families resided. Not all research. . . . So, to transformation. To transformative possibilities.

Only if we could convince our mother to pay attention to what we now do; perhaps through the eyes of her younger self. We would give her pause in her thinking about the value of research. We would, perhaps, come to see the worry and uncertainty, even fear, in her eyes fade away some as she came to understand Alzheimer's disease in a different way. We would give her insights that would transform elements of her care of her elderly stepfather who withered away under the debilitating diseases of dementia and Alzheimer's. Our work probably would help her understand the "normalcy" of her feelings, those of an only daughter struggling to care for an elderly parent, and offer insights into the possibilities of support and self-care.

We think that our new scholarship would challenge our mother's conceptions of work (ours and others) in the university "not [being] a real job" and that "It's cushy, easy, and all you do is sit and read!" She would become excited about the researching stories that are told—finely tuned narratives of experience. (Yes, she always liked a good story in which the inherent meaning was revealed in subtle ways.) This was a process and mode she would really understand. Kitchen tables were sites of storytelling and meaning making and work (but work predominated!). And she would be enthralled by poetic and theatrical renderings of research. She would probably even take notice of the visual articulations of inquiry that we pass her way; after all, her husband was an artist and she had some sense of art's transforming power even though she herself did not necessarily subscribe or succumb to art's possibilities. She would have some kind of bodily and emotive response to our researching. She might find it transformative. Later we provide a glimpse into some of the inquiry possibilities we, and colleagues, have realized (including an Alzheimer's disease project). As our orientation is influenced by family so also is it shaped by context.

"Location, location, location," the real estate broker howls in our ears. We are looking for a place to live in a city with sizzling residential house prices. We are tempted to look for out-of-the-way, unpopular, unique properties where we can reside quietly without concern about property values and the compounding of investment dollars. The broker, on the other hand, has greater financial investment aspirations for us, and we are reminded why we are researchers and not businesspeople. But, as far as researching goes, we know his words ring true.

The location, the site of work, can be transformational or it can be constraining, stifling, deadening. We work in an institution that has a history of welcoming alternative perspectives on researching. Close colleagues and others have supported our embrace of alternative researching processes and perspectives. They do not necessarily subscribe to our practice or views, but they are supportive; they are not concerned about a particular turf—"institutional real estate," as it were. Other colleagues around the world—figures of the virtual and real academic communities within which we move—also have provided example and inspiration, and these are essential to the development of intellectual and scholarly lives. We thankfully meet these colleagues at research conferences and occasionally at doctoral thesis oral examinations that are part of the Canadian tradition. In oral examinations these scholars reaffirm the directions of our mentoring work and the bold research of emerging scholars whose work—in process and form—is grounded in the fine arts. Indeed, we are inspired by the aspirations of graduate students who have pushed the boundaries of our own work.

Beyond the various forms of support received at our institution—although not from the "institution" itself but the people therein—we have relied on the spirit of experimentation and challenge that has come from at "home" and beyond. These calls for alternatives to traditional social science research, especially pertaining to educational inquiry, fundamentally urge scholars to critique the status quo in qualitative inquiry. This means:

- To question the prevalence of traditional modes of academic discourse, especially as found in educational dissertations and theses and, specifically, to explore the language of fiction, poetry, theater/drama, visual arts including film and video, as ways of advancing knowledge
- To try out alternative inquiry processes
- To embrace subjectivities and ambiguities
- To make scholarship more relevant and accessible to a public at large
- To honor research participants in fundamentally respectful and inherently ethical ways
- To demystify the processes and representational forms associated with traditional qualitative research.

art, s.b., skill in doing anything as a result of knowledge and practice; scholarship, learning, science; skill displaying itself in the perfection of workmanship (*sic.*); a pursuit or occupation in which skill in directed toward the gratification of taste or production of that which is beautiful; (as in fine art) those in which the mind and imagination are chiefly concerned; put together, join, fit (cf. harmony), hence artful.

inform, v., to give a thing an essential quality or character, to make it what it is, to pervade as a spirit, inspire, animate; to instruct, teach (a person), to furnish with knowledge.

informed, a., put into form; formed, fashioned; instructed, having knowledge or acquaintance with the facts; educated, enlightened. . . .

We hope that by now some of our notions about arts-informed research appear evident. Characteristics of such transformative inquiry work rest in the intersection of the researcher, the research focus, and the process and art form (including more traditional, complementary texts) that eventually comprise the public representations of research findings. Such works, clearly alternative in form or representation, are also as different in process, spirit, purpose, subjectivities, emotion, morality, and responsiveness as they are in passion, provocation, and pragmatics. So it is that arts-informed inquiry is more than just a method, an approach, a procedure.

As a way of providing a glimpse into the realm of arts-informed research, we provide some examples of research projects from our own work. The articulations of this work are varied and illustrate different elements of transformative inquiry. We do not offer the work as exemplars but rather as illustrations of process and possibility. Some are wholly imbued with an arts-informed perspective; others are characterized by more conventional information-gathering phases with arts-informed representations. We also provide examples of some colleagues' inquiries—recent graduates of the Department of Adult Education, Community Development, and Counselling Psychology at The Ontario Institute for Studies in Education, individuals whom Lorri Neilsen would call "scholartists" (in Neilson, Cole, and Knowles, 2001). Their adventurous "scholartistry," along with our own work, hints at the transformative possibilities when the arts are infused into social science research.

Teacher Educators Living in Paradox: The Limitations of Academic Discourse

During research conversations with professors of teacher education in a life history study, Ardra became acutely aware that some of the experiences being recounted were so imbued with emotion and poignant illustrations of the often dysfunctional relationship between academic institutions and individual faculty members that conventional forms of representing these experiences seemed inadequate. Frequently the teacher educators used graphic language to create images or metaphors to describe elements of their experience. Listening to them struggling to find words to adequately convey the passion and emotion felt when they talked about certain issues and experiences and, later, searching for appropriate words to convey interpretations revealed the inherent limitations of words, especially when crafted in conventional academic prose. In a search for a more authentic and meaningful representational form that would more closely render the aesthetic of lived experience, however partial, and afford readers better opportunities for their own resonant interpretations, Ardra turned to multimedia and to the tableau art form. She was drawn to the form as articulated by the American artist Edward Kienholz, who devoted a good part of his life (1927–1994) to producing large-scale, provocative works.

In conjunction with other life history researchers Ardra developed a three-part multimedia installation (Cole, Knowles, brown, and Buttignol, 1999a, 1999b, 1999c). The title, *Living in Paradox,* reflects the overarching theme that emerged through analysis. Each part of the three-part installation represents a predominant convergent theme that defines elements of the teacher educators' experiences as pretenured faculty members in university-based teacher education programs.

"Installation I: Academic Altarcations" suggests the paradox of sacrifice. The installation is a three-dimensional, life-size "altar." A simulated blackboard and other images of the university surround the altar: an ivy-covered wall, a men's room, a professor's office door, and adjacent message board. On the blackboard are words and phrases depicting the professors' emotion-laden perspectives about power differentials within institutions. Other issues and expressions are faintly printed as in previous, erased, chalk writing. Laid on the altar are symbols of the academy and tenure.

An electrically driven, bloodred conveyor belt carries, on small white satin pillows, symbols of personal sacrifice to the altar of the academy. These are sources of these professors' pains. Piled on the floor at the foot of the altar are more pillows. Background music of Gregorian chants is contrasted with audio-recorded chants composed from the professors' words—words that rationalize their sacrifices.

"Installation II: Wrestling Differences" is a commentary on the gendered nature of the academy. The installation is a simulation of a wrestling arena with World Wrestling Federation–syndicated, toy action figures pitted against a small female figure. She appears vulnerable yet defiantly resistant. The black emblem-embossed ring is set among a crowd of cheering onlookers. Poignant narrative excerpts of the teacher educators' experiences are projected onto the spot-lit and smoke-filled painted backdrop of the arena and inscribed on the bleacher-like supports of the installation.

"Installation III: A Perfect Imbalance" is less complex. It depicts the dual mandate of teacher educators' work that requires them to serve both the academy and the profession and that keeps their gaze focused on the fulcrum of their lives striving for balance. The installation is a simple balance scale weighted on either side with foam blocks that represent the many and varied roles and responsibilities of teacher educators' work and lives. On one side is a high tower of blocks each representing a different activity or role required of the professors of teacher education; on the other side is a single, multifaceted block that represents the activities the university deems most meritorious. This single item, a foam block, is weighted on the inside so that the scales will balance only when the entire pile of blocks identifying teacher educators' work is in place. The result begs the question of whether the scale is indeed balanced.

The information gathered through this project demanded an alternative representation. The installations represent a transformation from "conventional research" language and evoke a transformation through multimodal language. On each of several exhibits of this work teacher education professors and others have expressed strong resonance with it. The installation has evoked ranges of emotion from tears to expressions of affirmation, from unfettered anger to rational reflection, from silence to the recounting of new and similar stories of experience. We know this work has "spoken"' to many individuals.

LIVING AND DYING WITH DIGNITY: THE ALZHEIMER'S DISEASE PROJECT—INSTALLATIONS FOR COMMUNITY DEVELOPMENT

Ardra and Maura McIntyre share a common history. The personal and familial devastation that comes with a mother developing Alzheimer's disease pushed them into this inquiry. It is work from the heart. Their research program focuses on issues of care and caregiving for family members affected by Alzheimer's disease. It has two audiences—the general public and family caregivers of persons with Alzheimer's—and a commitment to knowledge advancement; education; and, community development. One purpose is to educate the public about the sociocultural dimension of Alzheimer's and the complex issues related to care. The second purpose is to help caregivers broaden their understandings of the disease and their experience of caregiving. The research is intended to honor the psychosocial dimension of family caregivers' experience with a specific aim of reducing isolation and promote well-being through opportunities for comfort, connection, and social support. Broadly, Ardra and Maura are interested in knowing more about how to better educate and prepare communities and society at large to respond to and care for those with Alzheimer's disease. Their primary emphasis is on the preservation of human dignity within the context of the disease.

Because the research is part of an agenda for community education and development and, therefore, concerned with issues of research relevance and accessibility, they employ alternative forms such as photography, narrative, three-dimensional imagery, and performance to enhance and aid in research communication. From data gathered from a variety of sources, including personal experience, they conceptualized, designed, and created a series of large, three-dimensional multimedia works, which represent significant themes and issues related to Alzeimer's disease and caregiving, all subsumed under the overarching theme of dignity. The exhibits also poignantly depict some of the characteristic problems and behaviors symptomatic of the illness. Several of the installations are completed; others are still under construction. Ardra and Maura are working to complete them and begin working with community and caregiver groups.

Their experience to date of displaying the representations has exceeded their expectations about the power of the exhibits to evoke the experience of caregiving. In one instance, word of the exhibit at an academic conference reached not only the official conference attendees but also support staff and other nonprofessionals who happened to be in the building. The exhibit audience was diverse beyond expectation. This is the power of the work as a three-dimensional palpable presence to evoke and access a wide range of responses. This is the kind of engagement arts-informed research can inspire.

"Dance Me to an Understanding of Teaching": Pedagogy as Performance

Ardra and Maura also were research partners in another arts-informed research project—a collaborative reflexive inquiry into Ardra's teaching. Working with data gathered in the conventional ways of interview, observation, and document collection (in this case, autobiographical writing), they composed and performed together at an international conference a three-act choreographed narrative involving text, musical selections, and dance. The project had two broad purposes: to better understand the relationship between teaching and learning, specifically as they relate to one's autobiography; and to explore the use of nonconventional forms for understanding and representing teaching and learning.[2]

Each of the three acts—Constraint, Conflict, and Creativity—is a metaphorical dance through the development and transformation of a teacher. Ardra is represented as young learner, teacher, then teacher educator, and Maura, her reflective mirror/alter-ego. Their reason for taking an artful approach to the reflexive study of teaching and learning is to comment on the inadequacy of conventional forms of inquiry and conventional linguistic forms of representation. They wanted to find a way to push at the boundaries created by an overreliance on linguistic forms to define and represent knowledge. Capturing and communicating the complex, contextual, and multidimensional nature of the teaching and learning processes as lived became a dynamic process.

Creating, composing, choreographing, and performing were all integral components of the inquiry and analysis process; the performance was much more than a presentation of "findings." In arts-informed inquiry, in particular, representation of the research text has a dual function or role. It is both communicative and evocative. It has both authorial intent to convey a substantive message and generative intent to engage the audience in a reflexive process. Put another way, there is an intention for the audience to "get it" and an intention for the audience to "experience it."

In each of the three acts of the performance, Ardra and Maura used a different text form to convey a particular tone, attitude, and theme. The different forms of text also convey a movement, from independence to conflict to interdependence, which captures the dynamic nature of development. Leaving the development unresolved at the end imparts a "truthful" quality and represents an ongoing and dynamic state of internal dialogue.

Like metaphor, music conveys so much more than words can say. Figuratively and literally, music has resonant qualities. Well represented research also has resonant qualities that attest to its rigor. In the performance, the text can conceivably be read through the music alone. The themes of constraint, conflict, and creativity are reflected in the musical selections. Combined with narrative text, however, the music adds a layer of meaning that both communicates and evokes feelings, associations, and thoughts differently for different people. The music makes the text come alive in a way that words alone do not, and the presence of music adds considerably to the resonant quality of the work for others.

The physical act of performing in costume is both an act of re-creation and creation. Performance of the research text is an embodiment and representation of the inquiry process as well as a new process of active learning. The performance and inquiry are transforming processes.

Place and Pedagogy: Collecting Information, Developing Understandings through Artistic Inquiry

Working first alone and more recently with Suzanne Thomas (2001a, 2001b), Gary has written extensively on the influence of place on pedagogy. This is not definitive, theoretical writing but is composed of book-length manuscripts, written in the spirit of American nature writers (John Muir's work being especially influential), in the form of a novel, and in journal form. With Suzanne he embarked on articulating, through two- and three-dimensional art forms, the experiences that informed the thinking behind pedagogical work in schools and the academy. In this work Gary and Suzanne expressly want to understand the relationship between embodied and experiential knowing and the subtleties of pedagogical practices. To do this they engaged in work that is at once "data" and representational form. They traced, through art making, narrative, and poetry infused with the theoretical understandings of other authors, the patterns and processes of their place experiences, place memories. Through this vehicle they explore questions of location and experience and how place perspectives and understandings influence fundamental pedagogical assumptions. The main form of work presented thus far in this project has been two large-scale paintings that incorporate multipaneled, multimedia (including paint on canvas and wood, narrative and poetic text on Mylar, photography, and found objects) representations of experience, place, and pedagogy.

STUDENTS' SENSE OF PLACE IN SCHOOLS: AN EMPHASIS ON PROCESS AND FORM

A serendipitous moment in an art gallery led to a research project involving art students in Toronto high schools.[3] Marlene Creates's photographic exhibition of people's experiences of place was highly evocative—charged with meanings associated with people located in very particular places and communities in Newfoundland, Canada. It depicted elements of life histories and was instantly recognized by Gary as artistically creative, visual, and textual life history research. Creates's model of inquiry consisted of assemblages that explore the relationship between human experience and landscape, nature and culture, and sense of place.

With Suzanne Thomas, Gary began to meld and fuse the artistic processes and perspectives of Creates's work into life history research projects (Knowles and Thomas, 2001a, 2001b). The intention was not to infringe on the creativity and uniqueness of the artist but, rather, to learn from and be responsive to the discipline of visual art as a way of informing alternative procedural and representational modes of inquiry. Artists' inquiries have the potential to enliven and inform scholarly work by:

> Enhancing researchers' conceptualizations of information gathering and the transformation of data and/or artifacts into representational forms. . . . Arts-informed perspectives, gleaned and imagined through the perceptive lens of visual artists, may enrich the complex rendering of personal stories and life histories in ways that are multi-textured, multi-layered, and multi-dimensional. The purpose of [the] project was to examine secondary school art students' experiences of "place" in schools. Our aim was to gain a sense of their contemporary lived experiences of schools . . . and to apply our understandings of Marlene Creates's artistic processes in our methodological approach. (Knowles and Thomas, 2001b: 210).

The "structure" and transparent processes of Creates's work was modified for the purposes of the students' work and was used as a framing guide: "This new, inquiry model . . . includes seven, linked elements which provided a multi-dimensional representation and format from which the student artists worked" and includes "several photographs, a narrative, a sketch, a two- or three-dimensional representation, and found objects" (Knowles and Thomas, 2001b: 211). The focus of the work was to get insights into students' experiences of school from their vantage point rather than those of adult researchers. There is little school-based research that places information gathering and respresentational responsibilities in the hands of students: "[The students] were at once information gatherers, portraiture artists, interpreters of experience and assumed full control over their individual inquiry processes" (Knowles and Thomas, 2001b: 211). Students' artistic "creations became at once 'data' and representations indicative of the inquiry focus. The artworks . . . in addition to conversations with the students . . . provided a comprehensive synthesis of 'data'/artifact gathering and analysis" (Knowles and Thomas, 2001b: 211, 213).

The researchers "'interrogated' the works for understandings of the students' individual and collective . . . experiences of school . . . [and] . . . 'asked questions' of the representations themselves, attempting to explore the meanings of the works, much as art viewers may do of a professional artist's exhibition . . . [doing so] with sensitivity, rather than a spirit of artistic connoisseurship. . . ." (Knowles and Cole, 2001b: 213). Gary and Suzanne brought their own experiences as artists and teachers to the conversations and public displays of the work. The project is in its second year, and twice the students' artistic research stories have been exhibited at a Toronto contemporary art gallery, the second time during the prestigious, citywide Contact 2001 photographic exhibition. Public displays also have been held in Louisiana and New York, and on all occasions extensive conversations about students' places in school and issues surrounding the conditions within public schools were foremost in the gallery discussions. Such exhibits, while costly to mount, create a different kind of public conversation, and the students' passions and honesty showed though. The students claim that their inquiry experiences were both empowering and satisfying, transforming in the broadest sense.

Emerging scholars, "scholartists" with whom we have recently worked, have pushed past the boundaries of their initial conceptions of doctoral thesis work and a university education and have produced inspiring works that have been personally transformative as well as having other transformative potential. Each researcher-artist has paid particular attention to the representation of his/her research. brenda brown (2000) has fully grounded her work in a particular art form, and her work has an internal consistency borne of congruence between process and form.

Telling without Telling, The Power of Poetic Fiction: Confronting Childhood Sexual Abuse

As brenda brown (a.k.a. Rowen Crowe) (2000) would acknowledge, the focus of her work had been with her for decades. She came to the university as a doctoral student knowing what she needed to do but unsure of vehicle for achieving her goals. She was passionate about her inquiry and worked tirelessly to complete the project. Arts-informed research allowed her to safely explore a sensitive topic. In her dissertation "Lost Bodies and Wild Imaginations: Expressing the Forbidden Tales of Childhood Sexual Abuse through Artful Inquiry," she presents for readers a poetic fiction "spun from the imaginary realm, held within theory, while engaging in a collaborative relationship with art, story and lived experience" (2000: ii). She goes on to tell of her work as "an exploration of what it's like to tell about childhood sexual abuse through artistic enterprise. Touching a touchy topic. It is also a deliberation on the process of telling. And an actual telling by Rowan, a reluctant story teller. . . . This is a testimony to lives lost and lives reclaimed, to the power of the imagination to return our histories to us and return these histories to their rightful place in the world" (2000: ii-iii).

Method and Metaphor: An Artful Exploration of Professional Practice

Maura McIntyre's (2000) work is grounded in arts-informed, life history inquiry (see Cole and Knowles, 2001) and is as much an exploration of methodological issues as it is a metaphorical articulation of lives lived. Her "detailed consideration of methodological issues is guided by the principle of creative, aesthetic and imaginative attention to process and relationship characteristic of arts informed research. . . ." (MacIntyre, 2000, p. iii). Maura uses the dissertation inquiry, entitled "Garden as Phenomenon, Method and Metaphor in the Context of Health Care: An Arts-Informed Life History View," to explore the notion of garden "in the personal professional lives of two women. . . . The garden as method . . . explores . . . our discovery of horticultural therapy as professional practice, and considers how the formalization of intuitive and common sense knowledge impacts . . . practitioner . . . work. . . . As metaphor this is a story which speaks to the influence of the garden . . . the particular people-plant connection, as a forum and mechanism for experiences of authenticity and connection" (2000: ii). Maura uses a variety of literary devices to invoke reader's engagement with her work.

MEN AT THE EDGE OF RETIREMENT: UPSTAGING THE EXPECTED, A TURN OF EVENTS IN A CONVENTIONAL LIFE HISTORY STUDY

Eric Miller (2001) was in the final stages of information analysis when Gary, his "supervisor," received a telephone call from him. Gary remembers Eric saying that the conventional approach to "reporting the data just doesn't work. These are colorful men, and I do not think I can do justice to them by using conventional forms." Gary was not surprised at Eric's response. Eric knew about arts-informed approaches and intuitively knew that he had to act differently. That a stage play script became the vehicle of representation was surprising given the more conventional beginnings of his project. "Closing Time, Men, Identity, Vocation, and the End of Work: A Stage Play as a Representation of Lives" is the title of his dissertation, which has a conventional approach to information gathering but diverges at the point of interpretation and representation. Eric wanted to "learn how the men characterized their careers and how they experienced the prospect and imagined consequences of vocational disengagement." While it had been his "intention to present the stories of the five men as separate stories" (140), he quickly determined otherwise. The stage play, *Closing Time,* is crafted around a fictional dialogue of five men and evokes the full range of emotional and intellectual issues with which they grappled. Within it Eric has crafted the responses and direction of the research act with the researcher fictionalized as a reporter, bringing these characters together for the purposes of writing a story about men at retirement.

WEBS OF MEANING-MAKING: METAPHOR AND STORY

A missionary kid in Africa, now adult, Lois Kunkel "wanted to understand how those 'missionary years' continued to influence [her] life and that of other [adult missionary kids]" (2000: ii). Lois's arts-informed dissertation, "Spiders Spin Silk: Reflections of Missionary Kids at Midlife," is an evocative literary work built with layer upon layer of metaphor and meaning. "West Africa, where I was raised, is the home of Ananse the Spider. The hermenutical framework for this inquiry is an Ananse story and the spider provides the metaphor around which the stories and experiences derived through the inquiry process are spun" (ii).

The account places the adult missionary kids' stories centerfold: "Telling stories and having their stories witnessed [referring to the participants in her inquiry] so that the teller can receive herself/himself in a new way is a form of arts-informed reflexive inquiry" (Kunkel, 2000: ii). In essence this work is about transformation of perspective that comes with telling.

Unlike processes of qualitative research as articulated in the 1980s, for instance, which were more discreetly defined and "regulated," eventually to be judged according to criteria grounded in positivism—reliability and validity, for example—the organic and fluid nature of transformational research speaks of a different set of values. Arts-informed research is based on a different frame, a different set of assumptions from traditional, qualitative inquiry processes. It is not highly structured or prescriptive, it is organic and emergent. It is about alternative ways of coming to know and of dispersing or sharing that knowledge.

Pervading qualities that comprise transformational, arts-informed work are "the aesthetic" and "the passionate." By aesthetic we mean consideration of the enduring principles of form and composition, of weight and light, of color and line, of texture and tone, as when working in the painterly arts, for example. (These concepts have parallels in other art forms.) The aesthetic element revolves around central principles upheld in a variety of art forms, be they dramatic, literary, performance, or music. The element of passion is evident at every phase of the research process and is as much grounded in the morality of purpose as it is in the morality of representation and even the pragmatics associated with seeking out alternative audiences, essentially the public at large. It is also evident in the doggedness required to bring to completion in the fullest sense possible, given energies and resources of researchers, work that is alternative and transformative in both vision and "reality."

We are speaking of research that, clearly, needs to be judged in ways different from how "traditional" research is assessed. We present some defining elements for making judgments about the "goodness" of transformative, arts-informed research:

- Intentionality. All research has one or more purposes but not all research is driven by a moral commitment. "Good" arts-informed research has both a clear *intellectual purpose* and *moral purpose.* Ultimately the research must stand for something.
- Researcher Presence. A researcher's presence is evident in a number of ways throughout the research account (in whatever form it is presented and, by implication, throughout the entire researching process). We especially infer a degree of self-consciousness in the procedural elements of the work. Moreover, we infer that researchers can learn from artists about matters of process. That is, the processes of the arts inform the inquiry in ways congruent with the artistic sensitivities and "technical (artistic) strengths" of the researcher in concert with the overall spirit and purpose of the inquiry. The researcher is present through an explicit *reflexive self-accounting;* her presence is also implied and *"felt";* and, the research text (the representational form) clearly bears her *signature or fingerprint.* Such research texts explicitly (although perhaps subtly) reveal the intersection of a researcher's life with that or those of the researched.
- Methodological Commitment. The arts-informed approach is guided by a set of articulated principles that are reflected throughout the inquiry process. As such the work reflects a methodological commitment through evidence of a

principled process and *procedural harmony.* These qualities evidence a deep reflexiveness about processes grounded in a set of coherent ideological principles. Among other things this implies the presence of a certain fluidity along with an organic, emerging process and openness to varieties of representational forms.

- Holistic Quality. From purpose to method to interpretation and representation, arts-informed research is a holistic process and rendering that runs counter to more conventional research endeavors that tend to be more linear, sequential, compartmentalized, and distanced from researcher and participants. A rigorous arts-informed "text" is imbued with *an internal consistency* and *coherence* that represents its seamless quality. Such a representation also evidences a high level of *authenticity* that speaks to the truthfulness and sincerity of the research relationship, process of inquiry, interpretation, and representational form.
- Communicability. Foremost in arts-informed work are issues related to audience and the *transformative potential* of the work. Research that maximizes its communicative potential addresses concerns about the *accessibility* of the research account usually through the form and language in which it is written, performed, or otherwise presented. Accessibility is related to the potential for audience receptiveness and response. Such representations of research have the express purpose of connecting, in a holistic way, with the hearts, souls, and minds of the audience—they are intended to have an *evocative quality* and a high level of *resonance.*
- Aesthetic Form. How research-informed insights are conveyed is as important as what insights are conveyed. Attention to the aesthetics of form is clearly important. Here we are concerned both with the *aesthetic quality* of the research account and its *aesthetic appeal.* By the former we mean how well the form adheres to a particular set of artistic processes and conventions. For example: Does the chosen form—say, of the novel—follow the conventions of that genre? Is the process of visual art making explicitly honored and adequately represented in the work? By the latter we mean how well the form "works" as a mode of communication.
- Knowledge Claims. Research is about advancing knowledge however "knowledge" is defined. As researchers, we make claims about what we have come to know through our work, and we do this in a variety of explicit and subtle ways. We reject any notions about the possibilities of an absolute and objective Truth that relieves the researcher of any responsibilities for making knowledge claims that are conclusive, finite, and universal. Any knowledge claims made must reflect the multidimensional, complex, dynamic, intersubjective, and contextual nature of human experience. In so doing, knowledge claims must be made with sufficient *ambiguity* and *humility* to allow for multiple interpretations and reader response.
- Contributions. Tied to the intellectual and moral purposes of arts-informed research are its theoretical and practical contributions. Sound and rigorous arts-informed work has both *theoretical potential* and *transformative potential.* The former acknowledges the centrality of the so what? question and the power of the inquiry work to provide insight into individual lives and, more generally, the human condition while the latter urges us, as researchers, to imagine new possibilities for those whom our work is about and for. We are not passive agents of either the state or the university or any other agency of society. We

have responsibilities toward fellow humans, our neighbors, and community members. Researching is never a faceless task!

The transformative potential of and by arts-informed research speaks to the need for researchers to develop representations that speak to audiences in ways which do not pacify or indulge the senses but arouse them and the intellect to new heights of response and action. In essence, and ideally, the educative possibilities of arts-informed work are foremost in the heart, soul, and mind of both the researcher and her purpose from the onset of an inquiry. The possibilities of such educative endeavors, broadly defined, are near limitless; their power to inform and provoke action are constrained only by the human spirit and its energies.

NOTES

1. All dictionary-like entries/definitions are composites, the result of consulting *both Webster's Universal Unabridged Dictionary* (Second edition, 1979) and the *Compact Edition of the Oxford English Dictionary* (1971).
2. The script of "Dance Me . . ." and a fuller discussion of the inquiry process and performance can be found in Cole and McIntyre (2001).
3. See Chapter 3 and Chapter 17 in *Researching Lives in Context: The Art of Life History Research* (Cole and Knowles, 2001). Another account of this work is found in C. Bagley and M. B. Cancienne's (2002) book *Dancing the Data* and in a CD that accompanies the publication. The CD contains images and narratives of the student researchers.

REFERENCES

Bagley, C. Cancienne, M. B. (2002). *Dancing the data.* (A volume in the Leslie College Arts and Learning Series.) New York: Peter Lang Publisher (in press).

brown, b. (2000). Lost bodies and wild imaginations: Expressing the forbidden tales of childhood sexual abuse through artful inquiry. Unpublished doctoral dissertation, University of Toronto, Toronto, Ontario.

Cole, A.L., and Knowles, J. G. (Eds.). (2001). *Researching lives in context: The art of life history research.* Walnut Creek, Calif.: Altamira Press.

Cole, A. L, Knowles, J. G., brown, b., and Butignoll, M. (1999a). *Academic altercations (first installation in) Living in paradox: A multi-media representation of teacher educators' lives in context (version II).* Multimedia (paint on canvas, timber, electric conveyor and conveyor belt, silk cushions, found objects, academic regalia, linen cloth, silver candle holders, music/recorded voices/narrative, academic and school clothing), three-dimensional (8′–0″ x 8′–0″ x 4′–0″) installation presented at the Annual Meeting of the Canadian Society for the Study of Education, Sherbrooke, Quebec, Canada, June 9–13.

Cole, A. L, Knowles, J. G., brown, b., and Butignoll, M. (1999b). *Wrestling differences (second installation in) Living in paradox: A multi-media representation of teacher educators' lives in context (Version II).* Multimedia (paint and crayon on canvas and plywood, photographic text-based images, action figures, plastic and elastic) three-dimensional (4′–0″ x 3′–6″) installation presented at the Annual Meeting of the Canadian Society for the Study of Education, Sherbrooke, Quebec, Canada, June 9–13.

Cole, A. L, Knowles, J. G., brown, b., and Butignoll, M. (1999c). *A perfect imbalance (third installation in) Living in paradox: A multimedia representation of teacher educators' lives in context (Version II).* Multimedia (paint and text on foam, balance beam/scales), three dimensional (1′–6″ x

1′–0″ x 6″) installation presented at the annual meeting of the Canadian Society for the Study of Education, Sherbrooke, Quebec, Canada, June 9–13.

Cole, A. L., and McIntyre, M. (2001). Dance me to an understanding of teaching: A performative text. *Journal of Curriculum Theorizing, 17* (2):.43–60.

Knowles, J. G., and Thomas, S. (2002). Artistry, inquiry and sense-of-place: Secondary school students portraying self-in-context. In C. Bagley and M. B. Cancienne. (Eds.). *Dancing the data.* (A volume in the Leslie College Arts and Learning Series). New York: Peter Lang Publisher. (in press)

Knowles, J. G., and Thomas, S. (2001). Insights and inspiration from artist's work, envisioning and portraying lives-in-context. In A. L Cole and J. G. Knowles. (Eds.), *Researching lives in context: The art of life history research* (pp. 208–14). Walnut Creek, Calif.: Altamira Press.

Kunkel, L. (2000). Spiders spin silk: Reflections of missionary kids at midlife. Unpublished doctoral dissertation, University of Toronto, Toronto, Ontario.

McIntyre, M. (2000). Garden as phenomenon, method and metaphor in the context of health care: An arts-informed life history view. Unpublished doctoral dissertation, University of Toronto, Toronto, Ontario.

Miller, E. (2001). Closing time, men, identity, vocation, and the end of work: A stage play as a representation of lives. Unpublished doctoral dissertation, University of Toronto, Toronto, Ontario.

Neilsen, L., Cole, A. L., and Knowles, J. G. (Eds.). (2001). *The art of writing inquiry.* Halifax, N.S: Backalong Books and Centre for Arts-informed Research.

CHAPTER 17

On Speaking Terms Again

Transformation of the Human–Earth Relationship Through Spontaneous Painting

Lisa M. Lipsett

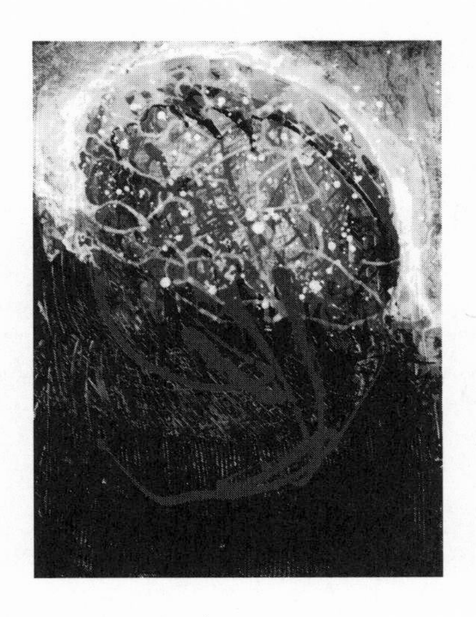

Figure 17.1 Earthroots

. . . I know I am made from this Earth, as my mother's hands were made from this Earth, her dreams came from this Earth and all that I know, I know in this Earth, the body of the bird, this pen, this paper, these hands, this tongue speaking, all that I know speaks to me through this Earth and I long to tell you, you who are Earth too, and listen as we speak to each other of what we know: the light is in us.

—Susan Griffin, *Women and Nature*

I am shocked into deep sadness at times. It is a grieving that I enter when I realize that the western-schooled human mind has been so closed, so systematically molded by the rational and the mechanistic that it is effectively not fit to live sustainably with other life-forms. The global ecological crisis facing us today is a mirror of this precarious state. How can we begin to undo, to open out to the wisdom that once came so naturally to us as children, that still comes so naturally to adults living in many indigenous cultures?

To be able to speak to each other clearly again, to be able to decode each other's messages, to be moved to act on behalf of the Earth in response to those exchanges, and to be able to make sense is the vision.

I have been painting spontaneously for three years now and have co-created hundreds of images. Over this period, I have experienced the fear associated with trusting that the process will be safe and nurturing. I have combated my urge to direct the outcome and analyze the resulting colors and forms. I continue to learn about the ways of the Earth through painting and am in awe of its power.

I may start a painting with the question: "Where am I at today?" and emerge with a beautiful green landscape or cosmic flow of colors. The painting is simultaneously me and the Earth. I can no longer separate the two. I love the feel of flowing clouds, flowers, and seas. I am clouds, flowers and seas. I have begun to fully experience myself as part of the natural order. Somehow I have come to know that the language of paints is a natural language. It is the language of all life. Colors run into one another to create new color and form. Sprinkled water droplets explode into starbursts while rivers of color flow from one side of the page to the other, as if transiting a landscape.

After many paintings, I now seem to experience colors in a more embodied way. I do not just see them and name them. Rather I feel them and am often moved by them. These vibrant landscapes have helped me to know that my emotions are an important part of the natural order. They help me to maintain a deep Earth connection. As well, fully experiencing and understanding my feelings makes me more attuned and feeling toward the natural world.

As a result I have become connected to this contemplative way of being in the world. It continually reminds me of its power. I feel part of the creative life force in a way that I never before thought possible. I dive into the colors and my soul takes flight. If I am able to open myself up by "letting go," I reap the brilliant shining treasure of Earth-connection. It seems that this "letting go" is universal and therefore available to anyone who is interested in experiencing a richer, more sustainable Earth connection. This work is an ongoing commitment to opening out, receiving and then giving back again.

By opening up to the Earth, new understandings are attained that are ageless and wise. We feel our connection with all living beings. Transformation of the human-Earth relationship is no small task in a culture that values reason over imagination, compartmentalized hierarchy over holism, action over contemplation, and domestication over wildness. Yet despite the perceived obstacles, many are screaming out that we need to once again become on speaking terms with the natural world (Roszak, Gomes, and Kanner, 1995). Human-Earth reconnection through spontaneous painting is my response to this call. Spontaneous painting over time nurtures our wild Earthy aspects and opens our hearts and minds to all Earth beings. By emptying ourselves out, we create space to let nature in. We become one with the wild spontaneous life force and learn a new way of listening. We co-create sustainable guiding visions, honor the Earth with our authentic creations, and find ourselves on speaking terms with the Earth once again.

Many believe that the consequences for not reestablishing such a sustainable human-Earth relationship will be devastating (Carson, 1962; Durning, 1992; Fox, 1990; Griffin, 1978; LaChapelle, 1988; Macy, 1991; Merchant, 1981; O'Sullivan, 1999; Purpel, 1989; Roszak, 1992; Shepard, 1982). The evidence is clear that nothing less than the survival of all life on Earth is at stake. Recent research by the World Watch Institute reveals that traditional environmental strategies have not allowed us to reverse the tides of deforestation, continued ocean dumping, widespread pollution, rapid climate change, habitat degradation, and continued inadequate access to clean food and water for the vast majority of living beings (Brown, Renner, and Flavin, 1998). It is clear that transformation of the current human-Earth relationship is imperative.

Simply being in the wilderness can be transformative. Wilderness immersion, whether it is a camping trip, a weekend hike, or sitting by a body of water, develops a renewed sense of connection and belonging to the Earth. We develop a sense of place by going out into wild spaces or back to sacred childhood places (Thomashow, 1996). We also cultivate a sense of spaciousness by exploring new wild places. We feel expansive, full of freedom and new possibility (Berry, 1988; Porteous, 1990; Tuan, 1979).

A sense of connection to the wild also develops when we connect to our local natural places. We can spend time observing the relationships between beings and reroot ourselves in the places we call home. Rediscovering the natural history of the places currently in our lives reconnects us to wilderness on a larger scale. This relationship to local places is crucial for the animation of our connection to the Earth in general. It is the linking of the psyche to the natural (Berry, 1988; Orr, 1992; Thomashow, 1998).

We also can engage in Earth-based rituals and exercises that expand our sense of self to include all living beings (Seed, Macy, Fleming, and Naess, 1988). We can go on vision quests and nature retreats (Clinebell, 1996). More drastically, we can move back to the land and live a simpler life in the spirit of Henry David Thoreau (1979). But is all this enough?

Many nature writers express despair and sadness in the face of the pollution and degradation of both their once-wild childhood places and their local bioregions (Carson, 1962). Since the experience of a place is so intertwined with who we are, the ravaging of landscapes is psychologically ravaging as well (Orr, 1992; Roszak, Gomes, and Kanner, 1995). As a result, we must learn to cope with the despair and heartache associated with the now degraded nature of places we once held sacred (Macy, 1991).

Also, wild spaces are becoming harder to find as more of the planet's surface is colonized by the telltale signs of the western consumer lifestyle. Over half the world's population now lives in cities and does not have regular access to the reconnecting affects of being on a mountaintop or in a forest (Brown, Renner, and Flavin, 1998). Even when we are able to retreat to the wild and simplify life, we carry the powerful vestiges of our socialization and our culture with us. "I am alarmed when it happens that I have walked more than a mile into the woods bodily, without getting there in spirit. In my afternoon walk I would fain forget all my morning pre-occupations and my obligations to society. But it sometimes happens that I can not shake off the village" (Thoreau, 1979: 600).

Further, many report that changed feelings and perceptions gained as a result of a wilderness experience wear off after a few days back in the routine of normal modern life (Devereux, 1996; Greenway, 1995; Harper, 1995). It seems we need a therapy of sorts that will open us up to the experience of what it is to feel

Earth-connection. Educating the mind does not seem like enough. "Only experiences that profoundly alter our view of nature and reconnect us with the divinity in ourselves and in the environment can empower people to commit themselves to the prodigious task before them. The therapeutic methods must be powerful enough to shift the ground of our being so that we experience the Earth in its living reality" (Mack, 1995: 284).

Therefore it seems that transformation of the human-Earth relationship cannot be attained solely from placing ourselves in natural settings. Although immersion in wild places is crucial for the development of a sense of connection, there must also be ongoing animation of a deeper connection to all Earth beings. This connection is arrived at only when there is an ongoing daily sense of wholeness or integration between ourselves and the Earth regardless of the setting we may find ourselves in (Harper, 1995). While being in the wilderness has the power to heal the body and mind (Clinebell, 1996; Roszak, Gomer, and Kanner, 1995), the body and mind also need to heal in order to fully and sustainably "be" in the wilderness (Fleischman, 1997).

Part and parcel of this healing is the rebalancing of our relationship to our own inherent wildness. It is this untamed or wild aspect of self that we find frightening and threatening and therefore, often avoid, control, and distrust it. However, when we distance from our wildness, we enter into a form of eco-alienation (Clinebell, 1996: 32). The human attempt to tame the wild has simultaneously tamed our deepest potentials.

"When Thoreau said in his essay on walking, 'in wildness is the preservation of the world,' he made a statement of unsurpassed significance in human affairs. I know of no more comprehensive critique of civilization itself, this immense effort that has been made over these past ten thousand years to bring the natural world under human control. Such an effort that would even tame the inner wildness of the human itself. It would end by reducing these vast creative possibilities of the human to trivial modes of expression" (Berry, 1999: 69–70). The alienation from the wild that results from our fervent attempts to tame and control it out of fear also leaves us with a lack of energy and desire to act on behalf of the planet. We have become impaired in our ability to access a guiding vision that will help us act in ways that adequately address the complex environmental problems of our time.

"Often we find that our attempts to fix things only end up by making them worse. Part of the impasse is that in dealing with an intricately interconnected network of patterns on the scale of the global ecology, neither our reasoning faculties nor our feeling faculties are equal to the job. The only capacity that our species has that is powerful enough to pull us out of this predicament is our self-realizing imagination. The only antidote to destruction is creation" (Nachmanovitch, 1990: 182). It is possible to develop the skills necessary for a transformed relationship with our own inherent creative wildness. This requires that we move beyond the fear-based need to control wild spontaneity and learn to live in the moment, all the while staying rooted in vibrant unpredictability. Animating, unblocking, and releasing our wild creative capacities allows us to open to a sense of Earth connection, animate new visions, and remain fueled and energized for the difficult tasks ahead. Once experienced, this transformed sense of connection can be put in the service of the planet in the form of creative right feeling, thought, and action. Self-interested action that is rooted in this intimate relationship is also in the Earth's best interest. Without a creative, sustainable, flexible vision, our actions will always be left wanting. We need artful creation for all life's sake. We need to connect to the mystery associated with experiencing wild creative spontaneity.

> To understand the human role in the functioning of the Earth we need to appreciate the spontaneities found in every form of existence in the natural world, spontaneities that we associate with the wild—that which is uncontrolled by human dominance. We misconceive our role if we consider that our historical mission is to "civilize" or "domesticate" the planet, as though wildness is something destructive rather than the ultimate creative modality of any form of Earthly being. We are not here to control. We are here to become integral with the larger Earth community. The community itself and each of its members has ultimately a wild component, a creative spontaneity that is its deepest reality, its most profound mystery. (Berry, 1999: 48)

The creative life force in everyone speaks in images, in music, in dance, in dreams, in trance and in mystical experiences. When fully engaged with it, we have a sense of timelessness, of total absorption and preoccupation. We are lost to its power, its motion, its flow. We access the essence of self and nature, the creative living force that binds all beings together.

The more we are able to nurture our own spontaneity, the more sustainable and ecologically sound are thoughts, actions, and feelings will become. Once this is accomplished, we become able to embrace cosmological transpersonal identifications where all beings are experienced as aspects of a single unfolding reality (Fox, 1990: 252). Fox likens cosmological identification to the branching pattern of a tree.

> If we empathically incorporate (i.e., have a lived sense of) the evolutionary, "branching tree" cosmology offered by modern science then we can think of ourselves and all other presently existing entities as leaves on this tree—a tree that has developed from a single seed of energy and that has been growing for some fifteen billion years, becoming infinitely larger and infinitely more differentiated in the process. A deep-seated realization of this cosmologically based sense of commonality with all that is, leads us to identify ourselves more and more with the entire tree rather than just with our leaf (our personal, biographical self), the leaves on our twig (our family), the leaves we are in close proximity to on other twigs (our friends), the leaves on our minor sub-branch (our community), the leaves on our major sub-branch (our cultural or ethnic grouping), the leaves on our branch (our species), and so on. At the limit, cosmologically based identification . . . therefore leads to impartial identification with all particulars (all leaves on the tree). (Fox, 1990: 255–6)

We enter into an "I-thou" relationship (Buber, 1937) where new habits can be chosen and new actions can be undertaken that are infused with a broad sustainable life-centered vision.

This spontaneous creative aspect seems to be easily accessible in childhood, bids a hasty retreat underground in the face of the development of schooled critical reason, but can be uncovered and accessed again with an openness and a letting-go of the fear that drives us to control its power.

The feelings of fear of both the unknown and the uncontrollable that surface for many when they are in the wilderness (Tuan, 1979) also emerge with art making of any kind (Cassou and Cubley, 1995). Letting go of fear and trusting the art-making process gets easier over time, just as letting go of fear of the wilderness and gaining a measure of confidence and security also develops over many experiences in the wild. Part of the process of letting go of fear involves accepting it and realizing that having control is an illusion. Paradoxically, once we let go to the spontaneous unrestrained wild aspects of self and the universe, we gain a sense of connection and security that is lacking when we try to plan and manipulate. We also become versed in the patterns that govern our ongoing relationship to the Earth and become better able to predict when we are in danger. We develop a healthy respect for that which we cannot control. We move from a position of fear-based alienation to one of secure ecobonding. Only then can we begin to remove the barriers that insulate us from the wild power of the Earth.

A sense of sacred connection comes from actively experiencing the wild creative self. It is the experience of the dance between: the inner and the outer, the mind and the body, the self and the planet. This dance takes place in the liminal space that is the creative act. We can come to better know ourselves and the wild in a joyful renewing manner when we enter this dance. Transformation of the self and the planet need not be agonizing and analytical. It can be a celebratory creation that affirms the inner wisdom and interconnection of all beings.

While painting spontaneously I am fully alive. What I do is natural. I set up the materials and let what is meant to happen that day, take place as best as it can. I cultivate the soil and then get out of the way as the seeds sprout on the page in the form of watery strokes. I close my eyes to choose the colors blindly and I do the same for the brush.[1] *Then I open my eyes and let my hand and the brush move the paint around the page at will. I need to be "present" in order to sense when the chosen color is finished so that the next color can be chosen blindly and applied spontaneously. I try to "let go" to the process and see where the paints will take me. I have no concern for the product created. The goal is to remain mindfully absorbed and to let go the need to arrive at any place in particular.*[2] *If my attention is focused on attempting to control or predict the finished product, then the contemplative state is temporarily lost. Therefore, there is a "being in the universe" aspect to spontaneous painting.*[3]

When I feel the painting is finished, I record how I have been feeling that day, the order in which the colors came up, any reactions I had to the colors or forms created, and then often a poem emerges around some noted aspect of the painting. The poem is free associative in nature and often provides deeper insights and wisdom. Often it initiates a dialogue between myself and the painted colors and forms.

Traditionally, in creative work of this nature, images often are placed in the service of the dreamer/artist much like life on the planet has been enslaved by the human. Therefore, I avoid analyzing the images for meaning. Instead, I interact with the images and experience them as living beings in their own right.[4] *By experiencing images in this way, I can begin the long journey back to reenergizing my own deep connection to all beings.*

The process of painting allows me to experience a sense of timelessness and universal connection while giving life-breath to aspects of my uniqueness.[5] *The product of painting—a form, a color, a texture, or a particular combination of the three—is like a record of the journey at this time. The resulting images are natural or wild creations that give color and form to the life force present during that session. Images recur and new ones emerge.*

I have begun to story both my self and the Earth with which I live. I have access to the source.[6] *By creating without rationality, criticism or an eye fixed on the quality of the end product, I have been able to enter a rich dialogue between myself and the Earth. Often we become as one and it is unclear whether I am in the Earth or the Earth is in me.*[7] *Both sensations are present simultaneously. Image creation is a process that captures my mind, heart, and soul. It has shifted my self-image to include all life-forms as I see my reflection in painted clouds, oceans, birds, sunbursts, stars, grass, trees, mud, and most recently moths.*

The following four paintings seem to capture the wild creative moment between stillness and life. Like the metamorphosis of a caterpillar to a moth or the creation of a spontaneous painting, there is a flurry of activity that does not happen systematically from the head to the abdomen. It is a simultaneous creation of the whole. It is characterized by a complete dissolving down of the caterpillar followed by an all-at-once, co-creative coming together of the moth. All of this is happening in the context of cocoonlike outward silence and stillness.

I have been exploring the words "still-life" and find they speak to me in many ways. A cocoon looks still and dead on the outside, yet inside there is incredible movement and creativity. Traditionally in the art world, a still-life involves painting or drawing carefully arranged plants, furniture, teacups, and dead animals. Like pinned butterflies, these renderings capture beautiful objects in an arrangement. In contrast, I see my spontaneous painting process as being one where I still myself in order to feel the motion, the wildness of life in that instant. Once finished I emerge with a still painting that is full of life.

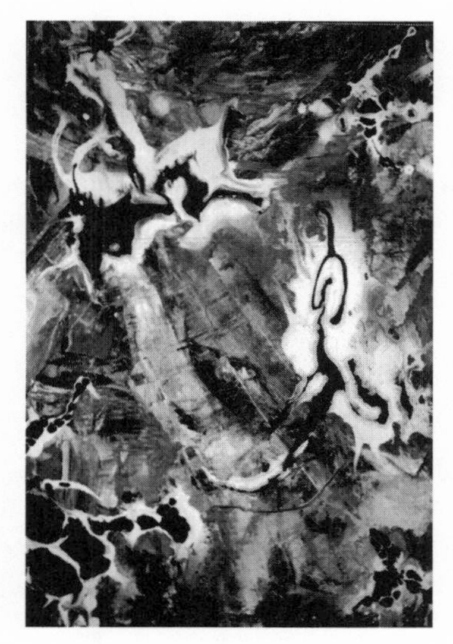

Figure 17.2 Still-life I

Today it's cold and snowy again
I too am in between seasons
No longer winter and not yet spring
No longer caterpillar and not yet moth

Bound up tightly in my cocoon
Waiting for the opening
For a beam of light to energize me
From the within and without
To move me into the onward

Movement, a sense of life-emanation

Is what I crave
Yet I must be alive to have such cravings
I am in suspended animation

Still, yet full of life
Alive, yet full of stillness
I am a moth-er-pillar
A still life

Through spontaneous painting I am able to create a sense of internal spaciousness while at the same time experiencing a deeper connection to the Earth. I can be in nature in a deeper more sustainable way. I do not need to take it in, acquire it, or co-opt it. I can move like a watery fluid between myself and all Earth beings. I am becoming Earthier and more balanced as a result. I am beginning to metamorph or transform myself while simultaneously being transformed by my ongoing connection to the Earth. I am slowly melding down into an Earth being again.

Through reflecting on groups of paintings over time, patterns have emerged that seem to hold together.[8] These are Earth-connecting patterns.[9] Space limits my ability to share them in detail, but I will briefly summarize the inherent Earth wisdom contained in each of the following six patterns: the spontaneous, the childlike, the embodied, the organic, the primitive/tribal and the wild. Each pattern connects us to the Earth in its own unique way; however, they are by no means separate or mutually exclusive aspects of Earth connection. One can enter Earth connection through each and immediately make the acquaintance of the other four. In my own explorations, I have focused on the spontaneous and the wild as they have been the most powerful connectors for me. Others find power through different entry points. Further, the qualities in each can be found in the other. I can feel the wildness in my spontaneity or embody the primitive at other times. Therefore, the teasing out of self-contained strands is a fluid activity of sorts and serves to provide focal points but not rigidly bounded categories.

THE SPONTANEOUS

Figure 17.3 Sunwhirl

The spontaneous aspect teaches us to clear our mind of thoughts and directives. We let go to the flow of impulse in the moment. The spontaneous helps us to connect to our instincts, our life spark, our unforced naturalness. It teaches us about the power of being in the moment and the co-creative aspect of all lived moments. It places us in communion with creative forces as our hand automatically moves for a color and creates a form that we could have never predicted. We learn how to breathe our way back into the flow of spontaneity and become mindful again when we move off track. Spontaneity promotes a lightness of being at the same time as it embraces a felt communion with all beings. While being spontaneous we are in the spin of things.

THE CHILDLIKE

Figure 17.4 Burst

The childlike aspect teaches us to remain playful, in awe and wonder of the mystery of it all. It helps us to take ourselves less seriously and encourages us to remove our guarded and suspicious adult cloak. We become able to appear foolish, silly and wide-eyed. We can spin in circles and follow that quiet still voice. This aspect keeps us in beginner's mind that encourages us to approach each new experience in the moment, with no preconceptions or looking to a particular outcome. We develop wonder and love for color and form in all beings and ponder where images come from. We can marvel at the universal nature of forms created by children the world over. We rekindle our trust and commitment to the creative process that often helps us to remember childhood nature experiences of deeply rooted Earth connection.

The Embodied

Figure 17.5 New Beginnings

The embodied aspect teaches us about opening up and being receptive to what the Earth has to teach on the material level. It teaches us about the fragility of life, the miracle of birth, and the power of the life force. It teaches us how to still ourselves and listen. It teaches us to trust our body and its inherent wisdom. We can move beyond "using" our bodies for certain purposes and instead co-create with all it has to teach us. The embodied aspect challenges us to let go and let be. It helps us to heal the body-mind imbalance and allows us enriched sensory experiences. We become better able to balance our thinking, feeling, sensory, and intuiting abilities. We can be with our breath in the moment. We can open to new ways of connecting in by clearing out blocks and filters. We can begin to know that images are our flesh and bones.

The Organic

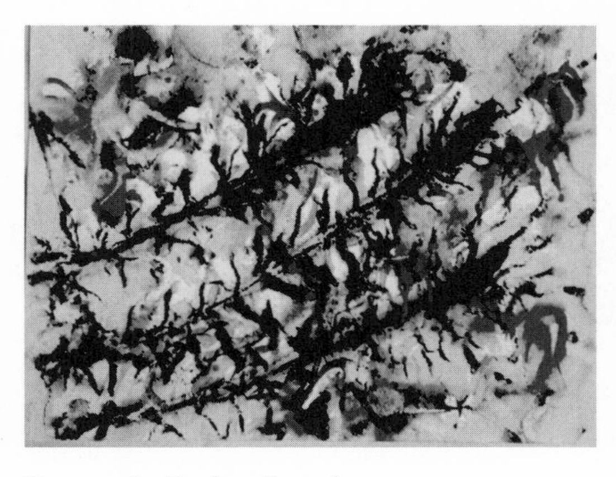

Figure 17.6 Verdant Growth

The organic aspect teaches us about natural cycles, about growth and decay repeated over and over again. It teaches us about rhythmic seasonal waxing and wanings, nourishment and patience. We learn that there is a time for all things and all things come in their time. We experience how sustainable growth cannot be hurried, that fetid decay fertilizes the soil for new growth, and that without nourishing soil nothing will grow to maturity. We begin to feel enchanted with the repeated mirroring of patterns and forms in our own work and in the natural world. We come to know that our own patterns are natural patterns. We develop trust in the life process and release the need to control the outcome of this process in order to feel safe.

The Primal/Tribal

Figure 17.7 Totem

The primal teaches about our tribal roots and our connection to all beings. It teaches about the historical need for humans to creatively express their connection to the Earth on walls, in pots, on the land, and in songs, dances, and rituals. It also teaches us about the co-creative collective nature of our relationship to Earth beings. It removes the delusion of separateness through sharing the timeless unchanged images and expressions of those who came before and lived sustainably with the Earth. We see historical patterns of color and form mirrored in our own work (e.g., Paleolithic cave paintings). We embrace ancient cultural art patterns as well (images from Mayan, Egyptian, Greek, African, Native American, Australian aboriginal cultures). We come to understand the importance of co-creating with a group (tribe) in an ongoing way in order to inform the process and provide support and encouragement. We begin to promote a culture of co-creation and nature connection.

The Wild

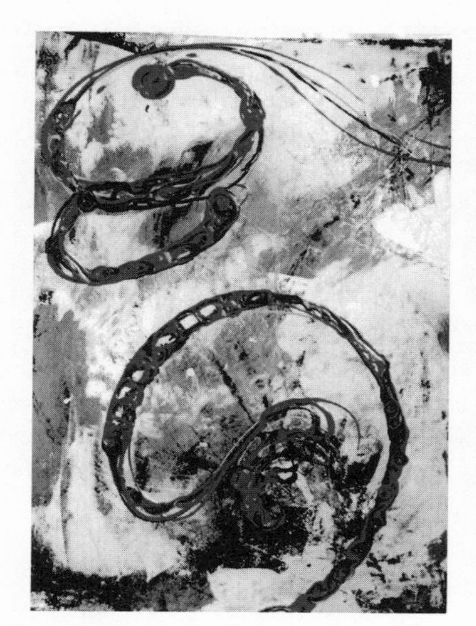

Figure 17.8 Cosmic Serpents

The wild teaches us to be resilient in the face of challenge and to be respectful of forces beyond our control, and to let go the desire to control wild forces out of fear. It teaches us how to remain grounded in the face of the unpredictable. We learn humility, sacrifice, and ultimately deep reverential respect. We also learn to persevere in pain and move through fear-invoking large, open spaces. We learn to stick with a process that leads to an unknown destination and fully embrace the unknown aspect of the wild. We become wilder ourselves. We begin to live in the present moment where nature lives and are able to be still in the face of complexity, the unknown, and the spacious. We begin to understand and get an overview on patterns that create and patterns that are destructive. We can experience metamorphosis and meld into wild sacred Earthy communion.

Conclusion

There are many passageways to Earth connection. Yet all seem to lead to the same place. It is the sacred place where one feels both freedom to explore and the sense of security that comes with belonging. It is the place where the visible and invisible worlds meld into one. It is the place where the creative capacities of the self are animated and simultaneously melded to Earth creation. Regularly visiting this sacred place animates our connection to the life spark and transforms our sense of self into an ecologically sustainable one. We find ourselves developing the skills necessary for a vibrant life in the "cosmic river." It is while in this river that we embody all Earth beings and we are once again on speaking terms with the natural world.

The cosmic river
Is the weaver of the web
That makes my heart sense quiver
Gives my soul its flow and ebb
It bathes me in its riches
It fills me with its charm
It sews me up with stitches
Made of watery stream-like yarn
It flows me around the rock face
It pulls me with undertow
It takes me into a dark place
Then teaches me how to know
The way to illumination
The space where the stars shine pure
It carves with determination
Its destination is sure
Its been flowing here for ages
Yet its liquid is always new
It pulls me deep in stages
With my love of painted hue
It freezes icy cold when the voice of reason shouts
The river takes its hold when the truth has been let out
That the river is our being,
The river is our core
The river is our seeing,
The river shows us more
Than what's floating on the surface
Than what's nagging on your mind
By streaming down the stir space
Buried treasure you will find

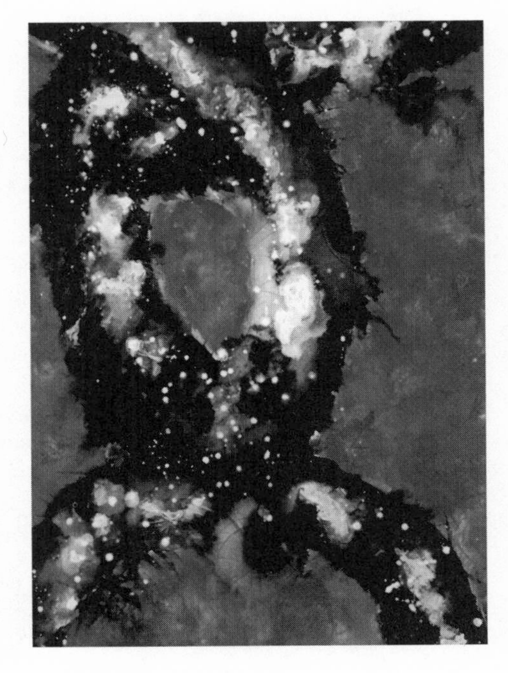

Figure 17.9 Cosmic River

NOTES

1. According to Nachmanovitch (1990: 51) "for art to appear we must disappear." We must let go of the mind, the personality, the judging eye and be lost to the flow. By closing my eyes I am helped into this state of disappearance.
2. The meditative state associated with the spontaneous creation of art is also akin to the flow experience described by Csikszentmihalyi (1996). The flow experience is characterized by a loss of self or ego and a deep concentration or preoccupation with what we are doing in the moment.
3. "Art is contemplation. It is the pleasure of the mind which searches into Nature and which there divines the spirit by which Nature herself is animated" (Rodin, 1983: 1).
4. It is detrimental to the life-enhancing message of an image to attempt to pin down a meaning or somehow "explain" what it is saying. James Hillman cautions against this tendency to define the meanings of animal symbols in his writings about dreams: "We must animate our images thereby giving a life soul back to them. In our eagerness for conceptual beings, we ignore the actual beast. Analysis and interpretation, even Jung's active imagination is done for the sake of the dreamer's soul not the soul of the animal image" (1997: 46).
5. Thomas Berry (1988) states that the universe is a communion of differentiated subjects and not a collection of objects. For beings to live sustainably on Earth, each must be able to animate its full uniqueness.
6. We are offered a connection to the source when we engage in spontaneous painting. The source is embodied, connects all of us, is revealed in creative play, and is the energy that first set the universe in motion. "The source is that deepest part of you, that part that remembers everything with absolute clarity. It is the body intelligence that exists without word. . . . As the collective memory of the creation of the universe and the history of evolution, the source is also home to the primordial imagery of myth, dreams and remembrances" (Gold, 1998: 9).
7. "So what the artist . . . is doing, fundamentally, is not recreating in the sense of making again what has been lost (although he is doing this), but creating what is, because he is creating the power to perceive it. By continually breaking up the established familiar patterns (familiar in his particular culture and time in history) of logical common sense divisions of me-not-me, he really is creating "nature," including human nature" (Field, 1957: 181)
8. In his book *The Self-Made Tapestry: Pattern Formation in Nature,* Philip Ball states: "You can't help concluding, once you begin to examine this tapestry, that much of it is woven from a blueprint of archetypes, that there are themes to be discerned with in the colorful fabric. Nature's artistry maybe spontaneous, but it is not arbitrary" (1999: 5).
9. After months of sitting with many small groupings of paintings, I read the following quote by Roszak and had an experience of crystallization that helped me to understand the paintings in a new light. "We begin to see how the urban-industrial reality principle represses much that is essential to the health of both person and planet: the primitive, the organic, the feminine, the child-like, the wild" (1992: 213).

REFERENCES

Ball, P. (1999). *The self-made tapestry: Pattern formation in nature.* New York: Oxford University Press.

Berry, T. (1988). *The dream of the Earth.* San Francisco: Sierra Club Books.

Berry, T. (1999) *The great work: Our way into the future.* New York: Bell Tower.

Brown, L., Renner, M., and Flavin, C. (1998). *Vital signs 1998: The environmental trends that are shaping our future.* New York: W. W. Norton and Co.
Buber, M. (1937). *I and thou.* Edinburgh: T. and T. Clark.
Carson, R. (1962) *The silent spring.* Boston: Houghton Mifflin.
Cassou, M., and Cubley, S. (1995). *Life, paint and passion: Reclaiming the magic of spontaneous expression.* New York: Jeremy P. Tarcher.
Clinebell, H. J. (1996). *Ecotherapy: Healing ourselves, healing the earth* New York: Haworth Press.
Csikszentmihalyi, M. (1996) *Creativity: Flow and discovery of psychology and invention.* New York: Harper Collins.
Devereux, P. (1996). *Re-visioning the Earth: A guide to opening the healing channels between mind and nature.* New York: Simon and Schuster.
Durning, A. (1992). *How much is enough?* New York: W.W. Norton and Company.
Field, J. (1957). *On not being able to paint.* Los Angeles: Jeremy P. Tarcher.
Fleischman, P. R. (1997). *Cultivating inner peace.* Los Angeles: Jeremy P. Tarcher
Fox, W. (1990). *Towards a transpersonal ecology: Developing new foundations for environmentalism.* Boston: Shambhala.
Gold, A. (1998). *Painting from the source: Awakening the artist's soul in everyone.* New York: Harper Perennial.
Greenway, R. (1995). The wilderness effect and ecopsychology. In T. Roszak, M. E. Gomes, and A. D. Kanner (Eds.), *Ecopsychology: Restoring the earth, healing the mind* (pp. 122–35). San Francisco: Sierra Club Books.
Griffin, S. (1978). *Women and nature: The roaring inside her.* New York: Harper & Row.
Harper, S. (1995). The way of wilderness. In T. Roszak, M. E. Gomes and A. D. Kanner (Eds.), *Ecopsychology: Restoring the earth, healing the mind* (pp. 183–200). San Francisco: Sierra Club Books.
Hillman, J. (1997). *Dream animals.* San Francisco: Chronicle Books.
LaChapelle, D. (1988). *Sacred land, sacred sex, rapture of the deep: Concerning deep ecology and celebrating life.* Silverton, Colo.: Finn Hill Arts.
Mack, J. (1995). The politics of species arrogance. In T. Roszak, M. E. Gomes, and A. D. Kanner (Eds.), *Ecopsychology: Restoring the earth, healing the mind,* (pp. 279–87). San Francisco: Sierra Club Books
Macy, J. (1991). *World as lover, world as self.* Berkeley, Calif.: Parallax Press.
Merchant, C. (1981). *The death of nature: Women, ecology and the scientific revolution* San Francisco: Harper & Row.
Nachmanovitch, S. (1990). *Free play: Improvisation in life and art.* Los Angeles: Jeremy P. Tarcher Inc.
Orr, D.W. (1992.) *Ecological literacy: Education and the transition to a postmodern world.* New York: SUNY Press
O'Sullivan, E. (1999). *Transformative learning: Educational vision for the 21st century.* London: Zed Books.
Purpel, D. (1989). *The moral and spiritual crisis in education: A curriculum for justice and compassion in education.* Grangy, Mass: Bergin and Garvey.
Porteous, J. D. (1990). *Landscapes of the mind: Worlds of sense and metaphor.* Toronto: University of Toronto Press.
Rodin, A. (1983). In P. Gsell. (Ed.). *Rodin on art and artists: Conversations with Paul Gsell.* New York: Dover Publications Inc.
Roszak, T. (1992) *The voice of the Earth.* New York: Simon and Schuster.
Roszak, T., Gomes, M. E., and Kanner, A. D. (1995). *Ecopsychology: Restoring the Earth, healing the mind.* San Francisco: Sierra Club Books.
Seed, J., Macy, J., Fleming, P. and Naess, A. (1988). *Thinking like a mountain: Toward a council of all beings.* Philadelphia: New Society Publishers.
Shepard, P. (1982). *Nature and Madness.* San Francisco: Sierra Club.

Thomashow, M. (1996). *Ecological identity: Becoming a reflective environmentalist.* Cambridge, Mass.: MIT Press.

Thomashow, M. (1998). *The ecopsychology of global educational change.* Environment Canada Environmental Education Online Colloquium, October 19–30. *www.ec.gc.ca/eco/education/Papers/thomasho.html*

Thoreau, H. D. (1979). Walking. In C. Bode (Ed.), *The portable Thoreau* (pp. 592–630). New York: Penguin.

Tuan, Y. F. (1979). *Landscapes of fear.* Minneapolis: University of Minnesota Press.

CHAPTER 18

Traces and Transformation

Photographic Ambiguity and Critical Histories

Amish Morrell

Introduction

In his essay "The Work of Art in the Age of Mechanical Reproduction," Walter Benjamin (1968) began to theorize the evolving relationship between reproducible images and abstract systems of social organization. He argued that techniques such as photography and lithography made it possible for reproduced objects to be placed in contexts determined by producers, distributors, and viewers and detached from their place of origin. Since then there has been a steadily evolving critique of the role of images in the production and reproduction of meaning and the organization of power in society. This chapter, through which I aim to contribute to this critique, is based on my own work in progress, in which I am using photographs to produce a social history of an inland community in Cape Breton, Nova Scotia. Through this project I address a set of questions that explore how to use photography as part of a transformative pedagogical and cultural practice.

I am interested in better understanding how photographs can be used to explore and represent memory and help us to better understand how we are located within the global forces of history. My research focuses on a community called Big Intervale that is part of the Margaree Valley in Nova Scotia. I look at how photography shapes peoples' memories of this place and attempt to produce a critical history of this community. My purpose is to develop a pedagogical method by which it is possible to become more critically aware of the social relations and historical conditions that shape visual meaning, and experience different ways of knowing, through the use of photographs. Part of my strategy for doing this involves recording absences. By documenting that which gives testimony to the past—roads overgrown by the forest or traces of what once were farms—my aim is to document traces of the past, helping the past to be seen not in time but through time. In a sense, it is an attempt to find a method for representing the unrepresentable or, more accurately, of understanding that which we cannot see or experience directly. By photographing traces, I aim to make visible the ultimate unknowability of the unrecorded past and raise questions about what this means for how we understand ourselves in the present.

The area upon which I base my work includes the small valleys and glens at the upper reaches of the Margaree Valley, where the Margaree River flows from a steep and uninhabitable canyon. It was first populated by Mi'kmaq First Nations people and then in the early 1800s by tenant sharecroppers expelled from the Western Islands and Highlands of Scotland as a result of the rise of British industrialism and its consequent demand for land for the production of wool and other raw materials. At the end of the nineteenth century and beginning of the twentieth century, many of these families emigrated to other parts of Cape Breton to find work in the coal mines or at the steel mill; to Boston or Maine where they worked as maids, carpenters, or lumberers; or to the prairie provinces where they worked as farm laborers. Throughout the twentieth century and until the present, people continued to leave, often to Ontario or Alberta as workers were needed in these provinces. Of the farms built by the earliest immigrants, often all that remains are the outlines of stone cellars or patches of meadow among spruce trees and alder bushes.

As the landscape itself becomes transformed, with roads being built to provide logging access to the surrounding highlands or buildings being renovated, torn down, or left to fall down, it loses the physical traces of its earlier inhabitants. As those who remember the past and the stories from those before them lose their memories and lives in hospitals and old-age homes far from this place, this place and its people are in danger of becoming orphans in time and in history.[1] As the people who remember also disappear, the context in which the lives of those before them was structured and made meaningful is rendered invisible. Consequently, it becomes increasingly difficult to access an understanding of the history of this area and the choices made available to the people who lived there.

Critical Theory and Photographic Practice

In terms of conceptualizing how to use photography as part of a transformative political project, it is important to consider work done by contemporary theorists looking at how photography organizes meaning and power within society. Many educators, researchers, and artists have looked at how photography can play a productive role in working toward increasing social equality, human well-being, and environmental sustainability. However, there are numerous structural limitations to successfully pursuing these ideals. Many of these barriers have been analyzed only recently by critics (Rosler, 1981; Sekula, 1986; Solomon-Godeau, 1991). While theorists have been able to describe and critique the role of photographs in the ideological production of everyday meaning, this is only a starting point for a socially transformative visual practice.

At the center of this problem is the ambiguous relation photography has to the material world. Photographs both evoke materiality and transcend materiality. They may represent people or places that we know, display emotions that we have experienced, or present products or lifestyles that we aspire to own or enjoy. At the same time they can carry us across time and space by representing places we have never seen, people who lived at different times, or experiences we have never known. They operate within preestablished structures of meaning to produce normalized ways of apprehending the world, or they sometimes can present us with contradictions that challenge our ways of thinking and seeing. As such, they are a powerful means of organizing meaning within our lives in ways that are largely unconscious. In being able to be reproduced and disseminated on a mass scale, photographs can be carried far from the specific relations in which they were produced and placed into other productive relations that may betray the ones from which they originated.

In the last half century, there have been considerable advances in theories of how we make sense of photographs and the relationship between photographic meaning and their social and political contexts. Marxian scholars frequently argue that images are the superstructural emanations of material relations. But what are the active dimensions of this process? How do images connect with structures? Visual theorist John Tagg (1988) writes that "[t]he photograph is not a magical 'emanation' but a material product of a material apparatus set to work in specific contexts, by specific forces, for more or less defined purposes" (3). Photographs represent a concrete and material reality, yet they are also a language and operate within abstract and fluid systems of discourse. According to structural linguistics, discourse is regarded as "a transindividual network or economy of [signifying] elements, conceived in ideal abstraction from the individual speech act" (Payne, 1997: 145). As visual sociologist Elizabeth Chaplin (1994) writes, there are certain codes, or rules of discourse, that shape this process of signification (1). Further, these codes and systems are constantly changing. In drawing from Michel Foucault, Donald Lowe (1982) writes that: "discourse is governed by unconscious epistemic rules or presuppositions, and that these rules as a whole change from one period to another. There is no universal logic of discourse; and knowledge resulting from one period is discontinuous" (1982: 9).

How do we locate the codes that Chaplin describes? How is discourse coordinated in order to produce certain structures of feeling and perception? Where are the sites of intervention? To understand this, it is useful to look at the processes by which meaning is coded through photographic images.

Roland Barthes (1977) delineates photographic meaning into three different types: the linguistic, the coded iconic (the connoted), and the noncoded iconic (the denoted) (36). The linguistic message encompasses the written or verbal text that often accompanies or is present in images. Linguistic messages serve either to anchor or to relay meaning of the image—or, as Barthes writes "to *fix* the floating chain of signifieds in such a way as to counter the terror of uncertain signs" (39). They serve to draw our attention to some preselected meaning that is associated with an image. The coded iconic message is a connotated form of meaning, in that it operates through implicit reference to something (i.e., a mental association) that exists outside of the frame of the photograph. Although the connotated message may be socially inscribed and collectively experienced, it is entirely subjective. The noncoded iconic message, or the denotative, refers to the physical objects or subjects contained within the image and is therefore objective. (For example, these could be a tree, a sunset, or a person.) These different types of signifiers operate together to constitute a rhetoric of the image that becomes what Barthes calls "the signifying aspect of ideology" (1977: 49). Photographs are ideological in that they operate through a vast network of associative meaning that is highly structured and largely preordained. It is the relationship between denotation and connotation that especially complicates photographic meaning. The connotative message uses the denotative message as a mask, subsuming our subjective associations of the photograph within its objective content and producing an illusion of truth. This process reifies visual meaning and makes it seem natural, facilitating an illusion of realism that obscures the constructed nature of photographic images (Barthes, 1977: 36).

In turn, realism has the social affect of denying the operations of broader social forces in shaping the meaning of photographs and ultimately denies our social agency. Work such as that commissioned by the Farm Security Administration (FSA) during the Great Depression to document social conditions in the American South and midwestern states served political functions that are not apparent in the images themselves. One of the best-known FSA images is Dorthea Lange's "Migrant Mother," a portrait of a woman sitting with her chin resting on her hand, framed by two children and an infant who sit with their heads turned away from camera. Martha Rosler (1981) argues that this kind of work serves as a way of assuaging the liberal consciousness. She writes that:

> [i]n the liberal documentary, poverty and oppression are almost invariably equated with misfortunes caused by natural disasters: causality is vague, blame is not assigned, fate cannot be overcome. Liberal documentary blames neither the victims nor their willful oppressors—unless they happen to under the influence of our own global enemy, World Communism. Like photos of children in pleas for donations to international charity organizations, liberal documentary implores us to look in the face of deprivation and to weep (and maybe to send money, if it is to some faraway place where the innocence of childhood poverty does not set us off in a train of thought that begins with denial and ends with "welfare cheat"). (73)

As Rosler points out, this does little to address the structural underpinnings of poverty and abdicates the viewer from any sense of responsibility. With this form of social documentary, there was little critical analysis of the role of images in mediating power.

Because our readings of images are subjective, ideology operates through a socially inscribed mechanism of connotation, or subjective association. Its effectiveness is guaranteed by the relationship between connotation and denotation. For the purposes of this chapter, I conceptualize ideology as the largely unconscious and prescribed frameworks and associations by which we organize our worlds (James and Hall, 1986: ix). It structures and activates preestablished ways of knowing and often denies the experience of not knowing. As such, it can be a substitute for and a way of organizing that which we do not know first hand. To understand the specific workings of ideology, it is necessary to focus on people's subjective readings of photographs. Why were these images made? What do they mean to their subjects? What do they mean to other people in the same community? By looking at what viewers find interesting, notable, or problematic, and by trying to understand the bases of these preferences, it is possible to link these specific acts of meaning making to broader structures of social meaning and organization. It is also possible to make visible the relations through which they are rendered meaningful and how people are differentially positioned within a set of viewing relations.

With photography, Victor Burgin (1982) writes, "[i]n the very moment of being perceived, objects are *placed* within an intelligible system of relationships (no reality can be innocent before the camera). They take their position, that is to say, within an *ideology*" (46). The meaning of the objects documented is already in action before the image is taken (47). With photographic communication, encoding begins the moment a subject is selected. But how do we transform the meaning of photographs? How do we create new associations and produce new narratives that can enable us to live our lives differently?

TOWARD A TRANSFORMATIVE PHOTOGRAPHIC PRACTICE

My work looks at photographs taken of Big Intervale over the past century that are held in people's private collections and in museums and public archives. In addition I use photographs that I take of present-day Big Intervale, many of which redocument the same places that are represented in the earlier photographs. Other images that I take serve as present-day illustrations of places mentioned in historical texts or that people describe in our conversations.

A critical source of information in my work is the stories that will be elicited through the viewing of historical photographs of Big Intervale. In my informal conversations with people there, I am interested not just in obtaining historical information but also in discovering what these photographs mean to their viewers. In looking at a collection of photographs taken in the 1910s, one person who was the subject of many of the images believed it would have been unlikely that someone living in Big Intervale would have owned a camera. This person thought that these particular photographs were taken by a person who had moved to the United States to find work and took these pictures on a return visit. Other photographs were thought to have been taken by professional photographers who came around offering their services.

Other images that I use are present-day photographs that I take during my research. Some of these are of places that arise in people's stories, which may have changed profoundly since they last saw them and to which they may no longer be physically able to travel. What do they think of these changes, and what memories and thoughts arise in seeing how these places have changed? Forms of information other than people's memories and stories also can shape the meaning of photographs. Sources such as church records, census reports, municipal or provincial petitions, and maps can provide information about the places and people that they describe as well as the institutions that they represent and how they shape our experience of a particular place. Just as the caption beneath a photograph determines the associations that we make with it and where we place its subjects within our imaginations, the texts or documents that we present alongside photographs will significantly shape their meaning. For example, it is obviously impossible to obtain photographs of Big Intervale at the beginning of white settlement, before photography had been invented, although photographs can be used in conjunction with other information to tell us about that time period. A strategy that uses photography to represent that for which there is no photographic record might be to present census information from an earlier time alongside photographs of the sites of the homes or workplaces as they appear now. If I could find the grindstone in the brook where the mill fell down around it and its timbers were swept away by the recurring spring floods, or remnants of the farm long ago reclaimed by a forest that was in turn exploited by a pulp and paper company it might be possible to better understand the changes of the past centuries in a way that neither photographs nor census data on their own could ever allow.

above: Big Intervale, circa 1970. Photograph by the late Norman MacLeod. Courtesy of William A. MacDonald.

above: Big Intervale, circa 2001. Photo by Amish Morrell, 2001.

By foregrounding our relationships with the photographs and their subject matter, it is possible to further shape a critical usage of photographs. As both image-makers and viewers we become an active part of the worlds we experience and represent and thereby are implicated in their production. To become part of a dialogical relationship in which we are reflexive of our own position(s) is to attend to a different notion of truth that is more responsive to those whom we represent. An example of where this has been done using photography is in collaborative work by the British cultural theorist Jean Berger and Swiss photographer Jean Mohr. In a collection of images titled "Beyond My Camera," they show an interesting and subtle way of doing photography that reveals the dialogical use of and the ambiguous nature of images (Berger and Mohr, 1982). They include, with Mohr's photographs, a running text on how they were made, what was Mohr's relationship with the subject in the making of the images, and how they were used. He tells us which images the subject liked, which ones evoked an unexpected response, which one they chose to display on their mantelpiece. In doing so, he presents his photographs in such a way that the context that makes them meaningful is visible. Both subject and photographer are visible within their respective social positions.

In viewing images, it is possible to commit a form of structural illusionism by projecting ourselves into the experiences of others. However, as film theorist Kaja Silverman (1996) writes, it is not necessary that we can imagine ourselves in the place of this "other." She describes this form of identification as "idiopathic," meaning that it is based on an incorporative logic that is a deep form of sympathy that completely absorbs and subsumes the other without any regard for their subjectivity or autonomy. Instead, in situations where there are substantive structural differences, Silverman advocates a "heteropathic" identification in which " . . . the subject identifies at a distance from his or her proprioceptive self. . . . The visual imago itself remains stubbornly exterior, like the original mirror reflection . . . at the expense of an imaginary bodily unity" (Silverman, 1996: 23). These viewing relations require each participant be viewed within their own context.[2]

To enable an encounter that falls into neither the trap of othering or idiopathic identification, it is necessary to use forms of representation that enable viewers to reflect on their own experience in relation to the subject. Images, film, and other art forms can allow us to apprehend or behold different social experiences, notably those of whom we may have previously identified ourselves in opposition to (Silverman, 1996: 93). Such forms allow us to better understand experiences that might otherwise be objectified or reified through formal discourse and academic concepts. The work of art—in my case, the photograph—can allow a subtler and less objectifying encounter with the topic under consideration. The aesthetic work mediates between the producer and the viewer and allows entry into another subjective space. It is here that the world of the "other" can be apprehended without it becoming subsumed within the identity of the viewer.

Much of what I wish to elicit through my research is not the known, but an encounter with the unknown. There are no photographs of the Mi'kmaq people who lived in Big Intervale and little available information about what happened to them when white settlers colonized the area in the early 1800s. There are only a few photographs predating 1900. There is no one still living who has firsthand knowledge of the area before 1910. There are definite limits to what can be known. Yet there are stories of traces and photographs of traces. Twenty years ago I saw a slide show

Figure 18.3 Site of the house of Angus MacInnis and Ann Robertson, who came from Inverness-shire Scotland and Wycogomah Nova Scotia, respectively. Their son, Duncan MacInnis, lived there until his death in 1916.
Photo by Amish Morrell, 2000.

by a man in Margaree who took up photography as a hobby after his retirement. He had images of stone circles in the meadows along the river that outlined the summer encampments of the Mi'kmaq. The circles are no longer there, and I do not know if even the pictures still exist. But I remember seeing the pictures, and this serves the same function as the pictures or as the circles themselves by enabling me to bring these memories to my understanding of the history of this place. This experience also exposes me to a way of knowing the past that is beyond the forms through which it is presented. History is explained through the markings left on the present by the past—and the frames that shape our knowledge of the past are made visible, enabling us to experience history in ways other than those that we already know.

This moves us toward a different way of conceiving how to use photography that facilitates a kind of cognitive ambiguity that is necessary in order to see differently. Jacques Lacan (1953) refers to the space between experience and language as the "real," or a world not yet coded. The real can also be, according to art theorist Hal Foster (1996), a rupture between perception and consciousness, or a break in taken-for-granted ideology (134). The real can be an experience in which the frames through which we view the world are rendered momentary visible in such a way that it destabilizes our preestablished notions of what is true. There can be something jarring about becoming aware of the mediums through which we often experience the world. This is the nature of the real—that it cannot be represented. Representation does not embody the real but can present the possibility of an encounter with the real. For this to happen we must somehow be aware of the frames, or relations, through which experiences are presented to us.

By framing photographs with people's stories, it is possible to address a particular problem endemic to photographic representation: that in being able to be removed from the context in which they were produced, photographs risk evoking meanings quite different from what the photographer originally intended. While the images themselves remain the same, the context in which they are viewed changes, and as a result their meaning changes. Without knowing what the lives of their subjects and makers were like, what they remember, what possibilities were available to them, and what decisions they had to make, their lives can only be abstract and disconnected from ours. As viewers, we risk losing our place in relation to the subjects of the images that we view. We can take photographs from the lives that they represent and those who made them and subject them to different uses. But we also can take photographs back to their origins, so that they can be used for the purposes of their makers and their subjects and for readers who hold a direct relationship to their subjects.

What we are looking for in a critical/transformative photographic practice is something much more elusive that defies structure altogether. Photography enables us to render aspects of the world nameable and, at the same time, experience something *unnameable.* With the method that I described earlier, I believe it is possible to trace how particular structures of feeling come into being and to understand the social meanings of images. But this is a deconstructive process, intended to make visible the fluidity of meaning and to demythify the workings of ideology. The aesthetic then can aid us in beholding the experiences of others, allow us to situate ourselves reflexively in relation to our histories, and imagine ourselves differently within the present.

Someone I have known for many years, and with whom I can credit much of my knowledge of Big Intervale insists on the importance of telling stories and of hearing stories. He describes it as "knowing where you came from."[3] These are the stories that surround photographs, as viewed by those whose lives they depict, and inscribe them with meaning specific to the context in which they were taken. Viewing images with the stories of both their subjects and their viewers is a way of bearing meaningful witness to and learning from the lives of others. This complicates our readings of images. However, it allows us to better understand the subjects' relationships to a place and to the world as well as our relations as viewers to what is being represented.

CONCLUSION

The photographic practice that I have outlined involves ambiguous images and images used in conjunction with text. It is important to disable the notion of there being any objective or socially disconnected photographic practice. As I stated earlier, we need to make visible the workings of photographs and how they shape the worlds that they represent. In being animated by those most intimate with their subject matter, the method I describe engages a different economy of relations than does most photography. The reproducibility of photographs enables them to be easily separated from the forces of their original production. However, this can be circumvented if the images are self-reflexively embedded in the social and historical contexts that have shaped their subject(s). The images, as they are presented in this chapter do not carry overt political or social messages. However, as I gather other images and the stories that surround these images, the social and political context of both their production and how they are read become more visible. Eventually this information will come together to produce a particular narrative of Big Intervale and those who have lived there. Also, by continually reflecting on how this project is conducted and making this reflection an integral part of the project, it is possible to make evident the conditions that shape our memories of the past.

The knowledge gained through photographs, like all forms of knowledge, is partial. However, the absence of photographic records does not have to prevent us from using photographs to better understand the past. Many of the photographs that I have taken show only the stone piles hidden in forests that had once been cleared to make fields or the trees that grow from where a house once stood. This is a way in which we can encounter lives that are not our own, or a past for which there is little record or memory, without projecting ourselves into those lives and thereby subsuming them. To do this is to enact Kaja Silverman's (1996) "heteropathic" viewing relations. In doing this it is possible to become more critically aware of how forms of representation shape what we know and enter the ambiguous space where there are few words or images to describe that which we seek to understand. It allows us to encounter what I refer to as the real. And it is in this ambiguous space that it is possible to learn how to know differently.

NOTES

1. Michael Ignatieff writes about friends of his whose families were killed in the concentration camps as being "orphans in time" (1987: 1). He extends this to refer to all those forced into emigration or exile due to economic reasons or political repression.

This is similar to the situations of those who came to Atlantic Canada as exiles from other places. After only several generations many of their descendents once again become economic exiles, moving to the United States (most commonly Boston) or to western Canada in search of work.

2. Claudia Eppert, drawing from the work of Emmanuel Levinas, also writes about this problem. She argues that a more authentic learning practice should "proceed not from the self but from one's encounters with the alterity of a unique other" (2000: 222). This "unique other" is "incapable of being masterfully 'known' or incorporated by the subject" (222).
3. I would like to note that there are certain dangers of inclusion and exclusion inherent in this phrase as it is commonly used. However, I think it serves as a good metaphor for a social/historical analysis.

REFERENCES

Barthes, Roland. (1977). *Image/music/text.* New York: Farrar Straus and Giroux.

Benjamin, Walter. (1968). *Illuminations.* New York: Schocken.

Berger, John, and Jean Mohr. (1982). *Another way of telling.* New York: Pantheon.

Burgin, Victor (Ed.). (1982). *Thinking photography.* London: Macmillan.

Chaplin, Elizabeth. (1994). *Sociology and visual representation.* London: Routledge.

Eppert, Claudia, (2000). Relearning questions: Responding to the ethical address of past and present others. In R. Simon, S. Rosenberg, and C. Eppert (Eds.), *Between hope and despair: Pedagogy and the remembrance of historical trauma* (pp. 213–230). Lantham, Md.: Rowman and Littlefield.

Foster, Hal. (1996). *The return of the real.* Cambridge, Mass.: MIT Press.

Ignatieff, Michael. (1987). *The Russian album.* Toronto: Penguin.

James, Donald, and Hall, Stewart (Eds.). (1986). *Politics and ideology: A reader.* Philadelphia: Milton Keynes.

Lacan, Jacques. (1953). Some reflections on the ego. *International Journal of Psychoanalysis, 34:* 11–17.

Lowe, Donald (1982). *History of bourgeois perception.* Chicago: University of Chicago Press.

Payne, Michael. (Ed.) (1997). *A dictionary of cultural and critical theory.* Cambridge: Blackwell.

Rosler, Martha. (1981). The Bowery in two inadequate descriptive systems. *3 Works.* Halifax: The Press of the Nova Scotia College of Art and Design.

Sekula, Allan. (1986). *Photography against the grain: Essays and works.* Halifax: The Press of the Nova Scotia College of Art and Design.

Silverman, Kaja. (1996). *The threshold of the visible world.* New York: Routledge.

Solomon-Godeau, Abigail. (1991). *Photography at the dock.* Minneapolis: University of Minneapolis Press.

Tagg, John. (1988). *The burden of representation: Essays of photographies and histories.* Basingstoke: Macmillan.

CHAPTER 19

Transformative Learning and New Paradigm Scholarship

MARY ANN O'CONNOR

NEW PARADIGM RESEARCH

It has been 100 years already since the first theory-shattering proofs in physics forced some western thinkers to abandon cherished assumptions about the nature of scientific "proof" or truth itself as well as assumptions about how scholarly work is done and what it can reasonably be expected to do. For those committed to empiricism and to the idea of scientific progress, it has been a bitter pill to swallow that the principal discoveries of the "new physics" so closely resemble the ancient discoveries of the pre-Socratic philosophers and of many even older wisdom traditions: namely, that the boundaries between subject and object are ambiguous, fluid, and mutually constituted; that causality explains far less than acausal connection; that processes and relationships offer more to understanding than products, reductive units, or fixed entitities.

To the extent the new research paradigm emerges from the new physics, it still carries the basic western orientation to modernity, which is that science leads the way. Scientists lay out the terrain as they go and create the definitions of what can be defined. Although new physics proofs are unlike the proofs of Newtonian physics, the ontological and epistemological legitimacy of hard science remains largely unchallenged. That is, within the western academy, hard science still holds a place of preeminence in terms of defining what knowledge is, and who gets credit for possessing it. When I say *what knowledge is,* I mean not only what contents count as knowledge, or how those contents are framed, but also how we can even agree about what it is to know. Indeed, we are now living in a time when knowing of any kind is contested, when the meaning of "to know" must be negotiated within particular contexts.

But the new physics is not the only source of the new paradigm, and it is a fascinating convergence of various sources that is reconstituting research. Postmodernism, at least in some of its varied forms, has effectively challenged the hold of science by embracing the complexity of our fundamental uncertainty. Foucault and many others have alerted us to the deeply political nature of all knowledge claims, demonstrating how particular constructions serve or neglect interests. Deconstruction has become a major method and tool of research, essential for revealing underlying structures of power, including structures of power within language, and for forcing people to question both personal and cultural assumptions. One result has been an impressive new intellectual tolerance for multiplicity of voice and perspective, an appreciation for diversity in principle if not yet in institutional practices.

In my view, creative approaches that both disrupt old boundaries and integrate disciplines in imaginative new configurations are the strongest and most promising challenges to the old research order. They are quite distinct from the range of postmodern approaches that favor disruption and/or deconstruction without any attempt at new integration. By "new integration" I most emphatically do not mean the reestablishment of boundaries in revised, fixed places, or the creation of a new formula for legitimate research. I mean the creation of research gestalts, a temporarily framed combination of both verbal and nonverbal texts that encourages a dialogue among the texts and among the researchers and readers/viewers for the purpose of learning more—and of feeling more—about the research question. This approach to research makes the researcher visible, not just as a defining set of attributes—white, working class, woman, and so on—but as a whole person, a creative vital personality.

In the next section, I attempt to explain my doctoral work, alternating explanatory narrative with poems from my dissertation.

FICTION AS RESEARCH

A poem begins my doctoral dissertation in adult education. My topic is learning in community, with a focus on the experiences of learning racism and antiracism. I am writing a two-part dissertation, the main part a novel in verse, the other part a more conventional analytical piece. I offer this description of my doctoral process not as a formula but as a provocative example of possibility within the new research paradigm. Each researcher must create her own process. I propose that if we are to survive on this Earth, we must create new emotions and new relationships to each other and to the natural world, and we must be prepared to take risks in our research to try to get at previously unexplored ways of feeling and relating. To a certain extent, this approach is bound to be an experiment in failure, but perhaps not entirely. A great deal depends on what we want to know and why. One of the things I want to know is how and why our academic culture so steadfastly avoids acknowledging our somatic and emotive ways of knowing and our place in the natural world and the connections that we have to each other as living beings. These connections do not in any way diminish the significance of relations of power. To acknowledge both sets of relations is a level of greater complexity, a requirement that we hold a larger tension.

Pure Sound

our bodies are pure sound. humming into one another,
singing into one another.
vibrations of the beginning—billion year old birth sounds—ring through us.
echoing us, us echoing
the endless distances and sudden proximities of space-time.
our sounds touch in waves and ribbons, pulses and undulations,
infinite resonances.
her sound as it touched him, his sound as it touched her—

those sounds moved through galaxies, sang between stars,
traveled out on points of light, and waves of possibility,
in the hope of being heard.

sound turns in us.
under our rivers, in depths that may feel soundless.
sound turns on a note.
one note, that people will almost certainly deny having heard.
this note, resounding in depths, is carried on silence, on absences.
this note is longing itself, the echo of space in space.

it can become a chorus, water whose motion is known as sound.
the music of river reaching deep, calling down the heavens.
the glittering stars that drop in answer.
they have said for ages these sounds belong to women.
calling down the heavens, the water choir, the chorus of water.

you can try to hold the sounds on your tongue, but they will not stay.
stars have music of their own, songs we can only hear in dreams.
and anyway people will tell you not to listen.
listen.

My research novel is about relationships in a small northern town on the banks of the Mississippi River during the early 1960s, a period when the civil rights movement was first coming to national and international attention. Music is the medium for encounter and exploration of cultures between two singers, a young black man from a stable family and a young white woman from an unstable and violent family. The ways in which the community—the networks of families, friends, neighbors, schools, churches, libraries—respond to the two singers forces both of them to examine their experiences of and participation in structures of race, racism, and racializing. Racist violence directed at each of them when their relationship is discovered causes body-mind ruptures that tear them from the land, from the music and their music-based ways of knowing, from their families, and from each other. What they learn is complex, nuanced, and full of pain. That is why I chose to write in the form of fiction, and particularly fiction in verse; it is the only way language will hold the complexity, nuance, and emotion. More conventional research language would kill the subtlety and eliminate most of the transformative potential.

An Unsung Land Is A Dead Land

I

Sing this: the land keeps us and we keep the land.
the river carries us and we carry the river.
history lives in us and we live in it, with every breath.

Dance this: the land keeps us. we keep the land.
the river carries us. we carry the river.
the histories are in our blood. the histories are in our breath.

Drum this: land keep us, we keep land. river carry us, we carry river. history lives in blood and breath. bleed justice. breathe justice.

sing this. sing this land.
dance this. dance this land.
drum this. drum this land.

II

beating heart of a drum, implacable.
beating heart of a drum will not surrender the land,
cannot surrender the land.
beating heart of a drum will sing each person's place,
each animal's place, and the place of every leaf and pebble.
drum heart knows the places.

humans do not make music all alone—no.
the land makes music with us.
wind and rivers make music with us.
through us, up through our bodies, up.

drum beat is earth beat,
the rhythms of the universe pulsing through us all.
vibrating into skin, past skin, into muscle and bone.
finally, into our spirits, the rhythm comes.

drums of the Iroquois, drums of the Apache,
drums of the Congo carried from Africa,
drums of the Ashanti.
drums of wars fought on the land.
heartbeats from the measured calls of men.
heartbeats from the naive pounding of boys.

War of Independence.
French and Indian War.
Spanish-American War.
Civil War.

ask the dry textbooks what happened to the hearts
that went deaf to the pulsing rhythms
of the song of the universe,
sung over the land, danced into the land, drumbeat into the land

come back. come back come back come back come

Contemporary intellectual discourse requires that a great deal of attention be paid to location, especially social location. I think this is entirely appropriate, so far as it goes. Race, class, gender, and other categories of exclusion/inclusion define very much about our lives, our bodies, our ways of being, our possibilities and perceptions. I would argue, however, that location must also include our material presence on Earth and our interdependence and interaction with all of the other presences of Earth. It is so interesting that critical theorists insist on the importance of material conditions but ignore our bodies and that postmodern theorists reify the Body but demonstrate little awareness of our bodies being material presences on Earth.

So I start my work with these poems about sky and Earth and river—because these are what we come from, where we come from, whom we come from. The journeys we may take in terms of social location are almost always journeys we take on Earth as well.

Rock That Weeps

mississippi river begins in rock that weeps,
the place where earth becomes water, and water earth.
is it land or lake—lake or river? it is a source.

way up north are the headwaters,
from which by both coincidence and mighty purpose,
a continental river first emerges.
who could know that from this weeping rock the waters
would find a course perfectly suited to the land?
that the land and the water would make each other: the water searching for a way to go,
the land opening a channel here, a rock bed there, a gentle slope, a steep canyon.

and then they carve together, land and water, making the path wider, deeper.
sometimes shifting course by unfathomable agreement.
joined by wind.
joined by every creek, brook, slough, and smaller river.
finding a way down, a chance placement of stones, a fold in the earth's crust, a flatland.

they sing the way to each other.
land sings the route to river, and river sings its way down to the delta.
every time another tributary enters, the music changes,
makes a new harmony or adds a few more altos.

And then there is the importance of place, of the land, as carrier of history. No researcher is ever, ever in a "neutral" or "sterile" environment. There is always a long and complex history alive in every place, a history to which our bodies link at the subatomic, intuitive, and emotive levels, an interconnected somatic history of place.

It Is Called the North

I

Mason-Dixon line gives us
separate versions of crime
land of the South
land of the North
A river doesn't care
where the land has been divided

II

Those neat rows of tombstones
so unlike the heaped or scattered bodies
on the torn-earth battlefields of the Civil War

earned this small town
a National Cemetery
whose population outnumbers the living

A vast city of the dead
surrounded by corn fields and alfalfa
A National Cemetery for soldiers
who died with the name of the same country
on their lips: United States
United—the North's agenda
States—the South's agenda
The issue was economic unity—not slavery
but the conflict turned on the uses
to which humans could be put, and plenty
of blond Iowa farm boys
fought and died to end slavery
while remaining committed racists
the logic of location dictating
their allegiance, the accident
of birthplace defining sides

This town was a base for the Northern army
the central swearing-in point
for the Iowa volunteers
and a major medical center
to which wounded soldiers were returned
for hot-iron amputations, leechings and septic death

By now the bodies are fertilizer
for the fields and forests nearby
for the shagbark hickory
and flickering silver maples

III

A river doesn't care
how the land has been divided
Every dead man's blood drains
from the soil and seeps into the water
we have all been drinking
for more than a century now

My research presentation is unconventional. My research process, however, involved many types of conventional research. I made several trips to the place where the story is set (a town in which I lived briefly during my childhood), took field notes and photographs, did interviews with residents, and studied the history of the area. I also went through the local newspapers for the period 1960 to 1963 and made copies of articles and opinion pieces dealing with race, civil rights actions, civil rights legislation, mixed-race relationships, murders and other violence based on race, and the March on Washington in 1963. In addition, I obtained other local library holdings such as an evocative/provocative collection of obituaries of African American community members that dated back to the 1890s and a *Community Self-Survey on Race Relations* published in the 1950s.

Excerpt from the Report on the Community Self-Survey of Human Relations, **1951**

a found poem, relined

In terms of its own reaction to the general circumstances
defining its status in the city,
the attitudes of the Negro family are generally favorable
toward the city and those relationships involved
with the neighborhood, school and public officials.
There is, on the other hand, a substantial feeling
toward practices of "racial discrimination"
which are associated with a variety of community situations.
The practices of places for eating and drinking in the city
constitute an important area of sensitivity and resentment.

The morale of the Negro family, as expressed
in the ambitions and aspirations held for its younger members,
is expressed in terms of very high and very low levels of expectation.
This kind of ambiguity and somewhat
unrealistic appraisal of opportunity
is rather generally characteristic of the responses
given by Negroes and other minorities to limited status.

In spite of a community situation of stability
and lack of conflict in race relations,
the position of the Negro group in the city . . .
is extremely marginal,
restricted
and limited in opportunity.

Once we know, quite literally, where we are, we can begin to explore our research question in more depth. It is the researcher's burden, responsibility, and exquisite pleasure to define the context of a research question. This could always be done a hundred different, "accurate" ways. All research has this enormously creative dimension, although most research still is presented in the most narrow and thought-constricting ways.

Laney: Young Woman in a Swing

her name was Lorraine, you may have known her.
may have known her as Laney.
may have known her as a strange one.
smart and distant and clear, most people afraid of her,
afraid for her: one way or another.

she was dozing in the oak swing on the front porch
of her grandparents' well-kept white clapboard house.
pale face up, knees up with her skirt tucked around them,
her long arms folded into her body like wings.

her dark curls dropped through the slats of the swing,
sweeping the porch floor.

she was sixteen. she felt her weight in the swing.
heard metal chains clink on metal hooks.
heard grinding at the rusted spot, the worn-out place.
a quavering lullabye, a creaking cradle of sound.
she heard the air as the chains swung through it.
a path cleared, an opening in the sound of the day.

she heard the air, you understand, the way most people breathe it.
her skin absorbed sound. her scalp her tongue her fingertips took in sound.
she lived with a steady reading of vibrations, listened with her body.
she heard music all of the time: in the trees just beginning to leaf
in the buzzing jazz of yellow jackets, in the movement of clouds.
she could hear danger in the heel-to-toe of one footstep.
in the turn of a key.
it was her only protection, this aural vigilance.
she was on guard for her self,
because it was certain all along: she would be unable to protect
her face, breasts, arms, legs—

but she could, if she moved very quickly, get her soul out of the way.

My research on the music aspect of the novel has involved taking voice lessons and studying vocal technique; interviewing singers; listening to many, many recordings; and reading widely in the musicology of blues, jazz, and rhythm and blues. My research on the civil rights movement has involved extensive reading of histories, personal accounts, and biographies as well as research into television coverage and documentaries. For contemporary analyses of structures of racism, I have studied critical race theory and the related law/policy approaches. I have read a great deal of material on antiracist approaches to education and have been deeply impressed by Black feminist books that offer alternative curriculum materials.

Let me give another example of how I work with fiction as research. In the novel's opening section about one of the main characters, Raymond, there are several poems about his family. One is about his father:

Raymond's Father's Barbershop

I

Howard came back from the war knowing
how to cut hair, trim beards, angle sharp sideburns.
he'd had to learn, once the Army realized the black soldiers' hair needed cutting too.
and the white barbers wouldn't do it, wouldn't let Negroes sit in the chairs.
he liked barbering well enough.
liked smells of shaving cream, hair tonic.
enjoyed the feel of lather on the boar's hair brush.
feel of men's bristly faces.
feel of his smooth metal scissors, his sure razor.

a decent job, he didn't come home smelling
of animal blood or tar or manure.
didn't have to bathe before greeting the family.

the rhythms of barbering: scissor clip, clip, snip, clip, snip,
strop slapping, vvvit, vvvit, vvvit, vvvit,
his whole body rocking to a solid work-rhythm.
he heard them all the time, the scissors and the strop.
heard them as the background rhythm of his own voice.
heard them in his carefully trimmed dreams:
the sound of not overreaching,
of taking what satisfaction he could.

II

some guys came nearly every day, took their customary places.
along the shelf of the front window: old timers, veterans,
ones waiting for work, a few past waiting.
Russell came limping in, his foot might as well
have been shot clean off for all the good it did him.
and Gerald toting his half-arm in the folded sleeve of his jacket.
they talked. *'Member how it was in France, man,*
those club owners askin' us in—
and those French women liked us—
Old Finney, man, he had 'em
two a night, every night without even tryin'—
Finney's grin widened and his belly chuckled
as he waved his hand back into the past.
war was good to me, what can I tell ya.
things ain't been so good since I come back here.
got some respect in France, got some appreciation.
there were nods all around, and head-tossing snorts of disgust.
hell, we was good enough to die for Uncle Sam.
good enough to do any job they needed doin' overseas.
Not good enough to have some respect—
and a steady job—here.
true. damn true.

Howard nodded too, laughed when the others did.
spoke along with them into the silences of shared war.
all of them conscious of not mentioning the unspeakables
that are the strongest bonds.
they were guilty-glad to be alive.
every day felt stolen from the dead.

This poem provides important historical information for the novel, at the same time that it helps us to understand Raymond's character and to anticipate what will happen later on in the story. It makes the connection between the experiences of Black Americans as soldiers in World War II and the civil rights movement. For many Black men, going to war raised their race consciousness. Boot camp, uniting men from various parts of the country, created the opening for a shared analysis of racism. When they went overseas, they experienced new awareness of the injustice they suffered at home and of the possibilities for things to be otherwise. Moreover, they experienced the anger and disillusion of having been full citizens when it came to fighting and dying for their country, only to find themselves second-tier citizens when they came home again. It is not surprising, then, that these ex-soldiers and the next generation, the children of the soldiers, became civil rights activists. It is historically accurate that white barbers would not cut Black men's hair and that Black barbers had to be trained. Later in the story Howard's skill with a razor, and the presence of the men in the barbershop, will save Raymond's life.

As I write, I try to make certain that I could document the elements of my fiction, as distinct elements and as sets of relationships. That is, I can "justify" every character's name, occupation, that follows are from the local newspaper. I also can document the communities' conflicting attitudes about race, the lines drawn around mixed-race relationships, the presence of Ku Klux Klan members in the area, the local news of civil rights actions in the South, the music stations and songs on the radio, how much "race crossover" there was in music listening and buying, the music people knew at church, and so on. In addition, I can document important outside influences on the community.

Up the River Came News

She read hurriedly before her grandparents noticed
where her attention was directed.
Everyday it was something:

Negroes Crowd Lunch Counter
Civil Rights 'A Carnival of Politics'
Violence Appears Over Lunch Counter Demands
Southern Demos Win Rights Bill Round
Arrest Threat Faces Lunch 'Sitdowners'
Begin Around-Clock Civil Rights Battle
Negro Arrests Soar
Dixie Senators See Revision As Way to Block Civil Rights

she needed to know this
she could feel it in the grim silences
of her teachers and other white people
see it in the faces of the lunchcounter protesters
hear it in the colored music they sang, hear it
in the bitter commentary of her grandpa
Goddam niggers—wantin' their "rights"—shit
I hope somebody shoots that King nigger—who
does he think he is—why the hell
does the newspaper even cover this -it ain't news
it's just goddam, lazy niggers wantin' more
his anger wrecked the newspaper,
filled the quiet living room with rage
made the furniture seem miniature,
made Laney feel small as a mouse
made everything shrink
grandma too

Reading Gandhi

Mild-mannered words of a skinny man in India
are coming to America, booming

across oceans, booming over the land
Vibrations rattle the arms
of those who read his book
shake the souls
of those who have been searching
Nonviolent protest
Nonviolent resistance
Nonviolent revolution
Courage is clarity about fear:
which danger is really greater

A hollow form of "civil rights"
has been here since Truman,
since the Supreme Court decision in *Brown*
But the spirit moves stronger with the vision
of nonviolence, of redemptive love
of an almost unbearable generosity
that whites only perceive as threat

I have had to do considerable research on the nature of research itself to locate myself in this contested terrain. As I have chosen to use both a critical and an arts-based orientation, I have had to develop my own theoretical account of the points of convergence of these approaches. A grounding in transformative learning theory, and particularly in the work of Edmund O'Sullivan (1999), allowed me to articulate a dialectical dynamic between what I have called a Sphere of Relations of Power and a Sphere of Relations of Natural Intimacy.

My way of working with all of this seemingly disparate material has been to use visual art as part of my reflection process. I have made collages of images of the Mississippi River, of the small town, and of the newspaper articles. I have made visual representations of my theorizing. I have painted landscapes, both representational and abstract, as a way to explore my perceptions and feelings. I have tried to draw my fictional characters and have been amazed to find how tellingly my drawings reflect the state of development of my written portraits. My field notes are in huge scrapbook albums and include photos, paintings, drawings, textual analyses of particular newspaper articles, collages, prose pieces, and poems. I go over this material, add to it, rethink it, agonize over it.

In my view, the emphasis on a uniquely individual construction *within* a network of relationships is the key element of new paradigm research. The "individual" here is not conceived as wholly separate, or isolated, or outside what she studies. There is little use to distinctions such as subjective/objective, or researcher/researched, or for that matter perceiver/perceived.

Another basic element in new-paradigm research is multimodality. By multimodality I mean the use of multiple means of inquiry, reflection, and reporting on research. These would include verbal communication that does and does not fit conventional research formats, such as life histories, poems, readers' theater, plays, songs, and fiction. These also would include nonverbal forms of communication, such as drawing, painting, collage, and other visual art, some performance art, photography, dance, quilting, and music making. The multimodal element is closely related to the individual-within-relationship character of the work because, again, it emphasizes the particular construction of an understanding, or of "knowledge," within a network of relationships.

In this brief chapter, I cannot tell the whole story of my novel in verse or of my research process for creating it. I can only give readers a feel for it all. When, toward the end of the novel, violence is done to Raymond and to Laney, the country as a whole is also confronting racist violence. In the next poem, the quotes are taken from the local Iowa newspaper coverage—that is, the quotes are what Laney and Raymond and their families would have read.

The Murder of Medgar Evers, June 12, 1963

I

just past midnight in north jackson mississippi
Medgar Evers was shot in his driveway
a bullet in his back.
he had known for months already
that death would come riding by—
definitely since that day in May
when he opened his front door
to a kerosene bomb.

II

the momentary silence of the night
once the assasins had sped off:
political movements circled like galaxies
in the eyes of the dying man
swirled like star dust

did he feel the wrenching memory of his mother's hands
the close resemblance of his daughter's hands
the touch of his daughter
did his skin remember as he died
remember all of the touches of love
to undo the quick hard case of a bullet

his wife's screaming voice, the calls of the neighbors
who clustered on porches in their nightclothes,
the radio of the police cruiser—
were these what he heard?
or was it the flutter of live oak leaves
the stillness of a hot night's holding
the stillness
and then, he entered it

The children his wife wept later to reporters,
we were up. And the children
came out and tried to talk to their Daddy.

III

From new york, Roy T. Wilkins
secretary of the NAACP, issued a statement:

In their ignorance they believe that by killing
a brave, dedicated and resourceful leader
of the civil rights struggle they can kill
the movement for human rights.
They cannot.

The way each one of us combines our investigations into life, our intellectual constructions, our past experiences, our imaginative worlds, and our formal research is highly idiosyncratic. And, I would argue, it should be idiosyncratic. A poet does not pretend to speak for anyone by herself, to know anything but the way *she* knows, and yet . . . and yet there may be much to learn. There may even be transformation.

REFERENCES

Abram, D. (1996). *The spell of the sensuous.* New York: Vintage Books.

Ai. (1999). *Vice.* New York: W. W. Norton.

Ai. (1993). *Greed.* New York: W. W. Norton.

Alaniz, Y., and Wong, N. (Eds.). (1999). *Voices of color.* Seattle: Red Letter Press.

Archer, J. (1993). *They had a dream: The civil rights struggle.* New York: Penguin Press.

Ayers, W., and Miller, J. L. (Eds.). (1998). *A light in dark times: Maxine Greene and the Unfinished conversation.* New York: Teachers College Press.

Barnhill, D. L. (Ed.). (1999). *At home on the earth.* Berkeley: University of California Press.

Bateson, G. (1972). *Steps to an ecology of mind: Collected essays in anthropology, psychology, evolution and epistemology.* London: Intertext Books.

Beatty, J. (1974). *Kiowa Apache music and dance.* University of Northern Colorado (Greeley), Museum of Anthropology, Occasional Publications in Anthropology, Ethonology Series, No. 31.

Belenky, M. F., Clinchy, B. M., Goldberger, N. R., and Tarule, J. M. (1986). *Women's ways of knowing.* New York: Basic Books.

Berman, M. (1981). *The reenchantment of the world.* Ithaca, N.Y.: Cornell University Press.

Berman, M. (1989). *Coming to our senses.* Seattle: Seattle Writers Guild.

Berry, T. (1988). *The dream of the earth.* San Francisco: Sierra Club Books.

Cixous, H. (1998). *Stigmata.* New York: Routledge.

Cohn, L. (1993). *Nothing but the blues.* New York: Abbeville Press.

Crenshaw, K., and Gotanda, N., and Peller, G., and Thomas, K. (1995). *Critical race theory.* New York: The New Press.

Dewey, J. (1980) [1934]. *Art as Experience.* New York: Perigree Books.

Diamond, B. and Cronk, M. S., and Rosen, F. (1994). *Visions of sound.* Waterloo, Ontario, Canada: Wilfrid Laurier University Press.

Donald, J. and Rattansi, A (Eds.). (1992). *"Race," culture and difference.* London: Sage.

Eisner, E. (1997). The promise and perils of alternative forms of data representation. *Educational Researcher* (August-September) 1997: 4–9.

Ellison, M. (1989). *Lyrical protest: Black music's struggle against discrimination.* New York: Praeger.

Escott, C. (1991). *Good rockin' tonight: Sun Records and the birth of rock and roll.* New York: St. Martin's Press.

Estes, S. (1999). *Simon Estes: In his own words.* Cumming, Iowa: LMP Press.

Fanon, F. (1963). *Black skin, white masks.* New York.

Farmer, J. (1985). *Lay bare the heart: An autobiography of the civil rights movement.* Fort Worth: Texas Christian University Press.

Fine, M. (1992). *Disruptive voices: The possibilities of feminist research.* Ann Arbor: University of Michigan Press.

Gablick, S. (1995). *Conversations before the end of time.* New York: Thames and Hudson.

Gallas, K. (1991). Arts as epistemology: Enabling children to know what they know. *Harvard Educational Review, 61:* (1).

Glesne, C. (1999). *Becoming qualitative researchers.* New York: Longman.

Goldberger, N., and Tarule, J., Clinchy, B., AND Belenky, M. (Eds.). (1996). *Knowledge, difference, and power.* New York: Basic Books.

Gioia, T. (1997). *The history of jazz.* New York: Oxford University Press.

Greene, M. (1995). *Releasing the imagination.* San Francisco: Jossey-Bass.

Greene, M. (1991). Texts and margins. *Harvard Educational Review, 61:* 1.

Heshusius, L., and Ballard, K. (1996). *From Positivism to interpretivism and beyond: Tales of transformation in educational and social research.* New York: Teachers College Press, Columbia University.

Heshusius, L., and Ballard, K. (1994). Freeing ourselves from objectivity: Managing subjectivity or turning toward a participatory mode of consciousness? *Educational Researcher* (April): 15–22.

Hines, J. (1998). *Great singers on great singing.* New York: Limelight Editions.

hooks, b. (1988). *Talking back.* Toronto: Between the Lines.

hooks, b., and West, C. (1991). *Breaking bread: Insurgent Black intellectual life.* Toronto: Between the Lines.

hooks, b. (1992). *Black looks: Race and representation.* Toronto: Between the Lines.

hooks, b. (1993). *Sisters of the yam.* Toronto: Between the Lines.

hooks, b. (1994a). *Outlaw culture: Resisting representations.* New York: Routledge.

hooks, b. (1994b). *Teaching to transgress: Education as the practice of freedom.* New York: Routledge.

hooks, b. (1995). *Art on my mind: Visual politics.* New York: The New Press.

hooks, b. (2000). *Where we stand: Class matters.* New York: Routledge.

hooks, b. (2001). *Salvation: Black people and love.* New York: HarperCollins.

Hull, G. T., and Scott, P. B., and Smith, B. (1982). *All the women are white, all the blacks are men, but some of us are brave.* New York: Feminist Press, City University of New York.

Jones, L. (Baraka, A.). (1963). *Blues people.* Edinburgh: Payback Press/Cannongate Books.

Jordan, J. (Ed.). (1997). *Women's growth in diversity: More writings from the Stone Center.* New York: Guilford Press.

Jordan, J., Kaplan, A., Miller, J., Stiver, I., and Surrey, J. (1991). *Women's growth in connection: Writings from the Stone Center.* New York: Guilford Press.

Kagen, S. (1950). *On studying singing.* New York: Dover.

Kilbourn, B. (1999). Fictional theses. Educational Researcher, 28(9).

Laura, R. S. and Cotton, M. (1999). *Empathic education: An ecological perspective on educational knowledge.* Philadelphia: Falmer Press.

Lawrence-Lightfoot, S., and Hoffman, David, J. (1997). *The art and science of portraiture.* San Francisco: Jossey-Bass.

Lorde, A. (1984). *Sister outsider.* New York: The Crossing Press.

Lorde, A. (1993). *The marvelous arithmetics of distance.* New York: W. W. Norton.

Lubiano, W. (1998). *The house that race built.* New York: Vintage/Random House.

Mezirow, J. and associates. (2000). *Learning as transformation: Critical perspectives on a theory in progress.* San Francisco: Jossey-Bass.

Morris, A. D. (1984). *The origins of the civil rights movement: Black communities organizing for change.* New York: Macmillan Free Press.

Morrison, T. (1987). *Beloved.* New York: Knopf.

Morrison, T. (1992). *Jazz.* New York: Knopf.

Neilsen, L. (1998). *Knowing her place: Research literacies and feminist occasions.* Nova Scotia: Backalong Books and San Francisco: Caddo Gap Press.

Olsen, T. (1965). *Silences: Classic essays on the art of creating.* New York: Dell.

Ondaatje, M. (1976). *Coming through slaughter.* Toronto: Vintage Canada.

O'Sullivan, E. (1999). *Transformative learning: Educational vision for the 21st century.* Toronto: University of Toronto Press.

O'Sullivan, E. (1990). *Critical psychology and pedagogy: Interpretation of the personal world.* Toronto: OISE Press.

Porter, L. (1997). *Jazz: A century of change.* New York: Schirmer Books (Simon and Schuster).

Porter, L., and Ullman, M. (1993). *Jazz: From its origins to the present.* Englewood Cliffs, N.J.: Prentice Hall.

Riches, W. T. M. (1997). *The civil rights movement: Struggle and resistance.* New York: St. Martin's Press.

Roediger, D. R. (1998). *Black on white: Black writers on what it means to be white.* New York: Schocken Books.

Roszak, T., Gomes, M., and Kanner, A. (Eds.). (1995). *Ecopsychology: Restoring the earth/ healing the mind.* San Francisco: Sierra Club Books.

Sacks, O. (1989/1990). *Seeing voices.* New York: Harper Perennial.

Spencer, J. M. (1996). *Re-searching Black music.* Knoxville: University of Tennessee Press.

Spencer, J. M. (1997). *The new Negroes and their music.* Knoxville: University of Tennessee Press.

Sudnow, D. (1978). *Ways of the Hand.* Cambridge, Mass.: Harvard University Press.

Sulter, M. (1990). *Passion: Discourses on Black women's creativity.* London: Urban Fox Press.

Tirro, F. (1977, 1993). *Jazz: A history.* New York: W. W. Norton.

Walton, O. (1972). *Music: Black, white and blue.* New York: William Morrow.

Ward, B. (1998). *Just my soul responding: Rhythm and blues, black consciousness, and race relations.* Berkeley: University of California Press.

Watters, P. (1971). *Down to now: Reflections on the southern civil rights movement.* Athens: University of Georgia Press.

CHAPTER 20

The Transformative Power of Creative Dissent

The Raging Grannies' Legacy

Carole Roy

Among the most subversive and powerful activities women can engage in are activities of constructing women's visible and forceful traditions, of making real our positive existence, of celebrating our lives and of resisting disappearance in the process.

—Dale Spender, *Women of Ideas*

The Raging Grannies have crashed official receptions, parliamentary hearings, Canadian Armed Forces bases' open houses, and meetings held by representatives of diverse levels of government to alert public opinion and authorities about environmental, peace, and social justice issues. With flair and style, they transform rage and powerlessness into humorous and creative protests. With the creation of the colorful, resistant, and seemingly enduring popular figure of the Raging Granny, some older women claim a public space of their own. They defy stereotypes and authorities alike with smiles, wit, daring, and imagination. In the *Canadian Theatre Review,* John Burns (1992) suggested that they "have reversed cultural expectations by empowering themselves within a society which belittles their experience and point of view" (21). Such a story of transformation requires attention: without stories of opposition, resistance, and transformation, we internalize patriarchy's ideology and pass its rules to the following generation (Lerner, 1997: 207–8).

. . .

You don't have anything
If you don't have the stories.
Their evil is mighty

But it can't stand up to our stories.
So they try to destroy our stories
Let the stories be confused or forgotten.
They would like that
They would be happy
Because we would be defenceless then.

. . . .

—Leslie Marmon Silko

Although feminists have been reclaiming different women's achievements and resistance from the past (Rowbotham, 1972: 16), Dale Spender warns us that "While we are prepared to put much energy into reclaiming women from the distant past, our record is not so good when it comes to preserving our more recent heritage. In fact, we have sometimes been careless about the way we have discarded that very heritage" (cited in Reinharz, 1992: 215).

Here, then, is an attempt to preserve the recent daring and creative heritage of the original group of the Raging Grannies from Victoria, British Columbia, a necessary inheritance in a world desperate for notions of political action that require both defiance of oppression and respect for life. Transformation demands fierce confrontation of oppression and deep imaginative caring for social justice: The Raging Grannies offer an example of innovative dissent and imaginative protest aimed at transformation.

I met the small group of women that was to become the core of the first Raging Grannies group in classes taught by a Salvadorean political refugee: learning Spanish was one of our efforts at continental consciousness and solidarity in the mid-1980s. I had recently returned to Victoria after walking for nuclear disarmament from New Orleans to New York and from Bonn to Vienna. At the time, I worked as a homemaker for the elderly and handicapped. These women were very active in the peace movement, and sometimes I joined them in their protests, sometimes also in their experimentation with street theater. In December 1985 and January 1986 I joined the Central America Peace March, an effort by 300 people from 25 countries initiated by Norwegian Women for Peace to express solidarity with the people of various countries in Central America under repressive regimes. This group of women studying Spanish were very supportive and organized a fund-raising dinner for the occasion. When I returned, Doran Doyle and I decided to organize a peace march for the conversion of Nanoose Bay Military Base and its underwater weapons-testing facilities used by the U.S. Navy. In the spring of 1986 we walked from Victoria to Nanoose Bay in five days. A few months later, in August 1986 I was part of the Motherpeace Civil Disobedience Action to reclaim the Nanoose military underwater weapons-testing ground for peaceful purposes. At the time, civil disobedience was controversial even among peace activists, but the women who became the Raging Grannies were very supportive. Two years later, some of them would commemorate that action with one of their own: The Grandmother Peace Action was also an act of civil disobedience. Although I was never part of the Raging Grannies, I am still in contact with members of the original group. They have been inspiring friends and I feel privileged to know them. I felt increasingly aware that their stories of creative protest were not systematically recorded and decided to work toward such a contribution. The daring and creativity of this group of women may provide inspiration to women in the future, as some women's resistance of the past encourages our own today.

Against the grayness of a wintry Victoria (BC) day in February 1987, a colorful group of elderly women climbed the stairs of the imposing BC Legislative Building with a basket containing a clothesline of female "briefs" (Howard, 1989: 14) to be presented at the provincial hearings on uranium mining. To their surprise, their new Granny persona was a roaring success with the crowd. Barred from entering the Parliament Building but encouraged by the crowd's response, they soon took their brand of satirical lyrics to captive audiences in movie line-ups on Tuesday for the well-attended cheap movie night and, more generally, took to the streets (Brightwell, 1988: 21). They have since spread like wildfire across Canada. Initially they were eight: Betty Brightwell, Doran Doyle, Lois Marcoux, Hilda Marczak, Bess Ready, Mary Rose, Joyce Stewart, and Fran Thoburn. Like many in the peace movement, they were white, well educated, mostly middle class (Lofland, 1993: 199). Teachers, a librarian, an anthropologist, a counselor, artists, and business women, they all had children and most, but not all, had grandchildren. Ranging in age from fifty-two to sixty-seven, some were World War II veterans, one was the wife of a military officer/defense scientist, and another was married to a refugee from Nazi Germany who joined the Canadian Army and is also a World War II veteran. Half were immigrant/refugee themselves or had parents who were so. Strong willed, they held a diversity of political views: Most disagreed with peace-through-strength but one maintained the importance of the armed forces (Ciriani, 1989: 3). Yet all agreed on the need to protest nuclear arms and have fun in the process (Acker, 1990b: 4).

The Victoria Grannies had an unending stream of innovative ideas for confronting what in their view needed change. In June 1988 the *Esquimalt Star* reported: "Saturday afternoon saw an unusual procession through Esquimalt heading for the gates of Dockyard. The Raging Grannies, who are against nuclear weapons, wanted to get into dockyard and lay bouquets of flowers on the guns of the ships stationed here. Not surprisingly they weren't allowed into the military complex and had to content themselves with singing anti-nuke songs from their carriage that had been halted by M.P.'s [Military Police]" ("Raging Grannies Sing," 1988: 2).

Riding in style in an inoffensive horse-drawn carriage, which contrasted with the danger of the nuclear submarine stationed at the base, they expressed dissent while offering the image of a cheerful and spirited group of older women. Dissenting with smiles can confuse people prepared to react to threats, as captured in an article about their protest: "Officials there [at the base] had to confer for quite a time about the request. . . . Finally the word came that the flowers couldn't be taken onto the base" ("Grannies Ride in Style," [nd]). Confronting people with unexpected avenues of expression creates a space where people have to question their roles, which is the aim of nonviolence. The Grannies' creative and unpredictable protests disturbed complacency and challenged established roles and nonquestioned assumptions. Invited to Ottawa by the federal government in 1990, they made officials nervous when they launched their Grannies' navy—paddling two dinghies on the dry concrete shoulder of the Centennial Flame on Parliament Hill, they warned of more mischief during the upcoming Commonwealth Games if U.S. nuclear vessels kept visiting Victoria's waters (Foley, 1990: 3). In spite of the humor, they assured people they were deadly serious. That ability to use any opportunity to broadcast their concerns was remarkable. As they returned from a massive protest at a Nevada nuclear test site, an article in the *Times-Colonist* read "Grannies agree to be contacts for draft evaders"—evaders, the Grannies suggested, who could say to immigration officials that they were visiting their grandmas (Brown, 1991: [np]). It is this ability of the Grannies to perceive the humor in any occasion and twist it into a catchy and meaningful line or action that made them good news.

Initially, they remained anonymous and refused to give their names ("The Raging Grannies," 1987: C8), "embarrassed by the antics they had to perform to publicize their cause" (Banner, 1988: 17). Granny Mary Rose said: "It is this very serious issue that causes otherwise conventional women, to be willing to make fools of themselves to bring attention to the risks nuclear ships and subs bring into our lives and those of future generations" (Rose, 1992: 1).

But their nerve increased, considering one of the songs they performed at the Fringe Festival in Edmonton in 1989: "When the Victoria Grannies roll across the stage, holding decidedly phallic versions of MX missiles while singing the old airforce ditty Roll Me Over, even the most loquacious politician is reduced to aghast silence. 'My husband nearly flipped,' says Granny Betty Brightwell. . . . But the Grannies weren't always this brash. 'It took us four tries to get out of the closet,' Brightwell admits" (Van Luven, 1989: A10).

Their daring was possible only because of a sense of togetherness: "We enjoy it when we are together. We could never do it alone" (Joyce Stewart cited in Banner, 1988: 17). Together they dared. They took risks. In August 1988 two of them, Lois and Doran, took part in the Grandmother Peace Action, a civil disobedience attempt to symbolically reclaim Winchelsea Island for peaceful purposes. Winchelsea Island is the nerve center of the Canadian Forces Maritime Test Range (CFMTR) at Nanoose Bay (BC) where the U.S. military tests its underwater weapons, bringing into the surrounding waters many nuclear-powered and/or nuclear weapons–capable U.S. warships and submarines. Arrested for trespassing on Department of National Defense property, they risked a $1,000 fine or one year in jail if found guilty of the criminal offense (Lightly, 1989: 15). Doran, who had "never been arrested before," felt she had "crossed out of the range of respectability" (cited in Howard, 1989: 14). In *Free Women of Spain,* Martha Ackelsberg (1991) suggests that crossing boundaries of "appropriate behavior" with a supportive group can empower and result in the questioning of the appropriateness of those boundaries in the first place (165). After the arrest, Lois was ostracized by the Girl Guide leaders of the group to which she belonged with her daughter: She had broken the law. Although she "felt rejected or pushed down or cast aside," Lois was never "sorry that we crossed the line." At the time, civil disobedience was controversial not only with the general public but even among peace activists.

Some of the Grannies had organized creative protests against the visit of nuclear ships to Victoria's waters before they became the Raging Grannies. On Thanksgiving 1985, they held a protest banging pots with spoons to alert the population ("Nuke Protest Today," 1985). On International Women's Day 1986, they went with brooms, brushes of all types, and vacuums to "clean up the base" ("Antinuclear Protesters Parade," 1986: A3). The principle of using things at hand would remain a thread with the Grannies' web of actions. On October 26, 1986, the Trident Submarine USS *Alaska* passed by Victoria on its way to the base at Bangor, Washington. With other groups, they spooked the passing Trident with a Dance Macabre, used "in the Middle Ages to dispel the Plague," in this case meant to be a "raising of the spirits of horror at the evil presence in our waters" (Malkinson, 1986). Doran said: "We measured out the length of the Trident in pantyhose tied together between the lamp posts at Bess' house. We measured out all these lengths of three football fields long [170 meters]" (cited in "Stretching a Point," 1986).

She recalls: "We had the co-operation of the fire brigade and people like that because we had Bar-B-Q pits to make flames and we played loud Dance Macabre and funeral march [music]. . . . It was around Halloween and that's why we were doing it that way. So that's bouncing off the festivals that were around" (Doyle, 1998).

Some had experimented with street theater: dressed in white lab coats, they were the NERT (Nuclear Emergency Response Team), carrying makeshift Geiger counters, testing puddles of water for radioactivity at malls and stores, explaining to bystanders that nuclear submarines were in the harbor and they were making sure there was no radiation in the water (Saunders, 1989). But "after Chernobyl occurred and we saw on TV the people going through the same gestures we had done in street theatre, running Geiger counters over posts, buildings, etc. . . . We decided we must do something else" (Letter by Sophia in Sophia Collection).[1] "Chernobyl . . . brought us to a halt with being fun and the play in the . . . common byways, the streets of Victoria and Esquimalt. And we went into despair almost at that stage. We had a big public meeting 'When can we eat our broccoli?'" (Doyle, 1998).

Shortly after these NERT actions they became the Raging Grannies. Doran Doyle came up with the name as she searched for a way to express her outrage at the presence of nuclear ships in the harbor. She recalls: "There is a long period where I just exploded inside because of my personal life and learning about this world, this beastly world, militarism. . . . It all came together reading Mary Daly about age. . . ." Doyle kept the inspiring quote:

> Rage is not "a stage." It is not something to be gotten over. It is transformative, focusing Force. Like a horse who streaks across fields on a moonlit night, her mane flying, Rage gallops onpounding hooves of unleashed passion. The sounds of its pounding awaken transcendentE-motion. As the ocean . . . its rhythms into every creature, giving sensations of our commonsources/courses, Rage, too, makes sense come alive again, thrive again, thrive again. Women require the contest of Be-Friending both to sustain the positive forces of moral Outrage and to continue the Fury-fueled task of inventing new ways of living. Without the encouragement of Be-Friending, anger can deteriorate into rancour and can mis-fire, injuring the wrong targets. One function of the work of Be-Friending, then, is to keep the sense of outrage focused in a biophilic way. (Doyle Collection)

There were hesitations about the name because of the stigma attached to anger and rage (Walker, 1998). But another Granny, Hilda Marczak, found a similar inspiration in Margaret Lawrence's writings: "It is my feeling that as we grow older we should become not less radical but more so" (cited in the Marczak Collection).

These feminist writers helped crystallize experience into concepts and had some influence on the definition of the Raging Granny character. Here anger and hope work together in a creative ferment, and the Grannies transformed fear into action. Doran Doyle (1998) also suggested a cultural dimension to rage: "To be raging in New Zealand or France or Britain is not so appropriate because people are more comfortable with expressing in a forthright way their feelings. Whereas in Canada I love the experience of how tolerant and how gentle people are but then, there are times when you think why don't they . . . get more aggressive."

The idea of rage is more jarring in a context where people are usually perceived as polite. That the Grannies originated in Victoria, retirement capital of Canada and a place of good taste and decorum (*Raging Grannies,* 1990), is also part of that larger context. But the coming of U.S. Navy nuclear vessels was a concrete threat and provided a focus for local action. James Bush, a retired U.S. Navy nuclear submarine captain, spoke out in Ketchikan, Alaska, and gave weight to their fears about the dangers of these ships: He advised people to ask for emergency plans in case of accidents and told them that the U.S. Navy had a policy of dumping possibly radioactive waste in the open sea. He warned that even though an accident was unlikely, it still could happen (cited in Farrell, 1991: 3).

Alienation from mainstream peace organizations was also at the origin of the Raging Grannies. Like the Greenham women who left the Campaign for Nuclear Disarmament, the largest British peace organization, because "it was male dominated and hierarchical" (Romalis, 1987: 93), both Doran and Betty mentioned sexism in the peace movement as an incentive to look for new ways. Involved with the Greater Victoria Disarmament Group, Betty said to a journalist that "We found we were becoming the fetchers and gatherers and we said 'No more of that!'" (cited in McCulloch, 1991). Doran echoes a similar experience: "Guys in the peace movement wanted us still to be bringing the coffee . . . not hearing us when we came up with things, but treated us as a bunch of older women . . . because it's the younger, more glamorous ones who would be heard. . . . It's all those different perceptions coming through that helped us to form as a group . . . o.k. they are not listening to us so we'll try it this way. We must have had these models from Greenham Common . . ." (Doyle, 1998).

Doran went to Greenham twice and experienced the concept and power of women-only actions. Others mentioned positive experiences with groups of women in their pasts. Their song *The Bad Old Days* conveys their changing views on women's roles:

Refrain:
We're the women who did the work
So men could get the credit
We said "Leave it all to us"
And wished we'd never said it.

Leave the dishes in the sink
You sit down and rest, dear,
I can finish clearing up
I can do it best dear.

Refrain and Kazoo

No, I don't mind staying late
I'll type another stencil
Can I bring your coffee now?
Let me sharpen your pencil.

Refrain

I'm sorry that the baby cried
I'm sorry that she wet you
I'm sorry she threw upon you
I'm sorry she upset you.

We're prepared to do the work
But we want more than credit
Equal pay for equal work
We'll sing until we get it.

—Marczak Collection

The Grannies were aware of gender and how it affected them. Not surprisingly, finding similar patterns in the "enlightened" peace movement sent them looking for other avenues of expression; they had explored the path of service to men enough for a lifetime. They were ready for something else. On occasion, they humorously suggested they were hoping for "Grumbling Grandpas" or "Raging Granddads" (Doyle cited in Howard, 1989: 15; Kavanagh, 1987: A16); but one only needs to try to imagine Raging Grandpas to know gender is a component of the Raging Grannies!

The Raging Grannies used their image of respectability to infiltrate where they were not welcomed. During the U.S.-Canada Free Trade talks, they crashed the reception for Pat Carney, then federal minister of trade, at the Empress Hotel in Victoria. They lined up and a smiling Pat Carney shook their hands, assuming they were part of a receiving line (Stuckey, 1989). Then they broke into a song written for the occasion called the "Free Trade Trot" (tune: "Playmates Come and Play with Me" with much Kazoo):

Oh Oh Oh Free Trade
Of you we're not afraid
Our fears have been allayed
This deal was heaven made.
Who needs a culture or an identity
When we have Dynasty on our TV

Oh Oh Oh Free Trad—the deal we've waited for
Our bucks will buy us more
Consumer goods galore
We may not know where our next jobs' coming from
But we'll have Calvin Kleins on all our
Bum, bum, bum, bum, bums.

—Marczak Collection

Acidly, Carney told them that their "singing was a hell of a lot better than [their] logic," (Stuckey, 1989), to the delight of the Grannies who appreciate a good line and readily admit to singing badly ("Cult Expands as Zeal to Save Planet Catching," 1989: B1). They made no pretensions of good singing: "we just start on any old note, and our singing is incredibly awful; we have people who can't sing a note" (Chodan, 1989: H3). "We pride ourselves on not being professionals, we're not musicians, we're just having fun" (Stuckey, 1989). They saw their appeal as being ordinary people with something to say (Kavanagh, 1987: A16). What was important for them were the issues, not the entertainment.

The Raging Grannies' defiance often was expressed through songs. Songs allow the expression of rage and outrage in public in a way speech does not. They used popular tunes and wrote satirical lyrics to comment on current events. In another song, they raise Canada's sovereignty:

Here in the land of the beaver,
We say we are nuclear free.
We want to be happy believers,
But tell us then how it can be?
There are nuclear ships in our harbour,
And the Tridents are out in the Strait.
We have tested the cruise, terrorized caribous,
Maybe we'll be the fifty-first state.
Nya nya nya nya nya nya, nya nya nya nya nya, nya nya.nya

When Mulroney and Reagan are talking
And they put on their friendship display
We wish Brian would say something shocking
And to Ronny emphatically say:
Take your nuclear ships from our harbours
Keep the Trident away from the Strait.
We will not test the cruise, terrorize caribous
And we won't be the fifty-first state.

(Brightwell, 1988: 22)

In "The Streets of Laredo," they linked militarization at home and abroad:

As we watched the news of what happened in Oka
And wait for the Gulf to explode on TV,
We see our brave leaders play cowboys and Indians
And wonder if this is how things have to be.
Our guns have been pointed at our land's First People,
Our planes are prepared for a Middle East War,
We're told we're the good guys and they are the bad guys,
But nobody says what we are fighting them for.
Maybe it looked good on the streets of Laredo
To reach for your gun when the going went rough.
But we wish our leaders had learned from their grandmas
To find a solution without playing tough.
We're sick of Mulroney, fed up with Bourassa,
We're not fond of Bush or of Saddam Hussein.
Who wants to return to the streets of Laredo!
So can that old Western and let's start again.

—Marczak Collection

Propaganda, they say, draws a line between us and them. In another song they link war and class:

Our sons in the desert are brave and are bold,
Our daughters can face any foe,
But the bravest by far in the Middle East War
Are Georgie and Brian and Joe.
When Saddam Hussein started being a pain,
Who hurried to save Texaco?
Who handed out gas masks to save gas supplies?
Georgie and Brian and Joe.
We're in a recession, we've got GST,
Where can all the unemployed go?
There's a place in the sun where they'll have lots of fun,
Say Georgie and Brian and Joe.
Remember the last war, remember Vietnam,
Remember who got picked to go,
To die in a ditch to make someone else rich—
NOT Georgie NOT Brian NOT Joe.

—Marczak Collection

War is, among other things, a class issue: The rich get richer, the poor get killed. Governments are quick to extend their hand and rescue large wealthy corporations as they withdraw that hand from the poor and "Let Them Eat Toast "(Tune: Quartermasters' Store):

There are kids, kids, lots of hungry kids
In the schools, in the schools
There are moms, moms, labelled welfare bums
By the governmental kools.

Refrain:
B.C. is rich our premier boasts
If they can't buy bread let them eat toast (2)
If we buy them buns, there'll be nothing left for guns
Company loans, or spacious homes.
Buying milk for tots will cost us lots and lots
Budget restraint is our first must!

It is folks like you who always fret and stew
Men like us, never fuss
If our budgets spent, we can slap on 7 per cent.
And it's you who'll pay, not us.

Refrain:
Canada is rich Mulroney boasts
If they can't buy bread let them eat toast (2)

—Marczak Collection

In the next song they link greed and war:

Money, power, living like royalty,
What a gold mine, dealing in weaponry.
Reactors and guidance systems
It's so hard to resist 'em.
We can't be blamed if they are aimed
At somebody's who's far away.
If we passed up all this gravy
To save some starving baby,
Then who would pay, for us to stay
In our God-given luxury.
What shall we do without the Commies
Spurring us on?
How shall we sell our guns and bomies
Now that the old foe is gone?
There's more money to be made
In war technology,
If we can just convince the leaders
We need it to stay free.
(Kazoo)
Food and shelter's such a boring
Way to spend our dough.
Thank goodness that we've still got—
the Iraqis—Helping the arms trade grow

—Marczak Collection

In "Radiation Country Gardens" they link local and global issues:

There is money to make
If from the ground you take
This substance called Uranium
Radiation abounds
For miles and miles around

Places where they mine uranium
The environment soon approximates the moon
When you're digging for uranium
They say we are alarmists
Technology is all
These are the same guys who engineered Bhopal.

—Marczak Collection

Songs served to collectively articulate and publicly express their point of view.

The media responded to their humor with their own. The Grannies were called "gray power platoon" when they volunteered for service at the Army Recruitment Center (Victoria) during the Gulf War, the "closest thing Victoria has to terrorists," or "Grannies who combat war with comedy." They were said to be unstoppable, outraged, outrageous, fun-loving cartoon Grannies, a ragtag group of grandmothers, a gaggle of Grannies, crooners, or social activists in sensible shoes. They "terrorize the stage" (Stuckey, 1989), "warble against war toys," and were "about as subtle as a chainsaw at a church social" (Ciriani, 1989: 3). A few images had negative connotations: They "rant all day" (Newman, 1996: 96), conveyed mindless rattle, while "a common fixture at peace and environmental rallies" (Goldberg, 1993: 13) dismissed them and the issues they espoused. But overall, the media were playful in their reports. Expressing dissent with humor generated an atmosphere where views were also aired with humor by reporters, cartoonists, and readers who disagreed.

The Grannies decided by consensus, a process that "honours the individual's voice, but makes governance slow. The structure . . . reflects their concern that traditional majority rule alienates the minority" (Burns, 1992: 23). Alison Acker, who joined the Grannies later, called it "cheerful anarchy" (cited in Burns, 1992: 23). Warren Magnusson, a University of Victoria political scientist, notes that there are no organizational guidelines they must all follow (Magnusson 1990: 533). Their leaderless, nonhierarchical, and unbureaucratic ways are familiar to many women's groups, says Lynne Jones (1983: 5). It allows greater creativity for individuals. While having fun is not part of the structure per se, it was a very important element: More than one said that they had fun, and if it was not fun they did not do it. This idea of fun as a sustaining element of struggle is also expressed by the Shibokusha, elderly Japanese peasant women who have been resisting the encroachment of the military over their lands since the end of World War II under difficult circumstances; they disrupt military exercises, sing, dance (Caldecott, 1983: 104) and say "it's quite fun and everyone is cheerful. If it weren't for that, we wouldn't keep it up for so many years" (Amano Yoshie, cited in "We Will Grow Back," 1982: 31). So "having fun" may be more important to resistance than traditional analysis has allowed.

The most remarkable impact of the Raging Grannies is the mushrooming of similar groups everywhere. Within a few years there were ten groups in BC and groups in Alberta, Saskatchewan, Manitoba, Ontario, Quebec, Nova Scotia, New Foundland, Texas, Wisconsin, even as far away as Greece, New Zealand, and Moscow (Chodan, 1989: H3; Ciriani, 1989: 3; Stainsby, 1989; Yandle, 1990; Acker, 1990b: 4; McCulloch, 1991; Gardner, 1993: 9; DeShaw, 1997; letter from Winnipeg Grannies Sophia Collection). The Grannies touched a nerve. Most who joined were older women. "Social activism can't be identified with any one generation anymore" (Howard, 1989: 14). "Women can feel good that they have a place when they get older" (Doyle, 1998). Overlooked by men or younger women in mainstream organizations, Burns suggests that older women became the norm within the Raging Grannies and could find their own voices (1992: 23).

The media often use "Raging Granny" as a generic term to denote persistence and humor. That the image of the Raging Granny stuck in the media and people's minds supports Doyle's (1998) assertion that "imagination is power." Their approach allowed women a wider range of emotions and turned rage into a positive transformative energy. Their verbal triumphs over the powers-that-be allowed other people to understand the power of active resistance. As for political effectiveness, they created a new space for politics. Their claim is not to authority but to being ordinary people with something to say: They turn their identity, usually a liability, into a resource. They raise consciousness by stimulating political debate (Magnusson, 1990: 536). They found a crevice in the media world, a crevice that paid off in the local political scene as it helped defeat MP Pat Crofton, who did not oppose nuclear ships. The Raging Grannies on Saltspring Island point to the defeat of a ferrochromium plant: The defeated industrialist called them "nuts" and other names, but he had to look elsewhere for his $42 million plant (Acker, 1990a).

The Raging Grannies were often invited into schools and seemed effective in that context as well. Many times students wrote to thank them and some offered reflections:

> I am very afraid of nuclear war. I don't like living everyday wondering if disaster is going to strike. Watching you perform made me laugh but also made me think. If I had the guts I would get up and express my feelings to the world.
>
> I'm very scared of nuclear war . . . and I'm glad you're trying to make people aware of what's happening.
>
> I enjoyed your performance . . . it opened my eyes to what I was trying to ignore. I haven't really been thinking about nuclear arms and was surprised when you explained that we are going to be buying those submarines. I thought we were getting rid of our weapons not getting more . . . you . . . opened my eyes. . . .
>
> When I first heard that you were coming I expected to see a bunch of radicals screaming about something. Your presentation gets a five star rating and I recommend it to other students.
>
> —Comments from Junior Secondary students in Marczak Collecton

The Raging Grannies are not about individual heroism but about the power of collective imagination. Doyle emphasized the collective dimension of creativity. Relationship is intrinsic to creativity: "Being able to play together as a group" made the Raging Grannies capable of tapping into a creative spring. They confronted dilemmas: Were they entertainment or intervention, concerned with outwardly directed goals or inward process? Given that activists are "living on the edge between forms of life in an alternative vision and the form imposed by the dominant order," (Magnusson, 1996: 101), such questions are not surprising. Doyle recognized the danger of taking themselves seriously in a way that endangered spontaneity and playfulness. She worried that notoriety and a measure of fame might lead the Grannies to lose their abrasiveness: "integrity means biting into something" (cited in Howard, 1989:

15). For political theater, success is often "as dangerous as failure to the purity of their politics," says John Burns (1992: 24). Because they managed to transform the stereotype into something positive, they are in demand by seniors' clubs; but singing in old-age homes may satisfy those Grannies interested in entertainment, yet "there is no challenge or thought-provoking disjunction in a group of older women singing to old people who have no voice in society" (Burns, 1992: 24). The original group were not entertainers but intervenors on a sociopolitical scene.

In her article on housewife activists, Harriet G. Rosenberg makes use of Eric R. Wolf's disagreement with "the notion that it is the most oppressed who mobilize": The underresourced poor or the ideologically constrained rich are not as well positioned or willing to challenge cultural assumptions or social structures because they have too much to lose (1995: 191). But those in between are strategically positioned for doing so. Magnusson calls the Grannies a "brilliant example of a group acting out their protests" who used their credibility as grandmothers to "undercut the legitimacy of military violence, corporate greed, and governmental insensitivity" (1996: 93–94). Also, solidarity can lead, as it did with the Grannies, to the development of new cultural forms (Cohen, 1979: 137). A sense of connection makes it possible to overcome a feeling of powerlessness that inhibits social change. The fluidity of such nonhierarchical organization can be scary for those who fear spontaneity and its necessary chaos. Yet that very ability to embrace the uncertainty of the creative process is what made the Raging Grannies noteworthy. They demonstrated what Vaclav Havel called deep hope, an orientation of spirit: Hope "is not the conviction that something will turn out well, but the certainty that something makes sense regardless of how it turns out" (cited in Cohen-Cruz, 1998: 65). Like the Greenham women who imparted a "sense of optimism regarding the value and role of women's collective action and solidarity" (Romalis, 1989: 173), the Raging Grannies imparted that sense about older women.

"Citizenship can grow out of conscious motherhood" (Kaplan, 1997: 40). Their willingness to "divest themselves" of the "artificial notion of decorum and dignity" (Walker, 1998) and make fools of themselves made the stereotype visible for all to see. This was something they shared with the Love Canal women protesters who similarly challenged stereotypes of motherhood through what Temma Kaplan (1997) described as spirited, dramatic, and colorful actions (38). Stereotypes keep a population regimented and controlled (Walker, 1998). The Argentinean Mothers of the Plaza de Mayo, who started from the tragedy of their individual motherhood and quickly expanded their commitment to human rights, chose to represent themselves as mothers because it made it harder for soldiers and police to attack them than if they had claimed their citizen's rights (Kaplan, 1997: 184, 186). They "translate the despair of personal searches for lost children into a profoundly radical understanding of the politics of terrorism" (Romalis, 1999: 11). "Mother Jones' army of women wielding kitchen pots and pans" is also an example of political motherhood. The dramatic use of symbols of women's domestic work show the connections between "the domestic, women's collective visions, and large political concerns." "The women's actions in all these cases are strategic and instrumental expressions of maternalist political discourses, rather than simply reflections of 'essential' womanhood" (Romalis, 1999: 11). The irreverence and subversive quality of the Grannies were due partly to their identity as older women with an "unmotherly" public rage.

Theirs is an example of the transformative power of the imagination; "imagination powers the processes of rational and creative action that . . . makes change possible" (Coult and Kershaw, 1990: 13). Creative imagination is the enemy of fragmentation and dehumanization; imagination is not diversion but an essential element of change as it challenges conditioned responses. Creativity is also expressed in humor: Sigmund Freud saw humor as rebellion and Jo Anna Isaak adds that it expresses freedom (1996: 14, 26). Satirists "tend to bloom in seasons of drought," and satire has been used to counteract the excesses of the Roman empire, the church, and European bourgeoisie: Melvin Maddocks (1987) calls it "literature's alternative to revolution." Transformation requires an ethic of risks; the ethic of safety is for those with privilege. For others, safety is the death of our ability to care. On the other side of silence is a raging roar (George Eliot cited in Belenky et al., 1986: 4) and a great laugh!

NOTES

1. Sophia is a fictive name as requested by the interviewee.

REFERENCES

Acker, Alison. (1990a). They're coming . . . Raging grannies' ranks swell. Doran Doyle Collection.

Acker, Alison. (1990b). Granny power: Cheeky protesters get message heard. *This Magazine, 23* (7): 4–5.

Ackelsberg, Martha A. (1991). *Free women of Spain: Anarchism and the struggle for the emancipation of women.* Indianapolis: Indiana University Press.

Antinuclear protesters parade. (1986). *Times-Colonist,* March 9: A3.

Banner, Chris. (1988). Lunch with the Grannies. *Southern Vancouver Island's Maritime Magazine,* January 31: 17. Marczak Collection.

Belenky, Mary Field, *et al.* (1986). *Women's ways of Knowing: The development of self, voice, and mind.* New York: Basic Books.

Brightwell, Betty. (1998). Interview by author. Tape recording. Victoria, BC, May 19.

Brightwell, Betty. (1988). I am a Raging Granny. *Briarpatch: Saskatchewan's Independent News Magazine, 17* (9): 20–23.

Brown, Nancy. (1991). Grannies agree to be contacts for draft evaders. *Times-Colonist,* January. Sophia Collection.

Burns, John. (1992). Raging Grannies. *Canadian Theatre Review, 72*: 21–24.

Caldecott, Leonie. (1983). At the foot of the mountain: The Shibokusa women of Kita Fuji. In Lynne Jones (Ed.), *Keeping the peace: A women's peace handbook* (pp. 98–107). London: The Women's Press.

Chodan, Lucinda. (1989). Raging Grannies combat war." *The Gazette,* August 26: H3.

Ciriani, Jean. (1989). Grannies all the rage: On visit to Duncan. *The Cowichan News Leader,* April 12: 3.

Cohen, Gaynor. (1979). Women's solidarity and the preservation of privilege. In Patricia Caplan and Janet M. Bujra (Eds.), *Women united, women divided: Comparative studies of ten contemporary cultures* (pp. 129–156). Bloomington: Indiana University Press.

Cohen-Cruz, Jan (Ed.). (1998). Part two: Introduction. In *Radical street performance: An international anthology* (pp. 65–66). New York: Routledge.

Coult, Tony, and Baz Kershaw (Eds.). (1990). *Engineers of the imagination: The welfare state handbook.* 2nd ed. London: Methuen Drama.

Cult expands as zeal to save planet catching. (1989). *Times-Colonist,* October 28: B1.
DeShaw, Rose. (1997). Goofy-hatted reconcilers in the land. *The Globe and Mail,* April 10. Sophia Collection.
Doyle, Doran. (1998). Interview by author. Tape recording. Victoria, BC, May 14.
Farrell, John. (1991). Nuclear submarine captain sounds nuclear alarm bell. *The Daily News Extra,* August 28: 3.
Foley, Dennis. (1990). Raging Grannies afloat: Nuclear ships on hit list. *The Ottawa Citizen,* April 21: 3.
Gardner, Alison. (1993). The Grannies are coming, watch out, watch out! *Maturity, 13* (3): 8–11.
Goldberg, Kim. (1993). Don't mess with the Grannies. *The Progressive, 57* (3): 13.
Grannies ride in style. Marczak Collection.
Howard, Keith. (1989). Those comic Raging Grannies. *The United Church Observer, 53* (3): 14–15.
Isaak, Jo Anna (1996). *Feminism and contemporary art: The revolutionary power of women's laughter.* London: Routledge.
Jones, Lynne. (1983). *Keeping the peace: A women's peace handbook.* London: The Women's Press.
Kavanagh, Jean. (1987). Vocal grannies are the rage in Victoria. *The Vancouver Sun,* December 17: A16.
Kaplan, Temma. (1997). *Crazy for democracy: Women in grassroots movements.* New York: Routledge.
Lerner, Gerda. (1997). *Why history matters: Life and thought.* New York: Oxford University Press.
Lightly, Marion. (1989). Raging Grannies: Singing the subs away. *Briarpatch: Saskatchewan's Independent News Magazine, 18* (1): 15.
Lofland, John. (1993). *Polite protesters: The American peace movement of the 1980s.* Syracuse, N.Y.: Syracuse University Press.
Maddocks, Melvin. (1987). Time ripe for satire. *Christian Science Monitor,* December 17. Marczak Collection.
Magnusson, Warren. (1990). Critical social movements: De-centering the State. In Alain-G. Gagnon and James P. Pickerton (Eds.), *Canadian politics: An introduction to the discipline* (pp. 525–41). Peterborough, Ont.: Broadview Press.
Magnusson, Warren. (1996). *The search for political space: Globalization, social movements, and the urban political experience.* Toronto: University of Toronto Press.
Magnusson, Warren. (1997). Globalization, movements, and the de-centered state. In William K. Carroll (Ed.), *Organizing dissent: Contemporary social movements in theory and practice: studies in the politics of counter-hegemony* (pp. 94–113). Toronto: University of Victoria Garamond Press.
Malkinson, Leah. (1986). UVic spooks dance for peace. *The Martlet,* October 30. Marczak Collection.
Marcoux, Lois. (1998). Interview by author. Tape recording. Victoria, BC, May 18.
Marczak, Hilda. (1998). Interview by author. Tape recording. Victoria, BC, May18.
McCulloch, Sandra. (1991). Victoria's Raging Grannies lead the battle against war. *Times-Colonist,* December 1. Sophia Collection.
Newman, Peter C. (1996). Most memorable 1996 absurdities. *Macleans, 109* (53): 96.
Nuke protest today. (1985). *Times-Colonist,* October 13. Marczak Collection.
Portelli, Alessandro. (1997). *The Battle of Valle Giulia: Oral history and the art of dialogue.* Madison, Wis. University of Wisconsin Press.
Raging Grannies. (1990). Narrated by Peter Downie. Written and directed David Cherniak. CBC Television Man Alive. Toronto. November 20
Raging Grannies Sing. (1988). *The Esquimalt Star,* June 15: 2.
Reinharz, Shulamit. (1992). *Feminist methods in social research.* New York: Oxford University Press.
Romalis, Shelly. (1987). Carrying Greenham home: The London women's peace support network. *Atlantis: A Women's Studies Journal, 12* (2): 89–98.

Romalis, Shelly. (1989). From Kitchen to Global Politics: The Women of Greenham Common. In Janice Williamson and Deborah Gorham (Eds.), *Up and doing: Canadian women and peace* (pp. 170–74). Toronto: The Women's Press.

Romalis, Shelly. (1999). *Pistol packin' mama: Aunt Molly Jackson and the politics of folksong.* Chicago: University of Illinois Press.

Rose, Mary. (1992). Raging Grannies at Sea. *Ocean River Sports Newsletter,* Winter:1–2.

Rosenberg, Harriet G. (1995). From trash to treasure: Housewife activists and the environmental justice movement. In Jane Schneider and Rayna Rapp (Eds.), *Articulating hidden histories: Exploring the influence of Eric R. Wolf* (pp. 190–204). Berkeley: University of California Press.

Rowbotham, Sheila. (1972). *Women, resistance and revolution: a history of women and revolution in the modern world.* New York: Pantheon Books.

Saunders, Joan. (1989). The medium is the message: Victoria's Raging Grannies. *Prime Life Magazine, 2* (8). Marczak Collection.

Silko, Leslie Marmon. (1977). *Ceremony.* New York: Viking Penguin Inc.

Sophia (pseud.). (1998). Interview by author. Tape recording. Victoria, BC, May 17.

Spender, Dale. (1982). *Women of ideas (and what men have done to them): From Aphra Behn to Adrienne Rich.* London: Pandora.

Stainsby, Mia. (1989). Raging Grannies: They sing up a storm for social conscience. *The Vancouver,* July 7. Doyle Collection.

Stretching a point. (1986). *Times-Colonist,* October 27. Doyle Collection.

Stuckey, Andrew. (1989). Raging Grannies "terrorize" conference stage. *Crows Nest Pass Promoter,* June 6. Marczak Collection.

The Raging Grannies. (1987). *Times-Colonist,* June 7: C8.

Van Luven, Lynne. (1989). Times not changing fast enough for the Raging Grannies. *The Edmonton Journal,* August 21: A10.

Walker, Moira. (1998). Interview by author. Tape recording. Victoria, BC, May 17.

We will grow back. (1982). *Connexions: An International Women's Quarterly, 6,* 3.

Yandle, Carlyn. (1990). Grannies' outraged. *The Peace Arch News,* February 28. Marczak Collection.

Index